Neutron Imaging

Dinesh K. Aswal · Partha S. Sarkar ·
Yogesh S. Kashyap
Editors

Neutron Imaging

Basics, Techniques and Applications

Springer

Editors
Dinesh K. Aswal
Bhabha Atomic Research Centre
Mumbai, India

Partha S. Sarkar
Bhabha Atomic Research Centre
Mumbai, India

Yogesh S. Kashyap
Bhabha Atomic Research Centre
Mumbai, India

ISBN 978-981-16-6272-0 ISBN 978-981-16-6273-7 (eBook)
https://doi.org/10.1007/978-981-16-6273-7

This Springer imprint is published by the registered company Springer Nature Singapore Pte Ltd.
The registered company address is: 152 Beach Road, #21-01/04 Gateway East, Singapore 189721, Singapore

Azadi Ka Amrit Mahotsav: Celebrating 75 years of India's independence

Foreword

We at BARC have been entrusted to carry out our activities with the highest quality of scientific and technological competence, be it in research, development or operation. Imaging technology is one such field where we have taken great strides keeping pace with the technological advancements in materials, detectors, controls and interfaces. Every field of research relies on imaging technologies at one point or the other. In this regard, neutron imaging plays the complementary role as compared to X-ray imaging due to their interaction mechanism. Neutrons can probe deep inside materials for either imaging or element-specific studies in bulk and play an important role in material research studies.

India was fortunate to have visionary scientific leadership of Dr. Homi J. Bhabha—an established mathematical physicist of international distinction—who chose to devote his time to the task of developing sustainable nuclear energy programme in India. Soon after Dr. Homi Bhabha started the atomic energy programme in India and Atomic Energy Establishment Trombay (AEET, later rechristened as Bhabha Atomic Research Centre) was created, the task to set up Asia's first research reactor Apsara, 1 MW swimming pool-type research reactor was taken up and completed in just about 16 months in August 1956. In fact, India became the first country in Asia to have indigenously designed, built and operated a nuclear reactor in the mid-1950s. Apsara reactor became the beehive of research activities for scientists working in various fields, reactor, nuclear, solid-state physics, radiochemistry and health physics. By early sixties, with the availability of 40 MW thermal reactor CIR (renamed CIRUS), the range of instruments to study various scientific problems began to grow and CIRUS was an important centre for neutron beam research activities till it was decommissioned in 2010. A 30 kW research reactor called KAMINI (KAlpakkam MINI) using U233 as fuel was commissioned at IGCAR Kalpakkam near Chennai. Incidentally, KAMINI is the only operating reactor in the world with U233 fuel and is also being used for neutron radiography.

Neutron beam research programme got a big boost in the mid-eighties with the start of 100 MW thermal reactor "Dhruva". Currently, Dhruva is the workhorse for carrying out neutron beam research and is the National Facility for Neutron Beam Research. The Mumbai Centre of UGC-DAE Consortium for Scientific Research

(UGC-DAE-CSR) has been promoting and supporting neutron beam research among researchers from various universities and other research organizations from all over India for more than two decades. BARC has played a pioneering role in the development of neutron radiography (NR) techniques in India and its applications in several key areas. In 2016, a new neutron radiography and tomography facility (NRTF) was commissioned at Dhruva to boost reactor use. BARC also has developed accelerator-based DD/DT neutron generators providing thermal as well as fast neutron imaging capability in laboratory, outside reactor premises. In order to popularize the use of neutron beam research and neutron radiography, BARC has collaborated with Mangalore University, Karnataka, India, in setting up an accelerator-based photo-neutron source.

I am delighted to see that group of scientists working on different aspects of neutron source development and imaging have come together in assimilation of basics, technological aspects of imaging, world scenario, case studies towards various applications using neutron imaging and studies carried out at BARC to publish a book on "Neutron Imaging—Basics, Techniques and Applications". I sincerely believe that this book would help in wider dissemination of knowledge of neutron imaging and its applications, thereby opening up new avenues in scientific and industrial research.

Dr. Ajit Kumar Mohanty
Director, Bhabha Atomic Research Centre
Mumbai, India

Preface

Neutron imaging, a powerful non-destructive technique for investigating the structure of materials, has gained significant importance in recent times for understanding the processes in advanced materials as well as in quality control of industrial components of nuclear, aerospace, defence, agriculture, food preservation, oil and gas, metal industries, etc. The history of neutron imaging began in 1935 when H. Kallmann and E. Kuhn generated first radiographic images of objects using a small neutron generator. The epoch-making discovery of nuclear fission in 1939 by Otto Hahn and Fritz Strassmann heralded an onset of a new era in the human civilization. Soon after this discovery, Fermi and co-workers created first self-sustained chain reaction, known as Chicago Pile 1 (CP-1), and it was the world's first nuclear reactor built underneath a football stadium of University of Chicago. The nuclear reactors are prolific source of neutrons and continue to be backbone of neutron imaging activities. Neutron imaging has advanced from traditional attenuation-based radiography/tomography to diffraction imaging, scattering contrast imaging, 4D tomography where 3D attenuation map is combined with spectral information from each voxel. These have been possible on account of the availability of high flux neutron sources such as high flux isotope reactor (HFIR) and spallation, and development of advanced detectors combining with advanced techniques such as wavelength-dependent imaging or time of flight. High flux neutron sources have yielded cold and ultra-cold neutron imaging possible, yielding higher contrast and invaluable information from bulk and also capturing dynamic phenomena at very short acquisitions down to microseconds.

Neutron beam research in India started almost immediately after Apsara reactor became operation in 1956 at Bhabha Atomic Research Centre (BARC), Trombay. The neutrons from this reactor were not only used for nuclear physics research, but also for condensed matter research, irradiation, neutron radiography and tomography, two phase flow studies and neutron detector development. By early sixties, with the availability of 40 MW thermal reactor CIRUS, BARC became an important centre for neutron beam research activities, viz. neutron radiography, tomography, phase contrast imaging and dynamic imaging. Neutron beam research programme got a big boost in the mid-eighties with the start of 100 MW thermal reactor Dhruva which was constructed entirely indigenously. It is a natural uranium reactor with peak thermal flux of 1.8×10^{14} n/s/cm^2 and India's main facility for neutron radiography and medical isotope production.

Partha S. Sarkar and Yogesh S. Kashyap would like to put in records their deep gratitude to Dr. Amar Sinha, who has been instrumental in initiating and advancing digital neutron imaging for various applications including tomography, phase contrast and real-time studies using both reactor and non-reactor sources in BARC, India. At BARC, Dr. Amar Sinha initiated neutron imaging activities, including tomography, phase contrast and real-time studies.

This book is aimed at introducing readers on advancement in neutron imaging during last few decades along with basic concepts of neutron imaging, neutron optics and relevant instrumentation. The book comprises 12 chapters. Chapter 1 deals with neutron properties, basic concepts of scattering cross sections and comparison of neutron and X-ray imaging. Chapter 2 discusses various modes of production of neutron sources and physics and design of suitable systems to adapt them for neutron imaging applications. In Chap. 3, basics of various mechanisms of neutron optics employed towards spatial beam shaping, spectral shaping and intensity increase are discussed. In addition, advanced neutron detectors and their use in specialized imaging experiments are also discussed. Chapter 4 gives a bird's eye view of various neutron imaging facilities (both reactor based and non-reactor based) across the globe and their utilization in a variety of applications. Chapter 5 presents the mathematics of neutron imaging in terms of system of linear equations and their conversion to obtain unknown attenuation coefficients of the object. In addition, methods for volumetric measurements using computed tomography and related reconstruction methods are discussed. Chapter 6 presents concepts and applications of phase contrast, polarized neutron and spin echo neutron imaging. Chapter 7 deals with the safety, regulations and standards towards implementation of neutron imaging activities in any neutron source. The basic requirements and implementation methodologies emphasize minimization of radiation doses to occupational workers and related limits set by regulatory bodies. Chapter 8 deals with application of neutron imaging in the testing of critical components of aerospace industry, such as turbine blades and pyro devices. Chapter 9 covers the applications of neutron imaging for the nuclear power industry: ranging from quality control of nuclear fuel and components to post-irradiation analysis of spent fuel and structural components. In particular, the neutron imaging investigation of hydrogen embrittlement, a phenomenon responsible for the degradation of metals in nuclear industry, is discussed in detail. Chapter 10 deals with application of neutron imaging in materials and engineering applications with an aim to study and optimize the process parameters, process efficiency, new material development in the areas of fuel cells, hydrogen storage materials and Li-ion batteries. Chapter 11 deals with application of neutron imaging in the areas of geo-mechanics, agricultural and food preservation industry. In these applications, the high attenuation of neutron for hydrogen in water is exploited to investigate the water flow/transport, water content, moisture or dehydration/drying. Chapter 12 describes applications of fast neutron imaging for security applications, i.e. detection of potential threat materials (weapons, contrabands, drugs, explosive, etc.) in large volume of transport trucks and cargoes. Fast neutrons allow inspection and imaging of large and thick samples composed of low-Z/high-Z materials or mixture of both.

We have tried to provide a broad overview of the basics of neutron imaging and its applications in selected important scientific investigations and industrial uses. Needless to say, many application areas may have remained uncovered. We believe that publication of this edition would simulate readers to explore new avenues of neutron imaging in variety of scientific and industrial applications. The contributions of Dr. Amar Sinha, who has initiated the neutron imaging activities at BARC, are acknowledged. The assistance of the Springer team during the various stages of the development of this book is greatly appreciated.

Mumbai, India

Dinesh K. Aswal
Partha S. Sarkar
Yogesh S. Kashyap

Contents

About the Editors

Dr. Dinesh K. Aswal is Director, Health Safety and Environment Group at Bhabha Atomic Research Centre (BARC), Department of Atomic Energy (DAE), Government of India, Mumbai. He has had several positions like Director, CSIR- National Physical Laboratory (CSIR-NPL), New Delhi; Director, Central Electronics Engineering Research Institute (CEERI, Pilani); Director, Science, Technology and Development Studies (CSIR-NISTADS, New Delhi); Chairman, National Accreditation Board for Testing and Calibration Laboratories (NABL); Secretary, and Atomic Energy Education Society (AEES), Mumbai etc. Dr. Aswal joined BARC through the training school batch of the year 1986 after completing M.Sc. in Physics (Gold medalist) from Garhwal University in 1985. He obtained his Ph.D. in Physics from Mumbai University and subsequently carried out post-doctoral research work at the Research Institute of Electronics, Hamamatsu, Japan. He is known for "Aswal model of inclusive growth" and made several outstanding contributions in the areas of metrology, molecular electronics, physics of organic films and their applications, thermoelectric power generators and gas sensors. He has had visiting professor/scientist positions at several international institutes/universities viz. Institut d'Electronique de Microelectronique et de Nanotechnologie (France), Sizuoka University (Japan), Commissariat à l'Energie Atomique (France), Weizmann Institute of Science (Israel), University of Yamanashi (Japan), University of Paris VII (France), Karlsruhe institute of Technology (Germany), and the University of South Florida (USA). He is a recipient of several national and international awards/fellowships, including Fellow of National Academy of Sciences, India (NASI); Academician, Asia Pacific Academy of Materials; Fellow, International Academy of Advanced Materials (Sweden); Distinguished Faculty Award of Homi Bhabha National Institute (HBNI); Materials Research Society of India (MRSI) Meda; Homi Bhabha Science and Technology Award; DAE-SRC Outstanding Research Investigator Award; etc. He has edited five books, published over 500 journal/conference papers.

Dr. Partha S. Sarkar is a Scientific Officer in Technical Physics Division at Bhabha Atomic Research Centre (BARC), Mumbai. He had joined BARC in 2000 after his

Masters in Physics from Indian Institute of Technology, Kharagpur, India. Subsequently he obtained his Ph.D. (Physics) from the University of Mumbai for his contribution to the field of radiation imaging techniques. He has been involved in developing radiation imaging techniques using X-rays and neutrons for non-destructive testing and evaluation towards nuclear, manufacturing industry, medical and societal applications. He has also been involved in developing systems towards medical imaging and field applications. He is presently involved in developing techniques and systems towards illicit material detection and special nuclear material detection towards nuclear safeguards. He is recipient of Young Applied Scientist and Technologist award from Department of Atomic Energy in the year 2009. He has published around 65 research papers in refereed journals including a book chapter involving Tomography with X-rays and neutrons. He is also a faculty member of Homi Bhabha National Institute (HBNI), Mumbai, as Associate professor in the discipline of Physical sciences.

Dr. Yogesh S. Kashyap joined Bhabha Atomic Research Centre (BARC), Department of Atomic Energy (DAE), Government of India, Mumbai as Scientific officer in 2001, after completing his Masters in Physics from Indian Institute of Technology, Kanpur, India. Subsequently, he obtained his Ph.D. (Physics) from the University of Mumbai, in the field of phase contrast X-ray and Neutron imaging. He was involved in installation & commissioning of Neutron imaging beamlines at CIRUS and Dhruva research reactor, Trombay and development of various industrial application of compact neutron generators. He has also made significant contribution in the development of X-ray imaging beamline at Indus-2, synchrotron source, India. He has worked on the development of multi-modal phase retrieval methods as a research associate at Diamond Light source, UK and developed in-situ near-field speckle based metrology system for quantification of mirror slope errors. He is recipient of Young Applied Scientist and Technologist award from Department of Atomic Energy in the year 2011 for development of X-ray phase contrast and neutron imaging techniques. He is also a faculty member of Homi Bhabha National Institute (HBNI), Mumbai, as Associate professor in the discipline of Physical sciences. He has contributed two book chapters and published around 90 papers in journal/conference.

Chapter 1
Introduction to Neutron Physics

Yogesh S. Kashyap

1.1 Introduction

Neutrons have played a very important role in many of the scientific disciplines by providing important breakthrough discoveries, thereby contributing to modern scientific and industrial progress. Neutrons interact with all known fundamental forces in nature and, therefore, are a powerful tool for investigating domains of quantum mechanics, nuclear physics and cosmology, etc. Although the existence of neutrons was hypothesized by W. D. Harkins and E. Rutherford in the 1920s, however, its confirmation was provided in 1932 by James Chadwick, through meticulously planned experiments. The late discovery of neutron can be attributed to its similarity with the gamma rays, which too are emitted from the nucleus and show the property of charge neutrality. In fact, Irene curie and Fredric Joliot were first to notice the important property of this radiation—creating recoil protons while going through hydrogeneous media. However, it was Chadwick who realized and proved that the energy of the recoil proton can be explained only if the incoming radiation has a mass comparable to the proton mass and named it as neutrons. In particular, the discovery of neutrons provided an important impetus in the development of nuclear energy using nuclear reactors, which have in turn contributed to slowing down of emission of greenhouse gases in the earth's atmosphere. Medical isotopes produced through the use of nuclear transmutation by use of neutrons in many of nuclear reactors have played an equally important role in disease diagnosis, treatment of cancers, sterilization of medical equipment, etc. Similarly, low-energy neutrons are also useful in the treatment of certain types of tumours through the use of techniques such as boron neutron capture therapy (BNCT). In parallel, neutron imaging and neutron scattering have been developed into a very important tool in modern science and engineering, and owing to unique properties of neutrons, neutrons make an

Y. S. Kashyap (✉)
Technical Physics Division, Bhabha Atomic Research Center, Mumbai 400085, India
e-mail: yskashyap@barc.gov.in

D. K. Aswal et al. (eds.), *Neutron Imaging*,
https://doi.org/10.1007/978-981-16-6273-7_1

important complementary probe to X-rays. Due to nuclear-dependent scattering cross section, neutrons are complementary to X-rays in terms of isotopic sensitivity and have other unique properties such as large mean free path in matter and sensitivity to magnetism. We briefly mention some important reasons that makes neutron a unique probe in the non-destructive study of matter.

(i) As neutrons carry no charge, they have a large penetration depth in most materials. Hence, they are ideal for probing industrial samples, metallic objects, biological materials and in situ studies under externally applied conditions of pressure, temperature, etc. Similarly, the large penetration depth of neutrons permits us to study experimental processes within a chemical reaction vessel. Similarly, neutrons play important role in detection of light elements such as hydrogen, lithium and boron, due to their high neutron absorption properties. Neutrons are a unique and helpful probe for studying hydrogen storage materials, organic materials and biomolecular samples or polymers because they are highly sensitive to hydrogen.

(ii) Neutrons primarily interact with matter either of two ways: through the nuclear force or through the magnetic moment. Neutrons possess an intrinsic magnetic dipole moment, and hence, they are sensitive to magnetic fields (external or generated by unpaired electrons). Using a polarized neutron beam, the magnetic moment of a neutron may be used to examine the magnetic characteristics of materials. Furthermore, because neutron–nucleus interactions are spin dependent, neutrons may be used to investigate the spin dependence of nuclear forces and detect nuclear spin.

(iii) The wavelength of a neutron depends on its energy, and for thermal neutrons (~1 Å), it is roughly comparable to the distance between neighbouring atoms in materials. This makes them an ideal probe for determining atomic and molecular structures using neutron diffraction. The use of neutron diffraction to detect residual stress in metallic samples such as zirconium, aluminium, steel and others is a powerful and important tool for micro-strain studies inside the bulk of the materials.

(iv) The energies of thermal and cold neutrons are of the order of few meV or μeV. This energy is comparable to the energy required for thermally activated random motion of atoms in solids and liquids, and therefore, neutrons are ideal probe for studying such diffusive motion of atoms. Similarly, thermal vibrations of atoms in a crystal lattice (phonons) or oscillations of the magnetic moments (magnons) and a coupling between them can be easily probed using neutrons. Inelastic neutron scattering methods are also employed in neutron-based vibrational spectroscopy to explore the vibrational modes in molecules. As a result, neutrons are a perfect probe for studying these matter characteristics.

(v) As the neutron interacts with the nucleus in a quasi-random manner, lighter atoms such as ${}_{1}H$, ${}_{3}Li$ and ${}_{5}B$ can be easily detected even in the presence of heavier atoms, and nearby atoms may also be distinguished from each other. Moreover, one can also vary the contrast in certain samples using isotopic

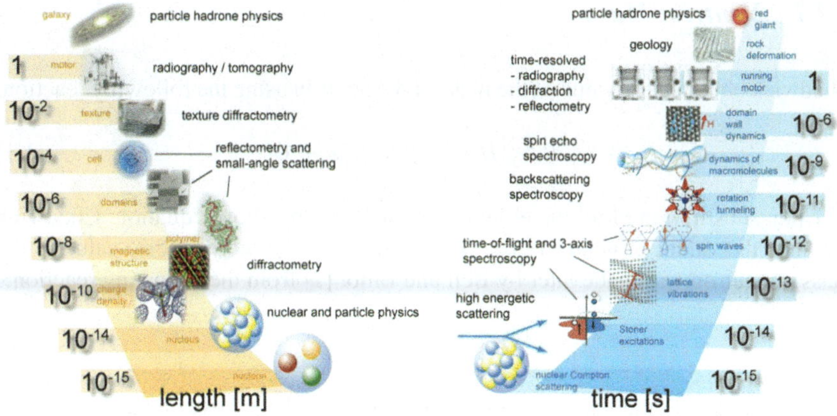

Fig. 1.1 Different length and timescales that can be probed using dedicated instruments using neutrons [1]

substitution techniques (e.g. deuterium for hydrogen, or one nickel isotope for another), and the same can also be used for study of complex ion formation process, partial structure factor determination and characterization of the internal structure of complex particles.

Figure 1.1 shows different neutron-based techniques and the length and time scales that they can probe. In the following section, we describe the fundamental properties of neutron and introduce the basic concepts in neutron imaging.

1.2 Properties of Neutrons

Neutrons are classified as hadrons, and they are subject to the strong nuclear force and belong to the family of baryons since they consist of three quarks. According to the standard model, the neutron is a neutral particle that is made of one up quark with charge +2/3 e and two down quarks with charge −1/3 e.

Proton, the other constituent of the nucleus, behaves almost identically under the influence of the short-range nuclear force (strong force) within the nucleus. Heisenberg proposed that the proton and neutron can be viewed as two quantum states of the same particle (charged and neutral) using the concept of isospin symmetry, and this formalism can be used to model the interactions of nucleons by the nuclear or weak forces.

Although there are many important properties of neutrons, we will focus on the following ones as they are most relevant for further development of the subject discussed in this book.

1.2.1 Neutron Mass

Chadwick obtained the value of the mass of the neutron using the following reaction:

$$^{11}_{5}B + ^{4}_{2}He \rightarrow ^{14}_{7}N + ^{1}_{0}n + Q \quad (1.1)$$

Upon substitution of all the atomic mass units in the above equation, Chadwick obtained the values of neutron mass as 1.0067 amu. The more precise value of the mass of neutron was made later by Bell and Elliot [2] from the following reaction:

$$^{1}_{1}H + ^{1}_{0}n \rightarrow ^{2}_{1}D + \gamma \quad (1.2)$$

The mass of the neutron obtained through the above measurement is 1.0008672 amu which was quite close to those obtained using a more precise measurement of 1.008664904(14) amu due to Bell and Elliot [3]. These measurements rule out any ambiguity that neutron could be composed of bound proton–electron pair, as postulated at the beginning of the twentieth century.

1.2.2 Neutron Charge

Since the beginning of discovery, zero electrical charge has been assigned to the neutron, as no measurable ionization is produced by neutrons in air. Experiments have been carried out by measuring neutron deflection as it passes through a transverse electrostatic field by Shull et al. [4] and gave an upper bound of the electric charge as 4×10^{-18} e (where *e* is the unit of electronic charge). Similar experiments were repeated at the Institute of Laue-Langevin in Grenoble, lateral displacement of neutrons was measured, and an upper limit $(-1.5 \pm 2.2) \times 10^{-20}$ e was obtained. More modern measurements have set an upper experimental limit on the charge of the neutron as -2×10^{-22} e reported by Olive [5]. In conclusion, therefore, one can safely use zero electrical charge as there is no experimental or theoretical evidence of any residual electrical charge on the neutrons.

1.2.3 Neutron Magnetic Dipole Moment

Even though the presence of an electric field has no effect on the neutron, studies have demonstrated that it may be significantly affected by an external magnetic field. The value of neutron magnetic moment was measured in a beautifully designed experiment by Alverez and Bloch [6] using the concept of nuclear magnetic resonance techniques and obtained the value which is quite close to the accepted value of −

1.91303 μ_n (in units of nuclear Bohr magneton). The negative sign emphasizes the fact that a neutron's magnetic moment and intrinsic spin point in opposite directions.

In the quark model of elementary particles, the neutron is made up of two "down" quarks of charge −1/3 e and one "up" quark of charge + 2/3 e. The magnetic moment of the neutron, therefore, arises due to the vector sum of the magnetic moments of the constituent quarks [7] and can also be calculated. Beg et al. [8] computed the ratio of intrinsic magnetic moments of neutron and proton to be −2/3, in good agreement with various experimental values to within 3%.

1.2.4 Neutron Spin

Neutrons have been found to behave like a Fermion, i.e. obey the Fermi–Dirac statistics, and have an intrinsic spin of 1/2. The interactions of the neutron's magnetic moment with an external magnetic field in a Stern–Gerlach type of experiments were used to confirm this hypothesis. In 1954, Sherwood et al. [9] employed neutrons in a Stern–Gerlach type of experimental set-up to record two spin states, thereby confirming that neutrons have an intrinsic spin of 1/2. It may be noted that neutron spin results in two refractive indices for neutrons in a magnetized media as demonstrated by Hughes and Burgy, who studied neutron reflection from ferromagnetic mirrors and observed angular distribution of reflection was consistent with spin 1/2 [10].

1.2.5 Neutron Lifetime

Unlike its counterpart proton, neutrons are not stable outside nuclei and undergo beta decay. These neutron lifetime measurement experiments can be broadly classified into two varieties, "bottle experiment" and "beam experiment". In a bottle experiment, ultra-cold neutrons are confined in a container, and undecayed neutrons are counted after a certain time has elapsed. In a beam experiment, researchers examine a stream of neutrons and count the number of protons produced by decay of neutrons. The neutron lifetime measurement has proved to be a difficult task primarily because that the data obtained with new experimental methods have been found to be in disagreement with earlier results [11]. There is nearly four standard deviations disagreement of 8.5 s between measurements of the neutron lifetime using the rate of neutron decay-producing protons measured in neutron beam experiments [12] (888.0 ± 2.0 s) and free neutron lifetime in bottle experiments [13] (879.4 ± 0.4 s). This lingering discrepancy is known as the "neutron lifetime puzzle", and it is still a matter of scientific debate; it is due to any unconsidered systematic effect or any new fundamental undiscovered physics.

1.2.6 Neutron Electric Dipole Moment

Although the neutrons have no electric charge, according to the standard model of particle physics, there exists a tiny separation of centres of the negative and positive charge distribution within the neutron, and this leads to a permanent electric dipole moment. Finding this neutron electric dipole moment would be a big breakthrough, in part because it could explain why matter in the cosmos did not simply vanish with antimatter at the beginning of time. It is worth noting that only if the centres of the negative and positive charge distributions inside the particle do not coincide, a finite electric dipole moment (EDM) can exist. So far, no neutron EDM has been found. The current best-measured limit for electric dipole moment is $(0.0 \pm 1.1) \times 10^{-26}$ e.cm[14]. Table 1.1 summarizes the discussion on various unique properties of neutrons.

1.3 Classification of Neutrons

As discussed earlier, neutron interacts with matter in a variety of ways, and the interaction probability is dependent on the energy of the neutron. Therefore, it is important to classify neutrons according to their kinetic energy. Moreover, neutrons can also be assigned a wavelength using the concept of wave–particle duality. The energy ranges described below are mainly concerned with differentiating how neutrons interact with materials for different applications.: nuclear fission, radiation shielding, neutron scattering, neutron imaging, etc. Note that the nomenclature and the energy ranges shown in the following definitions are not universal, and sometimes a group of energies can be recombined for ease of calculations (Table 1.2).

Table 1.1 List of some important properties of neutrons

S. No.	Important property	Value
1	Mass	1.00866491560(55) amu
2	Spin	1/2
3	Magnetic moment	$-1.91304273\ \mu_n$
4	Lifetime	885.7(8) s
5	Mean square charge radius	$-0.1161(22)$ fm^2
6	Electric charge	0 (experimental limits: $-2 \pm 8) \times 10^{-22}$ e)
7	Electric dipole moment	0 (experimental upper limit: 1.8×10^{-26} e.cm)
8	Electric polarizability	$12.5(2.5) \times 10^{-4}$ fm^3

Table 1.2 Classification of neutron according to the energy range [17, 18]

Neutrons type	Energy range	Wavelength [Å]
Ultra-cold	$\leq$ 300 neV	$\geq$ 500
Very cold	300 neV–0.12 meV	52.2–26.1
Cold	0.12–12 meV	26.1–2.6
Thermal	12–100 meV	2.6–0.9
Epithermal	100 meV–1 eV	0.9–0.28
Slow neutrons	1–10 eV	0.28–0.09
Resonance neutrons	10–300 eV	0.09–0.016
Intermediate	300 eV–1 MeV	
Fast	1–20 MeV	
High-energy neutrons	> 20 MeV	

1.3.1 Ultra-Cold Neutrons

Ultra-cold neutrons (UCN) have wavelengths greater than 500 Å ($E <= 300$ neV) and have several unique features. The first observation of the storage of ultra-cold neutrons was reported by Luschikov et al. [15]. This remarkable experiment was carried out at the Joint Institute for Nuclear Research at Dubna, USSR. It may be noted that for most materials, the optical potential of neutrons is less than ~300 neV, and, as a result, they are subjected to total external reflection at any angle of impact of the surface of most materials. This property enables neutrons to be stored in neutron tanks (or traps) to observe several fundamental characteristics [16]. The so-called trapped or bottled neutrons are mostly useful in understanding and addressing fundamental questions in physics, measurement of the neutron lifetime, gravitationally bound quantum states, neutron electric dipole moment, neutron beta decay studies, etc.

1.3.2 Very Cold Neutrons

Very cold neutrons have energies between ultra-cold neutrons and cold neutrons, typically corresponding to neV and sub-meV energies, i.e. from 300 neV–0.12 meV. These neutrons are obtained by use of D_2O ice, solid D_2 or CD_4 moderator cooled to 2–4 K, that is, a moderator at the temperature of liquid helium. In recent years, interest in the use of very cold neutrons has been increasing due to the possibility of obtaining better sensitivity in the detection of water in a fuel cell, water uptake in plant roots, etc.

1.3.3 Cold Neutrons

Cold neutrons have energies that are lower than thermal energies, often in the meV and sub-meV range., i.e. from 0.12 meV to 12 meV. The term, cold neutrons, refers to those neutrons whose wavelength exceeds the Bragg edge of polycrystalline beryllium (4 Å). Cold neutrons are produced in a neutron moderator using either liquid hydrogen or deuterium, supercooled gas, or solid (methane or ice) at temperatures ranging from 20 to 40 K. Cold neutrons are employed in a wide range of applications, including biomolecular structure investigations and neutron imaging, because they have a greater scattering and absorption cross section than hot neutrons. Moreover, because of high interaction cross sections, neutron optics such as mirrors, guides and polarizers can be more efficiently used to tailor the neutron beam properties for experiments.

1.3.4 Thermal Neutrons

Thermal neutrons are produced by slowing down more energetic neutrons in an uncooled moderator after they have been ejected from atomic nuclei during nuclear reactions such as fission, fusion and spallation. The high-energy neutrons constantly slow down and pick up energy, resulting in a Maxwellian distribution of neutron energies centred around the most likely energy (thermal energy). The term "thermal neutrons" refers to neutrons in a spectrum that are in thermodynamic equilibrium with moderator material at a temperature of 293 K, and corresponding mean energy is $E_{\mathrm{th}} = 25.3$ meV. In thermal reactors, this region of the neutron's energy spectrum is the most important and useful part of the entire spectrum. A large number of instruments, such as those for neutron imaging and scattering, use thermal neutrons. Many research reactors also include so-called hot sources for use in short-wavelength applications, using a radiation-heated block of graphite at about 2000 K temperature. For a 2000 K temperature, the mean energy is $E_{\mathrm{hot}} = 200$ meV.

1.3.5 Epithermal Neutrons

Epithermal refers to those neutrons which have energies higher than 100 meV, typically above 0.4 eV, the cadmium cut-off. Epithermal neutrons have a larger penetration depth as compared to thermal neutrons. Epithermal neutrons are preferred for application requiring large penetration depth such as boron neutron capture therapy and epithermal neutron imaging.

1.3.6 Resonance Neutrons

This usually refers to the energy range of 10–300 eV. This energy range corresponds to a nuclide's or element's resonance energy, usually uranium and its isotopes. As most of nuclei with which neutrons interact have a complex nuclear structures, it results in an enhanced interaction cross section or resonances in this energy range.

1.3.7 Intermediate Energy Neutrons

Typically, they have energy ranging from a few hundred electron volts (~100 eV) to one million electron volts (~1 MeV). Intermediate energy neutrons can be used to carry out neutron imaging applications, where greater depth of penetration is desirable.

1.3.8 Fast Neutrons

Generally, between 1 and 20 MeV, these are the energies of neutrons emitted by fission, accelerator and spallation sources. Another way of producing monoenergetic energy neutrons of 2.5 MeV and 14 MeV energy is to use D–D and D–T fusion reactions. In turn, this represents the upper limit of neutron radiobiology research and fundamental neutron interaction cross-section studies.

1.3.9 High-Energy Neutrons

These are neutrons having energy above 20 MeV. High-energy neutrons deliver a major portion of the dosage and effective dose in outer space, and they represent the area of greatest uncertainty in the biological effects of neutrons in space.

Figure 1.2 shows the Monte Carlo-simulated spectral intensity plot of neutrons as a function of energy for Intense Pulsed Neutron Source at Argonne National Laboratory for 100 K liquid methane for illustration purpose. The figure also shows a 293 K Maxwellian spectrum, which has a peak at an energy greater than the 100 K methane spectrum.

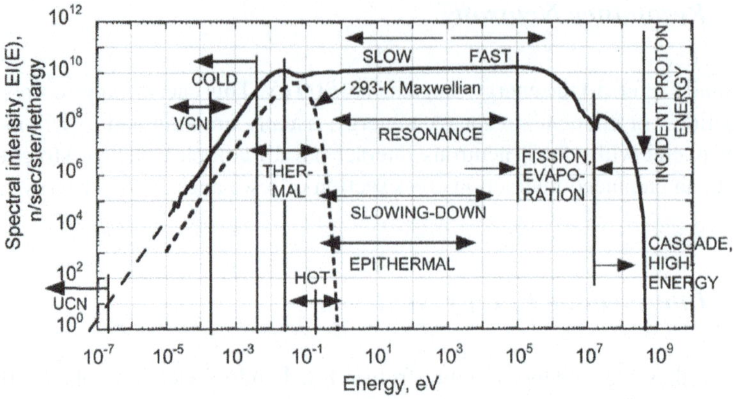

Fig. 1.2 Plot shows the spectrum of neutrons from 100 K liquid methane moderator of Intense Pulsed Neutron. *Source* at Argonne National Laboratory (reproduced from [19] with permission from Cambridge University Press)

1.4 Neutrons Matter Interaction

In the absence of charge and accompanying Coulomb interactions, neutrons interact with the nuclei of atoms by one or the both of the two mechanism: absorption and scattering (elastic scattering or inelastic scattering). Both of these interactions, absorption as well scattering, are highly dependent on neutron energy. Neutron absorption is due to neutron–nuclear interaction, and it usually leads to the release of gamma rays, charged particles, neutrons or fission fragments (Fig. 1.3). On the other hand, scattering could be due to nuclear interaction, magnetic interaction from unpaired electrons or Bragg scattering from the crystalline lattice and phonon and magnon excitation. As the wavelength and the energy of the thermal neutrons are comparable

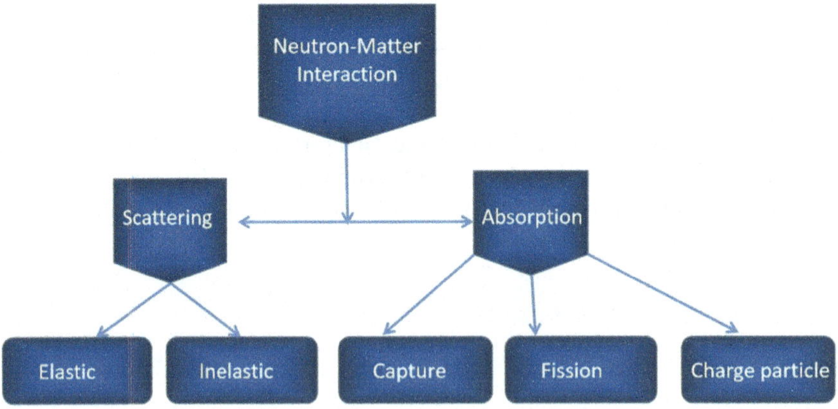

Fig. 1.3 Classification of scattering and absorption processes with neutrons

to lattice spacing and energy of various collective excitations in the crystals, they are useful in the study of crystal structures (including magnetic lattice), or phonon or magnon excitation. On the other hand for high-energy neutrons, nuclear interactions dominate over all other forms, and hence, they are more useful in the study of nuclear and particle physics, fast neutron activation analysis, etc.

The neutron matter interaction is best described through quantum mechanics in terms of wave function and probability density functions. This wave–particle duality is a fundamental idea behind quantum mechanics that describes the dynamics of subatomic particles, and therefore, one can sometimes refer to neutrons as particles and sometimes as waves. The parameter, namely scattering length (*b*), defines neutron interaction strength and is a phenomenological quantity. However, it cannot be calculated from fundamental principles, due to the complex nature of nuclear interactions.

The nuclear scattering lengths for the elements, as well as for individual isotopes, vary in a non-systematic manner with nucleus composition. The neutron scattering length of some elements is shown in Fig. 1.4.

As shown in Fig. 1.4, there is a markedly high difference in scattering length between hydrogen (^{1}H) and deuterium (^{2}H) ($b_H = -3.74$ fm and $b_D = 6.67$ fm). It should be noted that the scattering lengths of hydrogen and deuterium have distinct signs and hence used for contrast matching in many applications in soft matter research using neutron diffraction. Similarly, different isotopes of nickel show large variation in scattering length, and ^{58}Ni with the highest coherent scattering length is used for the fabrication of neutron optical elements such as neutron mirrors and guides. The quasi-random behaviour of scattering length of the neutron is in complete contrast with the more systematic scattering length of X-rays. It implies that neutrons can help in distinguishing isotopes or nearby elements in the periodic table.

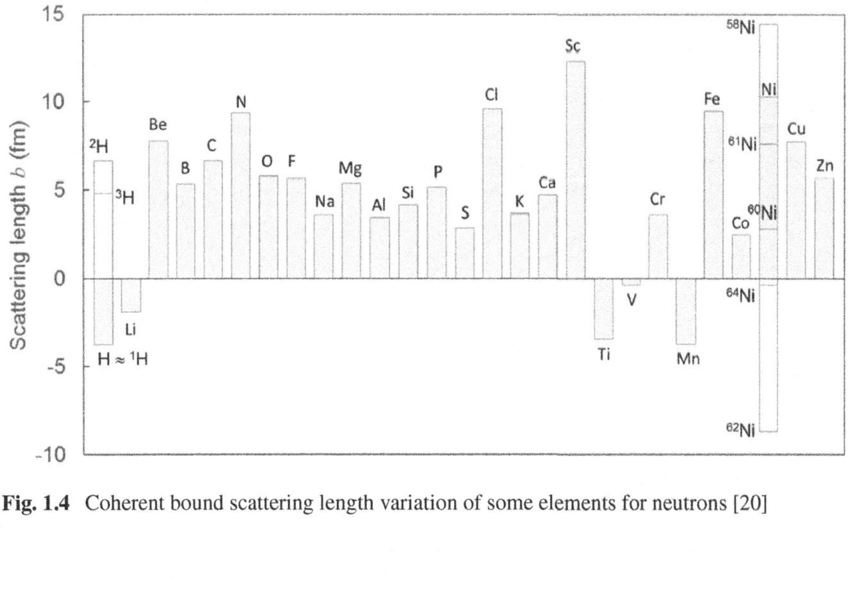

Fig. 1.4 Coherent bound scattering length variation of some elements for neutrons [20]

However, scattering lengths are not experimentally measurable parameters. Neutron scattering experiments measure the number of neutrons scattered by a sample according to the change in wave vector (Q) and the change in energy (E) of the neutron, described by the partial differential scattering cross section:

$$\frac{d^2\sigma_s}{\mathrm{d}\Omega\mathrm{d}E} = \frac{\text{no. of neutrons scattered per sec into d}\Omega\text{ in the energy range }[E,\ E+\mathrm{d}E]}{\mathrm{d}\Omega\mathrm{d}E \times \text{ incident neutron flux}} \tag{1.3}$$

The scattering intensity is divided by the neutron flux to ensure that σ is an intrinsic property, independent of the neutron flux at the particular experimental set-up. Integrating over all final energies, we reach the differential cross section

$$\frac{d\sigma_s}{\mathrm{d}\Omega} = \int \frac{d^2\sigma_s}{\mathrm{d}\Omega\mathrm{d}E}\mathrm{d}E \tag{1.4}$$

The total cross-section is found by a double integration, with dΩ as solid angle:

$$\sigma_s = \int \frac{d^2\sigma_s}{\mathrm{d}\Omega\mathrm{d}E}\mathrm{d}E\mathrm{d}\Omega \tag{1.5}$$

The neutron scattering cross section, σ_s, of a system is defined by its ability to scatter neutrons, and it has the unit of area. For a single nucleus, σ_s can now be seen as the effective area of the nucleus perpendicular to the neutron beam. The total cross section is a sum of two quantities: a term representing the probability of absorption by the nucleus (σ_a) and a term representing the probability of scattering by nucleus (σ_s). Figure 1.5 illustrates cross sections of some selected isotopes as a function of neutron energy in selected isotopes and hydrogen-rich plastic foils, showing that usually the cross section decreases as neutron energy increases.

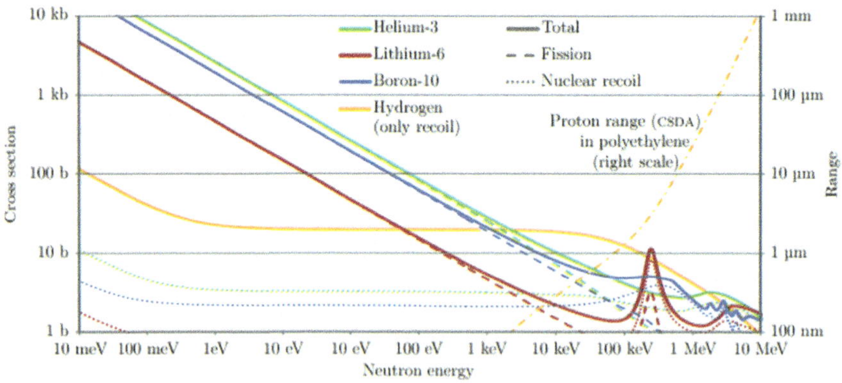

Fig. 1.5 Cross sections of various selected isotopes as a function of neutron energy and hydrogen-rich plastic foils [21]

In order to develop these ideas further, let us assume that neutron and matter interaction can be described by some finite range potential such that $V(r) = 0$ if $r > b$. Therefore, one needs to find out the solution of the time-independent Schrodinger equation,

$$\nabla^2\psi(\mathbf{r}) + K^2\psi(\mathbf{r}) = 0 \tag{1.6}$$

where

$$\begin{aligned} K^2(\boldsymbol{r}) &= \frac{2m}{\hbar^2}[E - V(\boldsymbol{r})]. \\ k^2(\boldsymbol{r}) &= \frac{2m}{\hbar^2}E. \end{aligned} \tag{1.7}$$

With the wave vector $\boldsymbol{k}$ outside the medium and wave vector $\boldsymbol{K}$ inside the medium (region of the potential), the incident particle being free, its wave function can be approximated by a plane wave

$$\psi = A\exp(-ik \cdot r) \tag{1.8}$$

where k is the incident wave vector and A is a normalization constant. After interaction with the matter medium, the amplitude of the outgoing wave can be approximated by

$$\psi = \frac{f(\theta, \phi)}{r}\exp(-ik \cdot r) \tag{1.9}$$

where $f(\theta,\ \phi)$ is the direction-dependent term and known as scattering amplitude. For completely isotopic scatter, needless to say, that it would be constant. The total wave function of the system consisting of the incident and scattered wave therefore can be represented as the sum of the incident and outgoing wave function as follows:

$$\psi = A\exp(-ik \cdot r) + \frac{f(\theta, \phi)}{r}\exp(-ik \cdot r) \tag{1.10}$$

Generally, we have the following relation between differential scattering cross section and scattering amplitude:

$$\frac{d\sigma}{d\Omega} = |f(\theta,\ \phi)|^2 \tag{1.11}$$

and total scattering cross section is given by:

$$\sigma_s = \int |f(\theta, \varphi)|^2 d\Omega \tag{1.12}$$

The total collision cross section, including both scattering and absorption, is given by the optical theorem

$$\sigma_t = \sigma_s + \sigma_a = \frac{4\pi}{k}\text{im}\,[f(0)] \tag{1.13}$$

The scattering amplitude can be expressed in terms of partial wave expansion. Consider a special case of scattering of the slow neutron from a fixed nucleus. When a neutron collides with this area, it scatters isotropically, that is, equally in all directions. Because the range of the nuclear interaction between the neutron and the nucleus is small in comparison with the wavelength of the neutron, the scattering is isotropic, and the nucleus appears to be a point scatterer. Generally, the scattering amplitude can be approximated by complex number, the scattering length b, which in general is complex quantity:

$$f(\theta, \phi) = -b + ikb^2 + O(k^2) \quad \text{where } b = b' - b'' \tag{1.14}$$

in which the k^2 term contains the effective range correction. When the higher-order partial waves are included, the k term also contains an anisotropic p-wave contribution.

Therefore, we can write

$$\sigma_s = 4\pi |b|^2 \left(1 - 2kb'' + o(k^2)\right) \tag{1.15}$$

$$\sigma_a = \frac{4\pi}{k} b'' \left(1 - 2kb'' + o(k^2)\right) \tag{1.16}$$

As it turns out that for most of the practical cases, $kb'' \ll 1$ and $b'' \ll b'$. Therefore, scattering and absorption cross sections can be as well approximated by the following simple expressions:

$$\begin{aligned} \sigma_s &= 4\pi |b|^2 \\ \sigma_a &= \frac{4\pi}{k} b'' \end{aligned} \tag{1.17}$$

In the above discussion, we have tacitly assumed that there are no resonances, which is usually true for thermal and cold neutron energies. In most practical cases, the interaction potential for s-wave scattering can be described in terms of scattering length b as follows (Fermi pseudopotential), which describes the effect of jth hard nuclei at the positions R_j.

$$V(\boldsymbol{r}) = \frac{2\pi\hbar^2}{m} \sum_j b_j \delta(\boldsymbol{r} - R_j) \tag{1.18}$$

Further, neutron scattering from several nuclei can be coherent or incoherent depending on whether the scattered waves from different scattering centres interfere or not. When considering elastic scattering from a crystal, the distinction between coherent and incoherent scattering becomes quite evident. Bragg peaks carry information about the relative positions of the atoms and results from coherent scattering of neutrons as a function of scattering angle. With the exception of isotopes with null nuclear spin (e.g. C, O), there are two possibilities in the interaction between a neutron and a nucleus: each atom's nuclear spin (I) couples with the neutron in either a parallel or anti-parallel manner. Since the neutron has a spin equal to 1/2, the spin of the ensemble, for each isotope, is $I + 1/2$ or $I - 1/2$ with the respective probabilities of $f_+ = I/2I + 1$ and $f_- = I/2I - 1$. Each interaction process corresponds to different scattering lengths, b_+ and b_-, respectively. We can define the average and the standard deviation that give, respectively, the "coherent" and "incoherent" scattering lengths:

$$\begin{aligned} b_c &= f_+ b_+ + f_- b_- \\ b_i &= \sqrt{\overline{b^2} - \overline{b}^2} \end{aligned} \tag{1.19}$$

Finally, we can define the neutron–nucleus interaction through the refractive index of the medium:

$$n(\boldsymbol{r}) = \frac{|\boldsymbol{K}(\boldsymbol{r})|}{|\boldsymbol{k}|} = \sqrt{1 - \frac{V(\boldsymbol{r})}{E}} \tag{1.20}$$

The mean interaction potential, or *optical potential*, for material is defined as follows:

$$\langle V(\boldsymbol{r})\rangle = \overline{V} = \frac{2\pi\hbar^2}{m} b_c N \tag{1.21}$$

where N is the mean number of scattering nuclei per unit volume and $b_c = \langle b\rangle$ is the mean coherent scattering length. Accordingly, the complex index of refraction can be defined as follows:

$$n = \sqrt{1 - \frac{\overline{V}}{E}} = 1 - \frac{\lambda^2 N}{2\pi}\sqrt{b_C^2 - \left(\frac{\sigma_r}{2\lambda}\right)^2} + i\frac{\sigma_r N\lambda}{4\pi} = n_r + in_i \tag{1.22}$$

The complex refraction index counts for absorption σ_a and incoherent scattering σ_i processes. $\sigma_r = \sigma_a + \sigma_i$ is the total reaction cross section per atom. If the imaginary part is small, therefore the index of refraction can be approximated as follows:

$$n = 1 - \frac{\lambda^2 N b_C}{2\pi} \tag{1.23}$$

1.5 Neutrons Imaging

Imaging science is a multidisciplinary field where two/three-dimensional pictorial distribution of certain object property is obtained. The most often useful property is the attenuation of the incoming radiation, as usually done in medical diagnostics, where X-ray radiography or a CT scan gives differential attenuation of the X-ray through the bones or tissues. These techniques work on extracting the relevant properties such as absorption coefficients which depend on the interaction of incoming radiation with the matter. Neutron imaging refers to a group of non-destructive testing methods, which exploit the transmission of neutrons through materials to probe the internal structure of objects, magnetic field distribution, etc. Owing to the large penetration depth in various materials, the development of neutron imaging started soon after the nuclear research reactor started operating. Only three years after the neutron was discovered, Hartmut Kallmann and Ernst Kuhn pioneered neutron imaging. Significant efforts by Thewlis marked the beginning of neutron imaging towards practical applications when he demonstrated it by utilizing a neutron beam from the British Experimental Pile (BEPO) reactor at Harwell. Initially, neutron imaging was carried out on the instruments used for neutron diffraction studies; however, it was soon realized that limited flux and small beam size did not permit the use of neutron for any practical sample. However, after J. P. Barton pioneered the divergent type of neutron collimators, leading to improved neutron flux and a larger field of view, thereby ushering a new era in the development of neutron imaging. Dr. W. L. Whittemore was the driving force behind the SNT-TC-1A personnel certification standard, which was first published in 1974 under the auspices of the American Society for Nondestructive Testing (ASNT). Until the 1990s, neutron imaging was carried out using film-based techniques, and hence, neutron radiography (two dimensional) was the only technique available to researchers. Since then, detection methods have improved drastically: digital neutron detectors and fast data acquisition and processing systems have been introduced, and neutron tomography has been implemented on most of the neutron imaging facilities. Moreover, introduction of new neutron optics like neutron guides and mirrors has also contributed to increased neutron flux at the sample positions. As a result of these innovations, and owing to the introduction of new imaging modalities, different from the conventional attenuation contrasts, new and challenging application fields have been established. Figure 1.6 summarizes different types of neutron imaging modalities along with the length scale probed by each one of them.

Broadly speaking, the current and the field of neutron imaging can be classified into four following major groups.

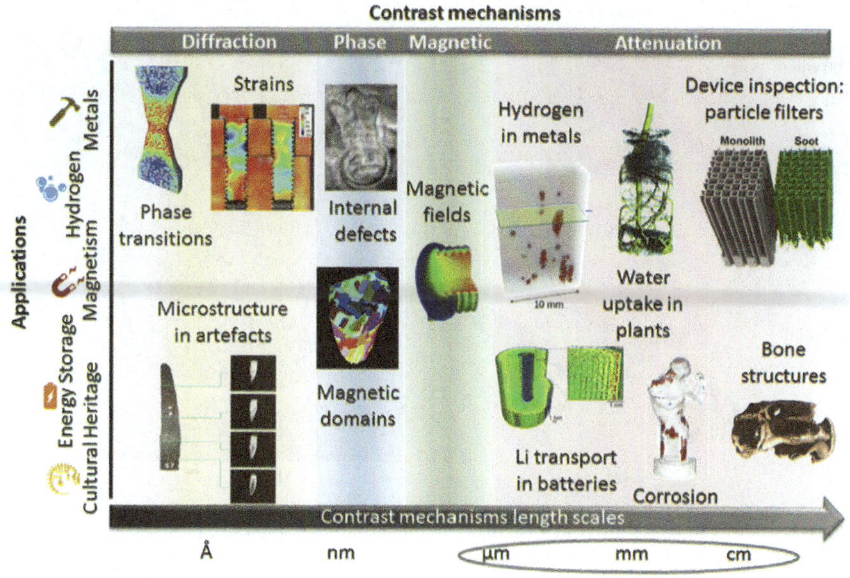

Fig. 1.6 Different contrast mechanism in neutron imaging (reproduced from [22] © 2018 Elsevier Ltd., with permission from Elsevier)

1.5.1 Attenuation-Based Neutron Imaging

This is the oldest scheme of neutron imaging, where an area detector (digital or film) measures the neutron beam transmission after it traverses through the object. The attenuation of the neutron beam can be easily modelled through Lambert–Beer law, which is given as $I = I_0 \exp(-\mu x)$, where μ is the neutron beam attenuation coefficient and includes all the ways neutrons beam is removed out of the path, due to various interaction mechanisms. This technique works best for those materials which have higher neutron absorption or incoherent scattering cross section. Another primary use of neutrons is in the structural investigation of high atomic number elements, where the X-ray fails to reveal internal or hidden details of the sample.

It may be noted that hydrogen (^{1}H) has the highest incoherent cross section, and therefore, this technique finds application where hydrogen detection or quantification is of primary concern. Moreover, one can collect the radiographs at different rotation angle, and complete three-dimensional distribution of the elements can be easily mapped. Figure 1.7 shows the three-dimensional distribution of hydrogen in the Zr alloy coupons, mapped using three-dimensional neutron tomography. The hydrogen concentrations were determined through use of calibrated samples.

As neutrons can penetrate in the metallic materials to larger depths, neutron radiography has also been widely used to map two-dimensional flow fields inside the metallic pipes as shown in Fig. 1.8.

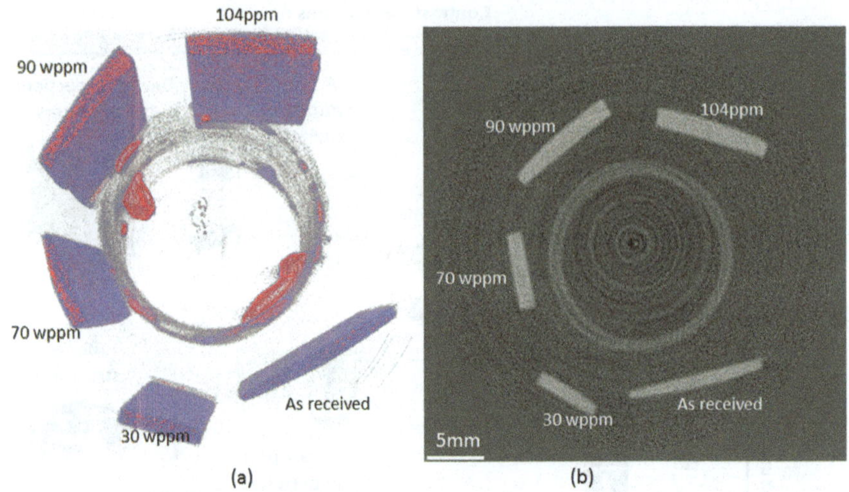

Fig. 1.7 Hydrogen distribution in Zr alloy using neutron tomography (reproduced from [23] © 2018, with permission from Elsevier)

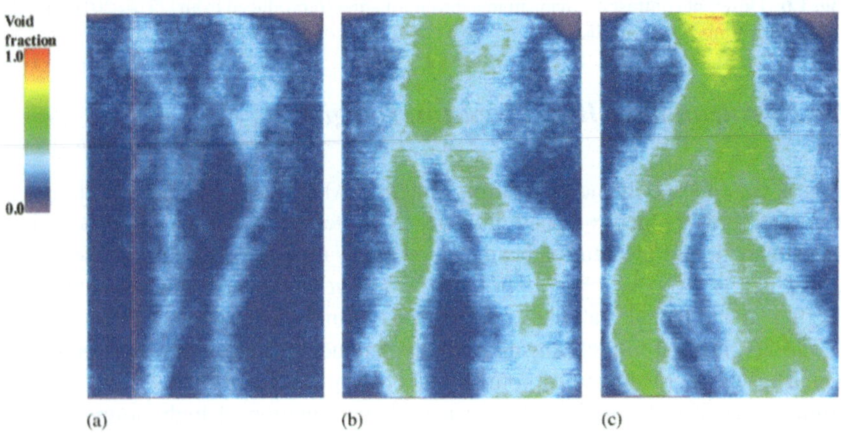

Fig. 1.8 Time-averaged void fraction profile taken by high frame rate neutron radiography (500 fps) for different gas flow rate. **a** $j_g = 1.4$ cm/s. **b** $j_g = 3.9$ cm/s. **c** $j_g = 12.3$ cm/s (reproduced from [24] © 2004 with permission from Elsevier)

The conventional neutron radiography also finds application in mapping the distribution of other elements such as lithium (Li-6), boron (B-10), cadmium (Cd-113) and gadolinium (Gd-157), as these elements have very large neutron absorption cross section. Figure 1.9 shows the dynamic neutron radiography was used to investigate lithium dendrite development during battery cycling. It shows dynamic distribution of lithium from the anode to cathode during charging and discharging cycles, induced by the internal short circuit due to lithium dendrite growth.

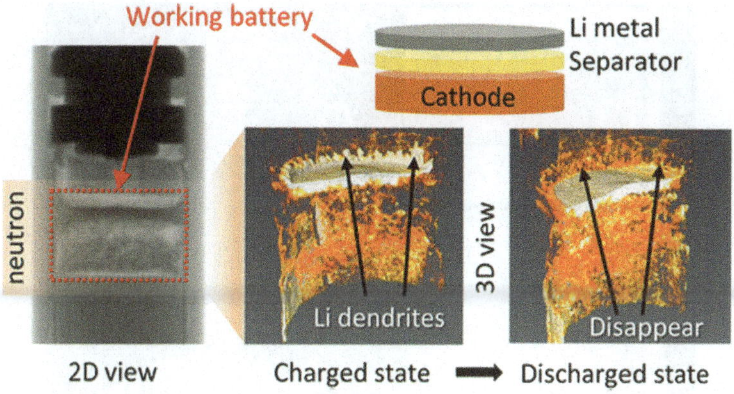

Fig. 1.9 Lithium distribution using in-operando neutron imaging (reproduced from [25] Copyright © 2019, with permission from American Chemical Society)

1.5.2 Phase Contrast Neutron Imaging

Conventional neutron-based imaging is based on so-called geometrical or ray optics. However, as neutron can also be treated as waves, we can associate a phase to the neutron wave, in analogy with the electromagnetic waves. This approach opens another avenue, namely the phase-based imaging techniques to study materials with low attenuation cross section, wherein the conventional attenuation-based imaging fails to reveal details within the sample under investigation. As the phase contrast effects are observed along with attenuation-based images, even the conventional neutron radiography improves in terms of detection sensitivity. A variety of interferometric and non-interferometric phase contrast techniques have been developed to measure either the slope of the phase or the curvature of the phase. Figure 1.10 shows an example of propagation-based phase contrast imaging using neutrons, and it clearly brings out the advantage of phase-based detection over the absorption-based one. Over the years, more sophisticated approaches which use crystal optics and near-field grating have also been developed to overcome the stringent requirement on the coherence as demanded by the propagation-based phase contrast technique. Another advantage of these phase contrast approaches is that we can obtain absorption, phase contrast and small-angle scatter contrast images from the same data set. The small-angle scatter contrast in these imaging modalities arises due to the presence of unresolved structures or structural inhomogeneities, below the point spread function of the system. Figure 1.11 shows tomographic reconstruction for a sample consisting of lead (black) and titanium (white) rods using grating-based neutron phase contrast imaging technique, showing the excellent sensitivity of the neutron phase contrast imaging in detection of materials with low neutron absorption cross section.

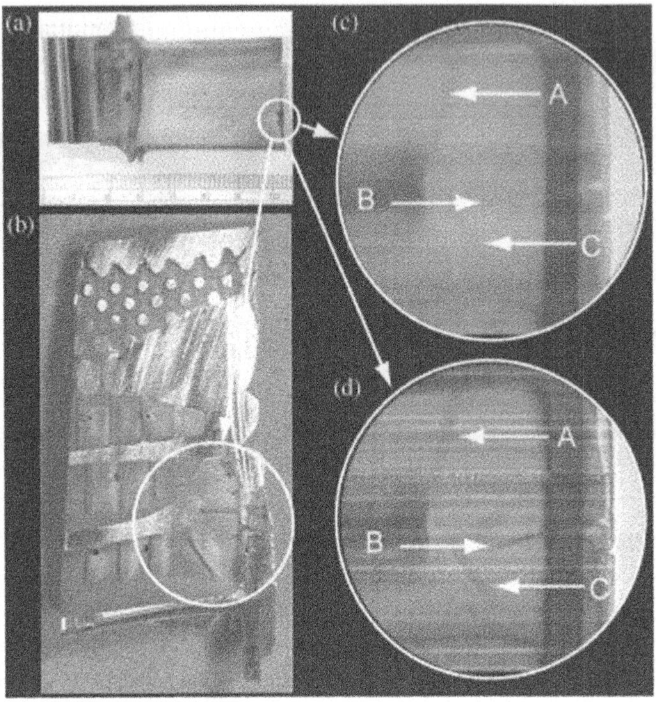

Fig. 1.10 **a** Encircled part of turbine blade was inspected for cracks, **b** destructive inspection of the turbine blade, **c** absorption-based neutron radiography, **d** phase contrast imaging of the same. The faint crack lines in **c** labelled with arrows *A*, *B* and *C* are greatly enhanced by phase contrast (reproduced from [26] © 2004, with permission from Elsevier)

1.5.3 Polarized Neutron Imaging

As discussed previously, although the neutron has zero electric charge, it does have a nonzero negative magnetic moment. The negative value of the magnetic moment implies that it is anti-parallel to spin angular momentum. This makes it possible for neutrons to interact with the magnetic moments of the atoms and, therefore, is used to study the detailed magnetic ordering in the solid materials. However, another application of neutron magnetic moment could be to obtain the spatial distribution of magnetic fields inside superconductors, visualization of DC or AC magnetic fields. Imaging with polarized neutrons is an important non-destructive technique to obtain the localized distribution of magnetic fields. For the polarized neutron imaging, a well-defined polarized neutron beam is initially selected using a supermirror or He^3-based polarizer. As the polarized neutron beam interacts with magnetic field of interest, the polarization vector undergoes precession, and the precession angle of the neutron polarization vector can be measured using an analyser. When used in combination of two-dimensional imaging detectors, and suitable data processing algorithm, the precession angles can be used to map the magnetic field direction

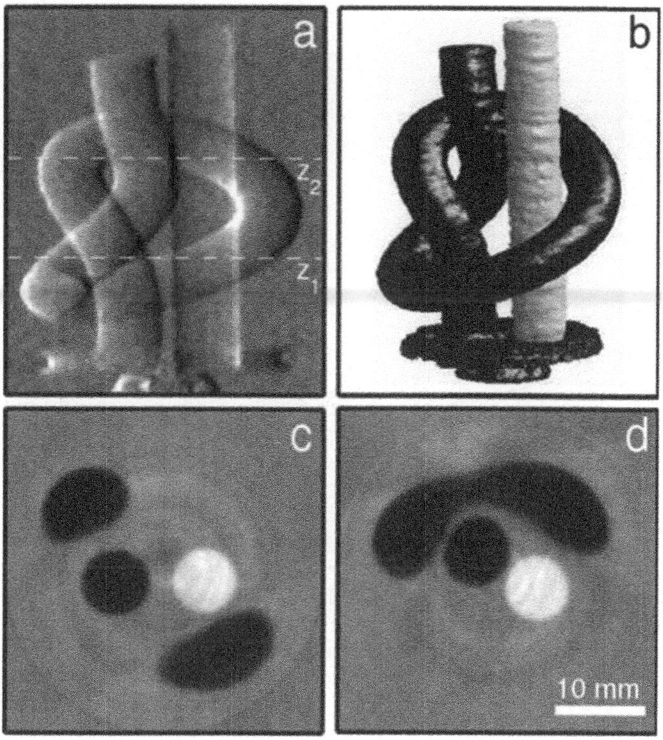

Fig. 1.11 Projection images and tomographic reconstruction of a sample made up of 6.35 mm diameter lead (black) and titanium (white) rods. **a** Image of a differential phase contrast projection. **b** A three-dimensional depiction of the sample's recovered coherent scattering length distribution. **c**, **d** two (out of 512) reconstructed tomographic slices (reprinted from [27] © 2006 with permission from the American Physical Society)

within the sample or given region of interest. The technique has been used to visualize modest magnetic fields (a few mT) with a low spatial resolution (100 μm) as shown in Fig. 1.12. In order to quantify the field integrals, one need to collect the data at multiple wavelengths, as a single measurement may have uncertainty of $\pm 2\pi$. However, quantification of spin rotation is more difficult if the fields are arbitrarily directed and not aligned perpendicular to the spin orientation.

Ramsey's technique of separated oscillatory fields, originally introduced for molecular beam resonance, has also been adopted for polarized neutron imaging. Two $\pi/2$ phase-locked spin flippers, oscillating at some frequency (ω), positioned in a homogeneous magnetic field region with the field strength B, flip the spins of a monochromatic, polarized neutron beam twice non-adiabatically. The neutron spin between the flippers precess with Larmor frequency (ω_0) in the plane perpendicular to the continuous magnetic field. After passing through the second flipper, neutrons pass through spin analyser before being detected by a two-dimensional area detector.

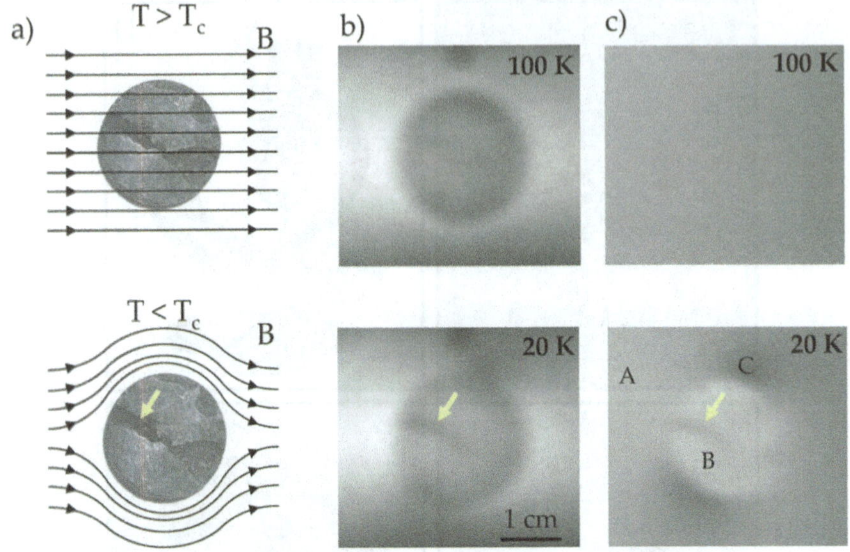

Fig. 1.12 Images of the magnetic field due to the Meissner effect in a YBCO pellet (superconducting) above and below the critical temperature ($Tc = 90$ K), displaying **a** schematic diagrams, **b** raw (unnormalized) radiographs and **c** radiographs normalized to a reference image. The reference was an image of the sample in zero field at 100 K. The arrow indicates a region where expulsion of the field is incomplete [28]

By scanning the frequency (ω) of the oscillating magnetic fields close to Larmor resonance (ω_0) and measuring the consequent spin polarization yields a so-called Ramsey oscillation pattern. By fitting these Ramsey patterns, a phase change at each pixel is obtained. Figure 1.13 shows an example of absorption and spin phase image using the proposed approach.

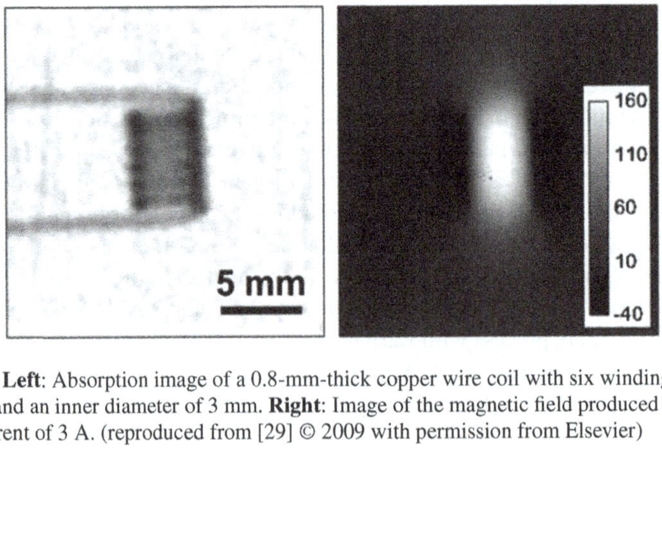

Fig. 1.13 **Left**: Absorption image of a 0.8-mm-thick copper wire coil with six windings, a length of 7 mm, and an inner diameter of 3 mm. **Right**: Image of the magnetic field produced by this coil with a current of 3 A. (reproduced from [29] © 2009 with permission from Elsevier)

The problem of precession of neutron spin in the external and sample fields can be corrected by a second, counter-oriented field to address the shortcomings of Ramsey spin-based phase imaging. This enables spin-echo polarized neutron imaging, in which neutron spin is precessed in the pre-sample flight route, transmitted through the sample and then precessed again in the post-sample flight path. A neutron spin analyser is used to monitor the number of spin precessions. This allows the use of low monochromatized neutron beam and quantification of field integrals which are also simplified due to spin-echo effects.

1.5.4 Bragg Edge Imaging

Bragg edge imaging, a wavelength-dependent neutron radiography, is a non-destructive characterization method and is used to obtain quantitative spatial information on the atomic and internal structures of bulk matter. In the traditional neutron diffraction experiments, Bragg diffraction peaks can be seen at wavelengths where the Bragg conditions are satisfied. It may be noted that the instruments at reactor sources record diffraction patterns as a function of diffraction angle, while pulsed sources record diffraction patterns as a function of wavelength at fixed diffraction angles. On the contrary, Bragg edge imaging is carried out in transmission mode and spectrum is recorded as a function of wavelength, and recorded spectrum displays abrupt changes in the transmission called Bragg edges. Bragg edges in the transmission spectrum are the characteristic of Bragg scattering from polycrystalline materials. For a specific crystal lattice family (hkl), with lattice spacing d_{hkl}, the scattering angle increases with the wavelength up to $\lambda = 2d_{\text{hkl}}\sin(\theta) = 2d_{\text{hkl}}\sin\ (\pi/2)$. The Bragg condition is no longer satisfied beyond this wavelength, resulting in a dramatic drop in the material's attenuation, known as the Bragg edge. Thus, the characterization of crystalline properties of materials such as lattice strains or phase fractions is possible, thanks to the analysis of the Bragg edges. Figure 1.14 shows an example of (110) lattice strain variation as obtained through the Bragg edge imaging technique under in situ load conditions.

Another useful imaging modality called white beam neutron topography, which is based on Bragg diffraction, has been used for visualization of defects such as dislocations and twins present within the crystal volume. In the neutron topography, the image of the object is recorded via Bragg diffraction by the object out of the incident beam, rather than the differential absorption of neutrons along the beam path, as in traditional transmission radiography. Neutron topography has proved useful for monitoring crystal growth and crystal quality and visualizing defects in many different crystalline materials. One can use both monochromatic and polychromatic neutron beam to carry out the measurements. Anti-ferromagnetic domains, chirality domains and other anti-ferromagnetic domains have all benefited from polarized neutron topography [31].

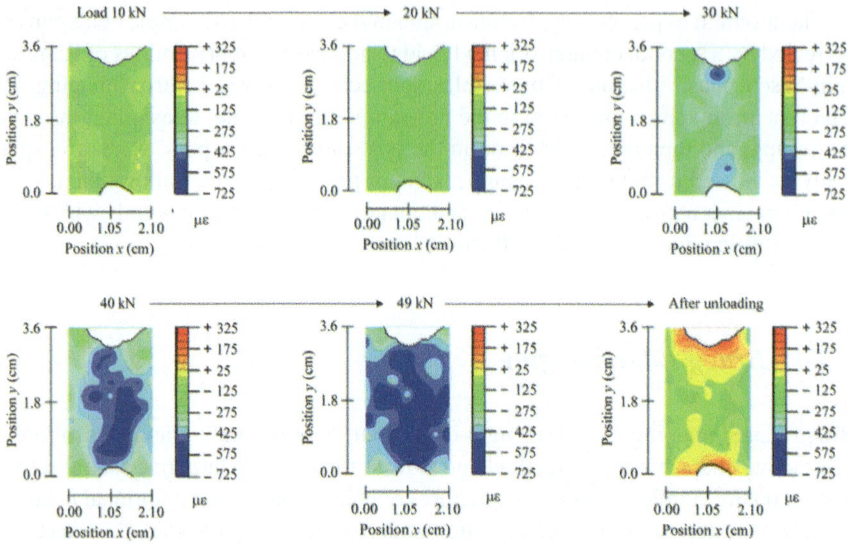

Fig. 1.14 In situ measurement of 110 lattice strain mapping during deformation due to external load using Bragg edge imaging in a 5-mm-thick ferritic steel plate (reproduced with permission of the International Union of Crystallography [30])

1.6 X-ray Versus Neutron Imaging

Soon after Wilhelm Rontgen discovered X-rays, it started being widely used in medical diagnostics and industrial non-destructive testing, due to the larger penetration depth of new radiation. Neutrons being charge neutral have even larger penetration depths in most materials and therefore offer a unique way to increase the probing depths. All traditional radiography technologies (such as X-ray or neutron) work on the same principle: radiation is attenuated when it passes through matter. The specimen is positioned in the path of the incoming incident radiation beam. After passing through sample, the unattenuated beam is detected at the detector, which records the fraction of the initial intensity transmitted by each point in the object. Inhomogeneities in the object or an internal defect (e.g. void, crack, porosity or inclusion) will manifest as a change in intensity of incident beam as recorded by the detector. Defect detection in conventional radiography is thus based on observing differences in radiation intensity as it passes through the object under examination. This occurs as in accordance with the Beer–Lambert law, which can be expressed more precisely as follows:

$$I = I_0 e^{-\int \mu \mathrm{d}l} \tag{1.24}$$

where I_0 and I are the incident and attenuated intensities, respectively, μ is the linear attenuation coefficient of the material (combining the interaction cross section and nuclear density), and dl is the path length through the sample.

As shown in Eq. (1.24), radiography provides integrated information about the amount of attenuation, which cannot be split into contributions from distinct elements and the amount of material corresponding to the path length. This is standard "inverse problem", though ill-posed, can be solved using the concept of radon transform and Fourier back projection techniques or through iterative reconstruction methods and is called "tomography". As a result, in order to obtain the distribution of the attenuation map, multiple radiographs (projections) for different angular orientations must be collected. The tomographic scan produces three-dimensional spatially resolved images (i.e. volumetric data) that show the distribution of attenuation coefficients in the sample volume.

Regardless of the type of radiation used, the broad strokes of the imaging method are the same. A source emits the radiation, which passes through a collimator. The collimator filters and restricts the passage of radiation ensuring that only parallel beams pass through. Beam collimation improves image quality, resulting in a sharper and more detailed image. Once the beam is collimated, it collides with the object to be imaged. However, there is one important difference: the attenuation X-ray cross section varies linearly with atomic number, whereas it is essentially random for neutrons (Fig. 1.15). This is attributed to the very different nature of their interaction within the matter: X-ray interacts primarily with electron cloud while neutron interacts with nucleus. Figure 1.16 shows the contrasting behaviour of X-ray and neutron radiographic imaging. The flower being rich in hydrogenous matter offers much better attenuation to neutrons as compared to the X-rays. Therefore, one can see the flower in neutron radiography even if hidden in a lead cask of 5 mm thickness. On the other hand, X-rays interact mostly with the lead matrix of cask, and flower

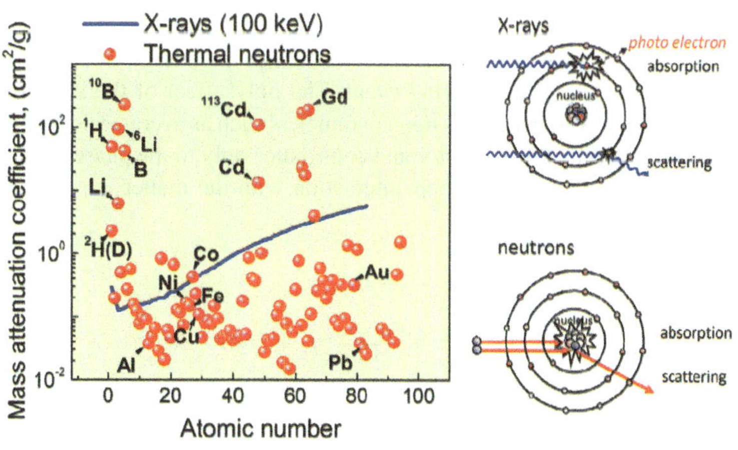

Fig. 1.15 Mass attenuation coefficient of thermal neutrons and 100 keV X-rays (reproduced from [32] © 2011, with permission from Elsevier)

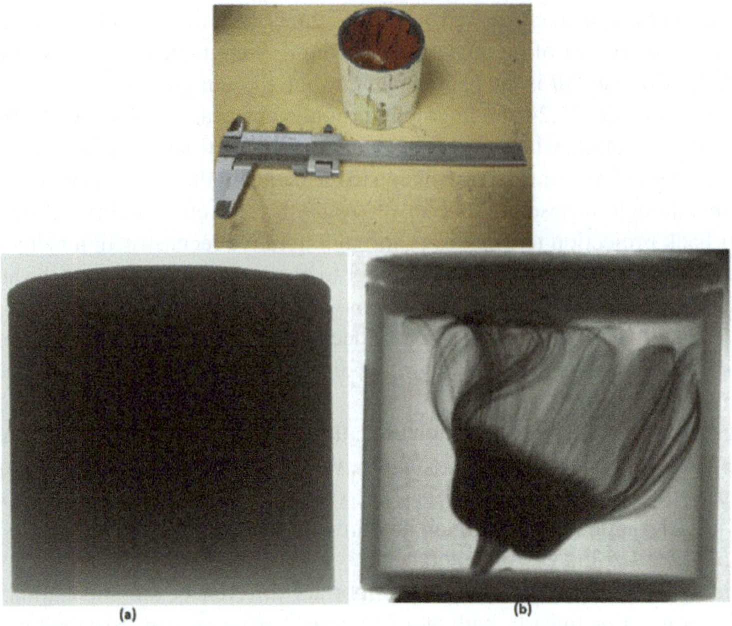

Fig. 1.16 Comparison of (**a**) X-ray and (**b**) neutron radiographs showing the sensitivity of neutron to hydrogeneous materials

being rich in hydrogenous matter, offers very little attenuation to X-rays and hence not visible in X-ray radiograph.

X-rays can undergo elastic or inelastic scattering while passing through a material medium. Photoelectric absorption and Compton scattering are the two major modes of inelastic scattering of X-rays. This results in loss of intensity in the forward direction, i.e. removal of photons along the beam path and hence both are treated as absorption phenomenon. In the elastic scattering, the energy of X-ray photons remains unchanged by the scattering event. The only effect of the scattering is a change of direction which depends upon parameters such as average electron density distribution and incident wave vector and contributes only to the phase shift. X-rays are electromagnetic waves, and their interaction with the matter is governed by a complex refractive index:

$$\begin{aligned} n &= 1 - \delta - i\beta \\ &= 1 - \frac{n_a r_e \lambda^2}{2\pi}(f_1 + i f_2) \end{aligned} \tag{1.25}$$

where $\delta = \frac{n_a r_e \lambda^2}{2\pi} f_1$ and $\beta = \frac{n_a r_e \lambda^2}{2\pi} f_2$.

The quantity $f = f_1 + i f_2$ is known as the effective atomic scattering factor. $f_1 \propto Z$ is responsible for the phase shift and is dependent upon the number of electrons, and $f_2 \propto \mu$ is responsible for the attenuation. In the energy regime where the photoelectric effect dominates, $\mu \propto Z^4$, implying that the higher the atomic number, the higher is the attenuation contrast.

It may be briefly recalled that quantum mechanics permits the neutron to be treated as a wave and hence corresponding wave equation is given by the Schrodinger equation. Taking a cue from the above treatment, one can compare neutron and X-ray interaction through the following definition of a complex refractive index for neutron:

$$\begin{aligned} n &= 1 - \delta - i\beta \\ &= 1 - \frac{\lambda^2 N}{2\pi}\sqrt{b_c^2 - \left(\frac{\sigma_r}{2\lambda}\right)^2} - i\frac{\sigma_r N\lambda}{4\pi} \end{aligned} \tag{1.26}$$

Thus, we can see that this approach of defining the refractive index is very similar to that of X-ray imaging. Thus, one can say that phase contrast imaging exploits the real part of the refractive index $(1-\delta)$ while neutron radiography exploits the imaginary part (β). In comparison with the refractive index of X-rays, which is generally slightly less than unity, the refractive index of neutrons can either be greater than or less than unity. This is due to the fact that the coherent scattering length for neutrons can be positive or negative. Moreover as shown previously, the nuclear scattering lengths vary in a very random way, and therefore, absorption cross section does not exhibit any systemic behaviour.

The resolution and penetrability of X-rays and neutrons are quite different. Neutrons are more penetrative; however, they have a lower resolution (~100 μm) and imaging speed (~*s*). However, neutron and X-ray images complement each other perfectly. Therefore, it is a good idea to probe the sample using both X-ray and neutron and carry out correlative X-ray and neutron imaging studies imaging. Figure 1.17 shows comparison of X-ray and neutron tomography and brings out their respective importance in non-destructive investigation of energy storage device such as batteries. As seen from these tomography data, structural details are readily visible in X-rays, while neutrons offer much better understanding of water kinetics of the chemical reaction. As shown in Fig. 1.18, correlative X-ray and neutron imaging is a powerful tool in non-destructive analysis where both structural and chemical information is needed. X-ray computed tomography enables identification of mechanical degradation processes while neutron tomography is useful in understanding of lithium diffusion process. This method is extremely useful because X-ray tomography provides mechanical information about the electrodes, such as structure and volume, while neutron tomography provides electro-chemical information, such as Li-ion transportation and intercalation, electrolyte consumption and wettability, among other things.

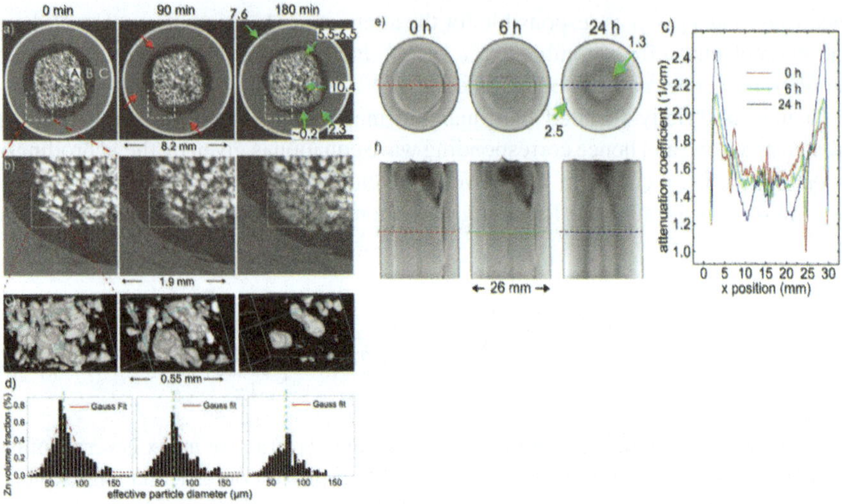

Fig. 1.17 Comparison of X-ray and neutron radiographs showing internal details of conventional battery. X-rays tomography shows up the structure details (**a–d**), while water migration is quite clear in neutron imaging (**e–f**) (reproduced from [33] with permission from Copyright © 2007 American Institute of Physics)

1.7 Summary

Neutrons, as free chargeless and spin 1/2 particles, have properties that make them unique tool to probe into a wide variety of materials. Neutron-based techniques complement a number of other diffraction (X-rays, electrons) and spectroscopy-based experimental methods. Table 1.3 summarizes a comparison of some properties of neutron and X-rays. The large mean free of neutron in most of materials and their ability to reveal the internal structure and distinguish between isotopes has made them an important tool in scientific and industrial research. Advances in instrumentation and refinements in the experimental techniques in the last two decades have led to the development of new neutron imaging modalities. Several new neutron imaging facilities with better neutron fluence and brilliance are under development. In particular, in recent years neutron imaging has been developed into powerful alternative to X-ray imaging techniques to studying the internal structure of the matter and their magnetic properties. The advanced methods phase contrast imaging, Bragg edge imaging and polarized neutron imaging provide more detailed information, e.g. visualization of the crystalline structure, magnetic domain walls or the crystal orientation. The neutron imaging techniques are also being used by industrial users, thereby expanding the scope of neutron imaging and making instruments available to the greater benefit of community.

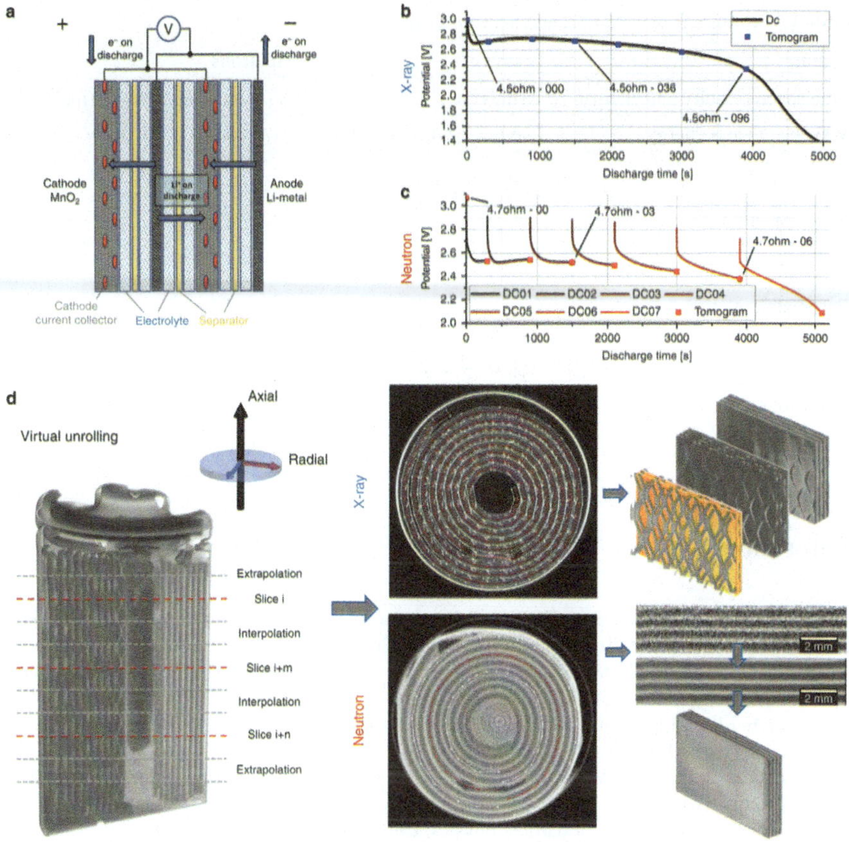

Fig. 1.18 **a** Schematic of Li/MnO2 CR2 primary cell from duracell **b** shows the constant resistance discharge curve for the CR2 cell, **c** depicts the constant resistance discharge curve, **d** shows reconstructed tomograms from neutron and X-ray CT scans, as well as examples of sections extracted after virtual unrolling of the reconstructions. The nickel current collecting mesh, which appears brighter than the Li_xMnO_2 active electrode material, is clearly visible in the X-ray images (reproduced from [34] with permission from Springer Nature)

Table 1.3 Comparison of X-rays and neutron's properties

Feature	X-rays	Neutrons
Free space velocity and energy	$V(\mathrm{m/s}) = c$	$V(\mathrm{m/s}) = 437.39\sqrt{E(\mathrm{meV})}$
Wavelength and energy	$\lambda(\mathrm{A}) = 12.4/E(\mathrm{keV})$	$\lambda(\mathrm{A}) = 9.05/\sqrt{E(\mathrm{meV})}$
Wave equation	$\nabla^2 E(\vec{r}) + \frac{\omega^2}{c^2} E(\vec{r}) = 0$ $E(r, t) = Ae^{i(\vec{k}\cdot\vec{r}-\omega t)}$	$\nabla^2 \psi(\vec{r}) + \frac{2mE}{\hbar^2} \psi(\vec{r}) = 0$ $\psi(\vec{r}, t) = Ae^{i(\vec{k}\cdot\vec{r}-Et/\hbar)}$
Dispersion relation	$k^2 = \frac{\omega^2}{c^2}$	$k^2 = \frac{2mE}{\hbar^2}$
Phase velocity	$v_{\mathrm{ph}} = \frac{\omega}{k} = c$ (wave packet profile is constant)	$v_{\mathrm{ph}} = \frac{\omega}{v/c} > c$ (wave packet profile changes with time)

References

1. https://mappingignorance.org/2020/06/10/neutron-sciences-as-an-essential-tool-to-develop-materials-for-a-better-life/. Accessed on 15 July 2021
2. Bell RE, Elliott LG (1948) Gamma-rays from the reaction ^{1}H(n, γ)D2 and the binding energy of the deuteron. Phys Rev 74(10):1552–1553
3. Bell RE, Elliott LG (1950) Gamma-rays from the reaction ^{1}H(n, γ)D2 and the binding energy of the deuteron. Phys Rev 79(2):282–285
4. Shull CG, Billman KW, Wedgwood FA (1967) Experimental limit for the neutron charge. Phys Rev 153(5):1415–1422
5. Olive KA (2014) Review of particle physics. Chin Phys C 38(9):090001
6. Alvarez LW, Bloch F (1940) A quantitative determination of the neutron moment in absolute nuclear magnetons. Phys Rev 57(2):111–122
7. Perkins DH (1982) Introduction to high energy physics. Addison-Wesley Advanced Book Program/World Science Division
8. Bég MAB, Lee BW, Pais A (1964) SU(6) and electromagnetic interactions. Phys Rev Lett 13(16):514–517
9. Sherwood JE, Stephenson TE, Bernstein S (1954) Stern-gerlach experiment on polarized neutrons. Phys Rev 96(6):1546–1548
10. Hughes DJ, Burgy MT (1949) Reflection and polarization of neutrons by magnetized mirrors. Phys Rev 76(9):1413–1414
11. Wietfeldt FE, Greene GL (2011) Colloquium: the neutron lifetime. Rev Mod Phys 83(4):1173–1192
12. Yue AT et al (2013) Improved determination of the neutron lifetime. Phys Rev Lett 111(22):222501
13. Particle Data Group (2018) Review of particle physics. Phys Rev D 98(3):030001
14. Schlenker M, Baruchel J (1986) Neutron topography: a review. Physica B+C 137(1):309–319
15. Luschikov VI et al (1969) Observation of ultracold neutrons. Sov Phys—JETP Lett 9:23
16. Zeldovich YaB (1959) Storage of cold neutrons. Sov Phys—JETP 9:1389
17. https://www.psi.ch/en/niag/neutron-physics. Accessed on 15 Sept 2021
18. https://www.nuclear-power.com/nuclear-power/reactor-physics/atomic-nuclear-physics/fundamental-particles/neutron/neutron-energy/. Accessed on 15 Sept 2021

19. Carpenter J, Loong C (2015) Elements of slow-neutron scattering: basics, techniques, and applications. Cambridge University Press, Cambridge. https://doi.org/10.1017/CBO9781139029315
20. Ederth T (2018) Neutrons for scattering: what they are, where to get them, and how to deal withthem. EPJ Web Conf 188:01002. https://doi.org/10.1051/epjconf/201818801002. Accessed on 25 August 2021
21. http://cds.cern.ch/record/1514295/plots. Accessed on 27 July 2021
22. Kardjilov N, Manke I, Woracek R, Hilger A, Banhart J (2018) Advances in neutron imaging. Mater Today 21:652–672
23. Shukla M, Roy T, Kashyap Y et al (2018) Development of neutron imaging beamline for NDT applications at Dhruva reactor, India. Nucl Instrum Methods Phys Res Sect A 889:63–68
24. Saito Y, Mishima K, Tobita Y, Suzuki T, Matsubayashi M (2004) Velocity field measurement in gas–liquid metal two-phase flow with use of PIV and neutron radiography techniques. Appl Radiat Isot 61:683–691
25. Song B, Dhiman I, Carothers JC, Veith GM, Liu J, Bilheux HZ, Huq A (2019) Dynamic lithium distribution upon dendrite growth and shorting revealed by operando neutron imaging. ACS Energy Lett 4(10):2402–2408
26. Jacobson DL, Allman BE, McMahon PJ, Nugent KA, Paganin D, Arif M, Werner SA (2004) Thermal and cold neutron phase-contrast radiography. Appl Radiat Isot 61:547–550
27. Pfeiffer F, Grünzweig C, Bunk O, Frei G, Lehmann E, David C (2006) Neutron phase imaging and tomography. Phys Rev Lett 96:215505
28. Dawson M et al (2009) Imaging with polarized neutrons. New J Phys 11:043013. https://doi.org/10.1088/1367-2630/11/4/043013. Accessed on 25 August 2021
29. Piegsa FM, van den Brandt B, Hautle P, Konter JA (2009) A compact neutron Ramsey resonance apparatus for polarised neutron radiography. Nucl Instrum Methods Phys Res Sect A 605:5–8
30. Iwase et al (2012) In situ lattice strain mapping during tensile loading using the neutron transmission and diffraction methods. J Appl Cryst 45:113–118
31. Baruchel J, Schlenker M, Palmer SB (1990) Neutron diffraction topographic investigations of "exotic" magneticdomains. Nondestr Test Eval 5(5–6):349–367. https://doi.org/10.1080/02780899008952978
32. Kardjilov N, Manke I, Hilger A, Strobl M, Banhart J (2011) Neutron imaging in materials science. Mater Today 14:248–256
33. Mankea I, Banhart J et al (2007) In situ investigation of the discharge of alkaline Zn–MnO2 batteries with synchrotron X-ray and neutron tomographies. Appl Phys Lett 90:214102
34. Ziesche RF, Arlt T, Finegan DP et al (2020) 4D imaging of lithium-batteries using correlative neutron and X-ray tomography with a virtual unrolling technique. Nat Commun 11:777

Chapter 2
Physics and Design of Sources for Neutron Imaging

Tushar Roy, Tarun Patel, and Nirmal Ray

2.1 Introduction

The earliest research on neutron radiography was carried out by Kallmann and Kuhn in Germany [1] in 1935 using an accelerator source with neutron yield of 4×10^7 n/s. Due to low output of the source, long exposure time (nearly 4 h per radiograph) was required. However, research work on neutron radiography was first published by Peters [2] in 1946. After the advent of research reactors, Thewlis and Derbyshire [3] demonstrated that much better quality neutron radiographic images were obtained using intense thermal neutron beam from reactor. Subsequent development and availability of research reactors opened up avenues for specific applications of neutron radiography [4–10].

Even today, research reactors remain by and large the main source of neutron imaging. However, nuclear reactors are not portable and are available only at few facilities in any country. Coupled with high cost of building a reactor, maintenance, safety and regulatory aspects, and liability of waste disposal costs resulted in the exploration of other sources of neutron production which could be used in the private industry or for field applications. This resulted in the use of radioisotope based neutron sources (^{252}Cf, ^{241}Am-Be and others) and accelerator-based neutron sources (using deuterium-deuterium or deuterium–tritium fusion reactions). These sources offer reduced costs and increased mobility (or portability) but at the expense of much lower neutron flux as compared to reactor sources and significantly reduced image quality.

T. Roy (✉) · T. Patel · N. Ray
Technical Physics Division, Bhabha Atomic Research Centre, Mumbai 400085, India
e-mail: tushar@barc.gov.in

T. Patel
e-mail: pateltp@barc.gov.in

N. Ray
e-mail: nirmal@barc.gov.in

D. K. Aswal et al. (eds.), *Neutron Imaging*,
https://doi.org/10.1007/978-981-16-6273-7_2

Due to poor image quality, the use of radioisotopes and accelerator-based neutron sources is restricted to few specific applications where constraints like portability, cost, or on-field use supersede high image quality requirements. Though spallation-based high fluence neutron sources are being developed at selected research centers across the globe, research reactor-based facilities still remain the workhorse in the neutron imaging domain.

2.2 Neutron Sources

Neutrons are produced in many different nuclear reactions like nuclear fission (both induced and spontaneous), nuclear fusion, spallation, photoneutron, photofission, and several other nuclear reactions. We discuss here briefly the different types of neutron sources that are suited for neutron imaging applications.

2.2.1 Research Reactor

The most widely used source for neutron imaging is a research reactor, where neutrons are produced due to nuclear fission. High neutron flux, highly collimated beam, and fast image acquisition make them almost the perfect choice for neutron imaging.

2.2.2 Isotopic Source

Neutrons can be produced from isotopic sources through three routes, namely spontaneous fission, (α–n) reaction and (γ–n) reaction. Spontaneous fission of ^{252}Cf produces 2.34×10^6 n/s per μg. However, a short half-life of 2.65 years coupled with low availability restricts its use for neutron imaging. Another class of isotopic sources popularly used as neutron sources use alpha-emitting isotopes like ^{241}Am or ^{240}Pu coupled with low Z materials like beryllium to produce neutrons via the (α, n) reaction. These sources, commonly named as Am-Be or Pu-Be, have a yield of the order of 10^6 n/s per Curie. Sb–Be combination is an example of a neutron source produced via (γ–n) reaction and 1 Curie of ^{124}Sb results in 2×10^6 neutron emission per second.

While there are many types of radioisotope-based neutron sources, they must be moderated and collimated to form a thermal neutron source for neutron imaging. This results in extremely low neutron flux when compared to either accelerators or nuclear reactors. Such sources have found limited application—primarily where portability is required. The other drawbacks of these sources include gamma radiation, high cost—both for initial purchase and subsequent replacement after the source decays—and disposal of isotopic source after use (radiological waste).

2.2.3 *Accelerator-Based Source*

With the advancement in accelerator technology, compact accelerators producing high current proton or deuteron beams have become very popular. When proton or deuteron beam is incident on a suitable target, it can produce neutrons via nuclear fusion, stripping, or other nuclear reactions. Among these, the most common neutron sources are nuclear fusion-based deuterium-deuterium (D-D) and deuterium–tritium (D-T) neutron generators. These neutron sources are based upon deuteron beams impinging upon deuterated or tritiated targets with relatively low accelerating potential and reasonably high current (~ 0.1–10 mA), with the resultant yield of 2.45 MeV or 14.1 MeV neutrons respectively at a neutron yield of the order of 10^8 to 10^{12} n/s. The high-energy neutrons produced are highly penetrating and can be used directly for fast neutron imaging. However, the high-energy neutrons that are produced can also, in principle, be moderated to thermal energy using suitable moderators to allow thermal neutron imaging. In practice though, neutrons of such high energy require large moderators and yield relatively low thermal neutron flux. The relative portability of systems based upon D-T generators makes in-field application quite possible. D-T tubes are being used largely for thermal neutron radiography for specific field applications since exposure times are quite large.

Neutrons are also produced from accelerators based on other nuclear reactions like p-Li, p-Be, d-Be, etc. but are not commonly used for practical neutron imaging applications and are mostly restricted to research purposes.

Another route for producing neutrons is through the generation of photoneutrons using electron accelerators. High energy (~ 10 MeV or higher) and high current electron beam is incident on a tungsten or tantalum target to produce Bremsstrahlung radiation. The X-ray photons are then incident on a beryllium target to produce photoneutrons. Similar to previous cases, the neutrons have to be moderated to thermal energy and the resulting thermal neutron flux is low. The electron accelerator based photoneutron sources are neither portable nor inexpensive. They find limited application in neutron radiography for on-site industrial applications.

2.2.4 *Spallation Source*

In spallation reaction, energetic particles like proton or deuteron (~ 1 GeV) incident on high Z-target such as mercury, lead, uranium, lead–bismuth eutectic produces neutrons. An intense spallation neutron source can be used as neutron source for neutron imaging. The total neutron yield of a pulsed spallation source is higher than a conventional research reactor. Also, short pulses and low backgrounds between pulses make it an attractive choice for energy-resolved or time-resolved neutron imaging.

2.3 Nuclear Fission

Nuclear fission, discovered in 1939 by Otto Hahn, proved to be a great source of energy in the twentieth century that could be tapped and efficiently converted into electricity. Nuclear fission is the splitting of a heavy nucleus into two lighter nuclei (or fragments) of unequal masses and releasing large amount of energy along with the production of neutrons and gamma rays. The discovery of nuclear fission and the subsequent demonstration of self-sustaining nuclear chain reaction in Chicago Pile 1 (CP-1) experiment by Enrico Fermi and his team in 1942 (as part of the Manhattan Project) proved that controlled nuclear fission was possible and it led to the birth of nuclear reactors.

In a nuclear reactor, the rate of neutron production and the rate of neutron losses (both through absorption and leakage) are balanced to reach a critical state [11]. In nuclear fission of ^{235}U, an average of 2.4 neutrons are released. In order to sustain chain reaction, at least one neutron should be available for fission in the next generation, accounting for the losses due to absorption and leakage. Mathematically, neutron multiplication factor k is defined as the ratio of neutrons available in the present generation to the neutrons available in the previous generation. To maintain criticality, k should be equal to 1.

2.3.1 *Reactors as Source of Neutrons*

Research reactors are abundant sources of neutrons where a large number of neutrons per unit area per second (flux) are readily available. Neutron flux at the entrance of a neutron collimator can be of the order of 10^{13} to 10^{14} n/cm^2-s. After collimation, typical neutron flux at the image plane is generally of the order of 10^7 to 10^8 n/cm^2-s. This allows to achieve a high collimation ratio of more than 100 and fast data acquisition (exposure time of few seconds). The neutron flux at the imaging plane obtained from a reactor source is typically 100–1000 times higher than the neutron flux available from typical radioisotope or accelerator-based neutron sources. Moreover, neutron energy spectrum from a reactor neutron source covers a vast range (few MeV to 20 MeV) and the neutron beam can be converted to cold neutron source or ultra-cold neutron source (using liquid hydrogen) as well as hot source (using graphite). It is also possible to use filters to cut-off undesirable energy neutrons (for example, sapphire single crystal is used to filter out fast neutrons) or monochromators to choose a narrow band of energy (or wavelength) or guide tubes to eliminate unwanted radiations (gamma) from the beam.

2.3.2 *Moderation*

Most research reactors are thermal reactors which are either light water reactors (LWR) or pressurized heavy water reactors (PHWR) with ^{235}U as the fuel. The fission of ^{235}U is initiated with the capture of a thermal neutron followed by the emission of fast neutrons (defined by the fission-watt spectrum). In reactor parlance, thermal neutrons are defined as those neutrons which are in thermal equilibrium with the surroundings at 300 K. The Maxwell–Boltzmann distribution is used to describe the energy spectrum of thermal neutrons. The most probable energy of thermal neutrons is 0.025 eV which corresponds to neutron velocity of 2200 m/s. The fast neutrons produced in the fission reaction need to be slowed down or moderated to thermal neutrons to continue the fission chain reaction. To achieve this, moderator is used, in which, the neutrons undergo scattering with the nuclei of the moderator atoms and lose energy till they are in thermal equilibrium with the motion of the atoms around them. For a material to be an efficient moderator, it should have a high scattering cross-section and low absorption cross-section. In addition, the energy loss per collision should be high so that energy of the neutron is reduced in minimum number of collisions which in turn will reduce the overall size of the moderator. Some of the main candidates for moderator materials are light water (H_2O), heavy water (D_2O), graphite and beryllium.

2.4 Design of Reactor Based Neutron Imaging Sources

Most work on neutron imaging is performed with either thermal neutron (defined as neutron with energy of about 0.025 eV) or cold neutron (defined as neutron with energy typically 5 MeV or less). While cold neutrons penetrate materials less compared to thermal neutrons, the differential attenuation between two materials may increase, depending on the precise energies and materials involved. This can result in substantially improved contrast between two materials. However, thermal neutrons are still the preferred choice for neutron imaging due to two reasons. Firstly, neutron within this energy range can exhibit useful attenuation feature and has higher penetration than cold neutrons; secondly, thermal neutron can be easily obtained. Subsequent discussion on reactor-based neutron imaging will be limited to thermal neutrons.

Figure 2.1 shows a schematic configuration of neutron radiography from a reactor neutron source. Neutrons coming from nuclear reactors usually diverge in direction. Therefore, it is necessary to collimate the neutrons to generate a sharp radiograph with high resolution. The collimated neutron beam passes through the sample and the transmitted intensity of neutron beam is recorded on the detector.

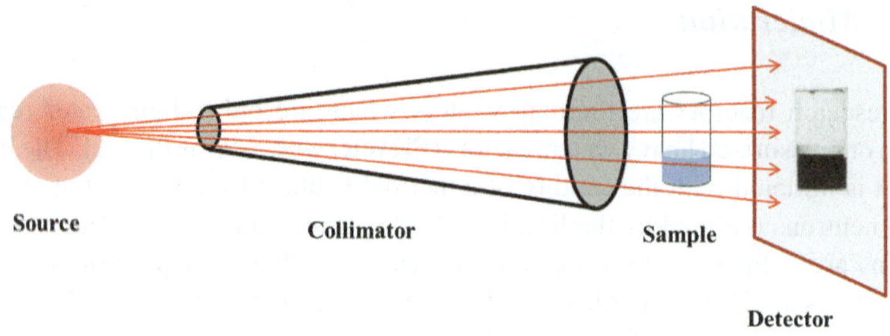

Fig. 2.1 Schematic configuration for neutron radiography

2.4.1 Collimator and Its Role

A neutron beam of large divergence comes out of the reactor beamport (usually with a circular shape). A collimator is used to limit the divergence of the incoming neutron beam and also serves the purpose of shielding scattered neutron and gamma radiation coming out from the reactor. To accomplish this, an aperture (an opening that will allow neutrons to pass through it surrounded by neutron-absorbing materials) limits the neutrons entering the collimator. Neutrons that are not traveling the length of the collimator in the desired direction are absorbed by neutron-absorbing materials (for example, boron). The degree of collimation is determined by the L/D ratio of the collimator where L is the distance (or length) from the input aperture of collimator to detector and D is the diameter of the input aperture. Neutron imaging experiments require neutron beams with a large L/D (small divergence of the neutron beam) over a large surface (large field-of-view).

The neutron collimator is mainly of two types—(a) Soller collimator, and (b) Divergent collimator.

(a) **Soller or multitube collimator**

This collimator uses multiple parallel tubes or plates to extract parallel neutron beam to the test specimen (Fig. 2.2). Those neutrons which do not travel in parallel with the collimator axis will hit the side of the tube or plate and are then absorbed allowing only neutrons traveling in parallel with the tube axis to reach collimator exit. The

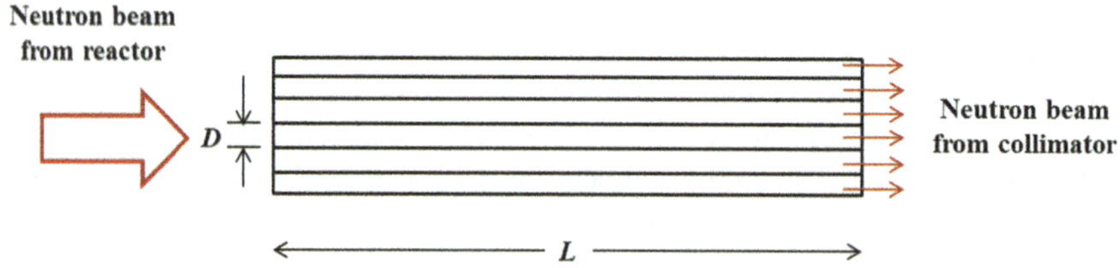

Fig. 2.2 Schematic of Soller collimator

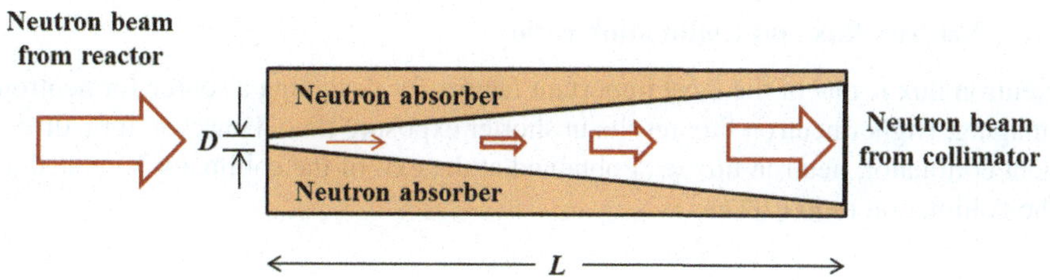

Fig. 2.3 Schematic of divergent collimator

main drawback of Soller collimator is that the pattern of parallel plates or tubes may be seen on the image.

(b) **Divergent collimator**

This collimator uses an input aperture (small hole of diameter ~ few mm) at the input end (towards the reactor source) of the collimator and diverges at the other end (Fig. 2.3). Neutrons are allowed to enter only through the aperture and the neutrons entering the collimator from the remaining area are absorbed using neutron absorber (like boron). The collimator is also lined with neutron absorber (like cadmium) to absorb the scattered neutrons from the surrounding areas other than those coming through the aperture.

Divergent collimator is easy to construct and can also be used with non-reactor-based sources (like radioisotopes and accelerator) for neutron imaging. The drawback of divergent collimator is that the image sharpness may not be as good as Soller collimator.

2.4.2 Image Quality Parameters

The design of the collimator is an integral part of setting up of neutron imaging facility, as it performs beam shaping and also critically guides the image quality parameters. For example, the overall neutron flux will largely determine the exposure time or available spatial resolution of the system. Similarly, the scattered neutron content is vitally important to the sharpness of the images. While there are many parameters that need to be established and some tradeoffs have to be adopted, the constraints guide the final design. These constraints vary from facility to facility and hence the collimator design variation. Some of the constraints are beam-tube dimension, maximum reactor neutron flux, resolution requirement, exposure time (speed) requirement, space availability, etc.

Following are some of the important image quality parameters.

(a) **Neutron flux and Collimation ratio**

Neutron flux is one of the most important factors for designing a source for neutron imaging. Higher neutron flux results in shorter exposure time. However, for a divergent collimator, neutron flux φ_{exit} obtained at the exit of the collimator is related to the collimation ratio L/D as

$$\varphi_{\text{exit}} = \frac{\varphi_{\text{source}}}{16(\mathrm{L/D})^2} \tag{2.1}$$

where φ_{source} is the neutron flux at the input of the collimator.

A tradeoff exists between neutron flux, collimation ratio, and exposure time. Higher collimation ratio will produce sharper image at the cost of reduced flux thereby increasing exposure time, whereas higher neutron flux will produce high contrast image at the expense of reduced collimation ratio and hence reduced image sharpness.

(b) **Cadmium Ratio**

The cadmium ratio is a measure of the epithermal neutron content in the neutron beam. It will impact the effectiveness of the beam's penetration power in certain materials such as nuclear fuels. Cadmium ratio is measured by neutron activation of bare gold foil and cadmium-covered gold foil. Cadmium cover measures total flux of all neutrons above the cadmium cut-off energy of 0.4 eV. Cadmium ratio is estimated from the ratio of the total neutron flux to thermal neutron flux.

(c) **Contrast**

Contrast is the ability to differentiate two areas with different materials or densities. It is characterized by the degree of grayness between different areas in a radiograph. Higher contrast means that even small difference in density values can be detected.

In terms of image, the image contrast is defined as

$$\text{Contrast} = \frac{I - I_{\text{min}}}{I_{\text{max}} - I_{\text{min}}} \tag{2.2}$$

where I is the average image intensity, I_{min} is the minimum image intensity and I_{max} is the maximum image intensity in the measurement area of the image.

(d) **Geometric unsharpness**

Geometric unsharpness for relatively thin samples close to the image plane is given by

$$U_g = \frac{t}{\mathrm{L/D}} \tag{2.3}$$

where t is the sample thickness. To reduce geometric unsharpness, L/D ratio should be increased. Lower the geometric unsharpness, higher is the image sharpness and hence better the image spatial resolution.

(e) **Neutron to gamma ratio**

Neutron to gamma ratio is defined as the ratio of neutron flux to gamma dose rate. It quantifies the effect of gamma dose on image quality. As most of the neutron detectors are also sensitive to gamma, higher gamma background tends to produce blurring in the image, thereby reducing image quality. For a good quality neutron image, it is recommended to reduce the gamma background such that $n/\gamma \geq 10^6$ n/mR - cm^2. This condition ensures that the gamma contribution to image will be small relative to that from neutron.

(f) **Modulation transfer function**

Mathematically, modulation transfer function (MTF) is defined as the magnitude of the optical transfer function which is the normalized Fourier transform of the impulse response of the imaging system.

$$\mathrm{MTF}(\omega) = \frac{F(\omega)}{F(0)} = \frac{\int \mathrm{LSF}(x)\cos(2\pi\omega x)\mathrm{d}x}{\int \mathrm{LSF}(x)\mathrm{d}x} \tag{2.4}$$

where $F(\omega)$ is the Fourier transform of line spread function (LSF) and $F(0)$ is the total area under the line spread function

The desired MTF of an imaging system can be calculated by using a step function input (Edge Response Function), differentiating to obtain impulse response (Line Spread Function), and then finally transforming to frequency domain by Fourier methods.

2.4.3 *Shielding*

The experimental area of a neutron imaging facility has to be properly shielded for both neutron and gamma radiation coming from the reactor beamport. For gamma shielding, the energy distribution of the gamma rays must be carefully considered, as well as the intensity. Careful design of the collimator will ensure that the gamma rays generated by collimator components are of low energy so that they are readily shielded. Primary gamma radiation from the reactor core in a radial beamport will require additional shielding for the high-energy component. The most common material used for gamma shielding is lead or high-density concrete.

Like gamma radiation, neutrons undergo scattering and absorption interaction with matter. These interactions form the basis for methods used to shield neutron radiation. In general, for fast (high energy) neutrons, scattering interactions are more likely than capture interactions. The energy of neutrons is first reduced through scattering interactions (thermalization/moderation) using low Z materials like polythene, wax, etc. The thermal neutrons can then be absorbed through neutron capture interactions in neutron absorber material like boron and cadmium. Practically, neutron shielding materials combine both moderator and absorber together. Some examples of neutron shielding materials are borated polythene, borated wax, etc.

2.5 Non-reactor Based Neutron Imaging Sources

Neutron can also be produced from several non-reactor-based reactions, for example, radioisotopes and accelerator-based reactions. Radioisotopes-based reactions are the spontaneous fission in heavy actinides, (α, n) and (γ, n) reactions. Accelerator-based reactions, such as fusion and spallation also serve as neutron sources. These different radioisotopes and accelerator-based neutron sources are described in the following subsections.

2.5.1 Radioisotope Based Neutron Source

Nuclear reactor is the most popular source of neutrons. But, they have limitations for field applications due to the lack of portability, high cost, regular maintenance, production, and disposal of transuranic elements. On the other hand, neutrons are emitted by spontaneous fission in actinides, for example, 30 mg of ^{252}Cf can provide neutron of strength 10^{11} n/s with a half-life of 2.2 years [12]. Furthermore, alpha particles of energy of few MeV are emitted by some actinides such as, ^{210}Po, ^{238}Pu, ^{239}Pu, ^{241}Am etc. These alpha particles in reaction with light elements such as Beryllium (^{9}Be) can produce neutron and are widely used as isotopic neutron source. The details of the different alpha based isotopic neutron sources are listed in Table 2.1. Amongst the alpha based neutron sources, ^{238}Pu-Be and ^{241}Am-Be are widely used for strong isotopic neutron source due to their availability from spent nuclear fuel. The alpha-Be neutron sources have similar energy spectrum, and have small variation in energy due to the initial energy of alpha emitted by actinides.

Table 2.1 Characteristics of alpha-Be neutron sources [11]

Source	Half-life	Alpha energy (MeV)	Neutron yield per 10^6 primary alpha particles (Theoretical)
^{239}Pu-Be	24,000 y	5.14	65
^{238}Pu-Be	87.4 y	5.48	79
^{210}Po-Be	138 d	5.30	73
^{241}Am-Be	433 y	5.48	82
^{244}Cm-Be	18 y	5.79	100
^{242}Cm-Be	162 d	6.10	118

Antimony (^{124}Sb) emits β^- with gamma energies 1691 and 2091 keV with half-life of 60 days, and the corresponding branching ratios are 0.484 and 0.057 respectively. Beryllium has threshold energy of 1.66 MeV for photo neutron reaction. Thus, gamma-ray from Sb can produce photo neutrons from ^{9}Be. It has been reported that a 10^3 Ci Sb–Be sources can produce 10^{10} n/s [12]. But, the isotopic source has limitations due to the lower flux compared to reactor which result in poor image quality of object.

2.5.2 *Accelerator-Based Neutron Source*

Particle accelerator based neutron sources, also well-known as neutron generators, have a long history and their continuous developments have evolved since the last few decades due to their potential uses in various fields such as scientific research, industry, medical, security, nuclear safeguards, etc. Accelerator-based neutron generator has many advantages over nuclear reactors and radioisotopes. The most important advantage is that it can be turned on or off as per the requirement. This feature has improved the safety of its operation, transportation, shielding, and application. Furthermore, the neutron yield can be varied by adjustment of the accelerating voltage and ion beam current. Moreover, the accelerator-based neutron generators can also be operated in pulsed mode. This feature has opened up many areas of applications where timing information of neutron production is very important to improve the signal-to-noise ratio in measurements. With the present technologies, neutron generators are able to produce neutrons in the range of 10^2 to 10^{14} n/s. The size and the design of neutron generator can also be customized for specific applications.

There are many reactions that can be used to produce neutrons using charged particles. Most common approach is fusion of light particles. The most commonly used reactions for neutron production using accelerators are listed in Table 2.2 [13].

These reactions have low threshold energy and higher cross-section compared to other reactions. Figure 2.4 shows cross-section as a function of incident particle energy for several of the neutron-producing reactions considered in this chapter. Neutron yield mainly depends on the energy-dependent cross-section of the reaction. In practical, it also depends on the condition of the target and its thickness.

Table 2.2 List of nuclear reactions used for accelerator-based neutron production

T(D, n)^{4}He:	${}_1^2H + {}_1^3H \rightarrow {}_2^4He + n + 17.59$ MeV
D(D, n)^{3}He:	${}_1^2H + {}_1^2H \rightarrow {}_2^3He + n + 3.27$ MeV
^{9}Be(D, n)^{10}B:	${}_1^2H + {}_4^9Be \rightarrow {}_5^{10}B + n + 4.36$ MeV
^{7}Li(D, n)^{8}Be:	${}_1^2H + {}_3^7Li \rightarrow {}_4^8Be + n + 15.03$ MeV
^{9}Be(p, n)^{9}B:	${}_1^1H + {}_4^9Be \rightarrow {}_5^9B + n - 1.85$ MeV
^{7}Li(p, n)^{7}Be:	${}_1^1H + {}_3^7Li \rightarrow {}_4^7Be + n - 1.64$ MeV

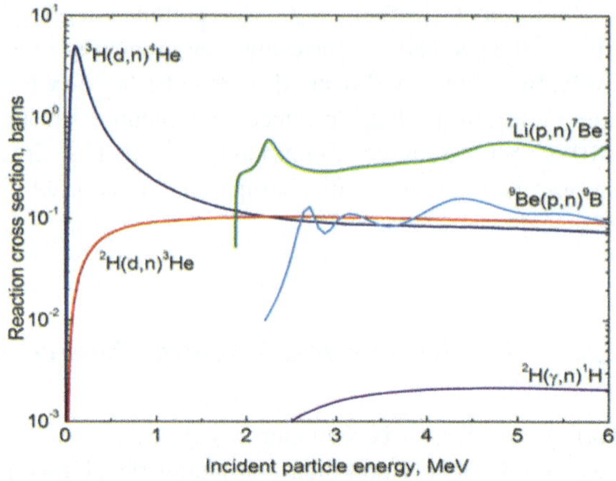

Fig. 2.4 Reaction cross sections for accelerator-based neutron producing reactions (reproduced from [13] with permission from author)

Theoretically, the variation of the neutron yield per μA incident beam current with incident particle energy is shown in Fig. 2.5 [13, 14].

Among the fusion reactions, T(D, n)^{4}He is the mostly used reaction for producing neutrons at low energy accelerator as it has a broad resonance with a maximum value of 5 barns at deuteron energy of 107 keV which is the highest cross-section among all the above reactions. It is an exothermic reaction with a very high Q-value (17.59 MeV) which produces almost isotropic and mono-energetic fast neutrons (14.1 MeV).

D-D reaction has two exothermic reactions D(D, n)^{3}He and D(D, p)T with Q-values 3.27 and 4.03 MeV respectively. Both the reactions are equally probable. At lower incident energy (<300 keV), it produces almost 2.45 MeV mono-energetic neutrons. But neutron yield of this reaction is roughly 100 times less compared to

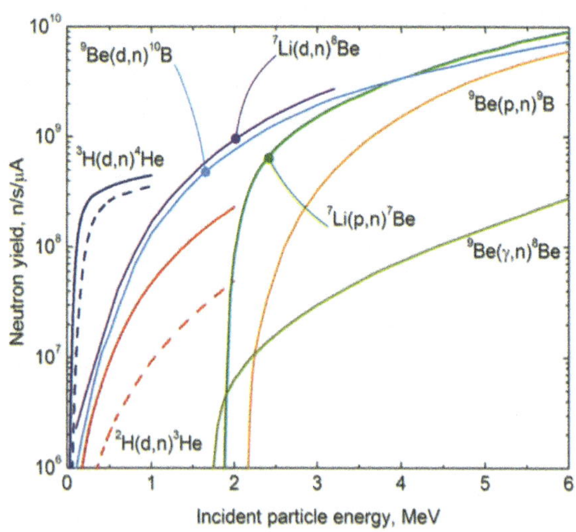

Fig. 2.5 Total neutron yield variation for various fusion reactions (reproduced from [13] with permission from author)

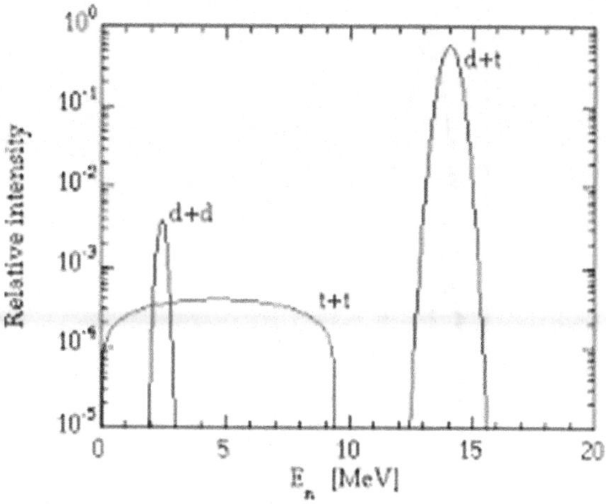

Fig 2.6 Normalized neutron spectra for DD, TT, and DT neutron generator (reproduced from [15] with Open Access)

DT reaction at this energy. On the other hand, deuterium target has a longer lifetime and is very cost-effective compared to Tritium and also eliminated the safety related to handling of Tritium. Figure 2.6 shows the normalized neutron spectrum of DD, TT, and DT reactions [15, 16].

^{9}Be(D, n)^{10}B and ^{7}Li(D, n)^{8}Be are both exothermic reactions and have almost identical cross-sections. However, ^{7}Li(D, n)^{8}Be has a high Q-value (15.03 MeV) which is approximately 3.4 times higher than Beryllium (4.36 MeV). In comparison with Lithium, Beryllium target has the advantage of high melting point (1287 °C), excellent thermal conductivity, and is chemically inert. For incident deuteron energies between 1 and 2 MeV, neutron outputs of these reactions are comparable with DT neutrons at low energy. Figure 2.7 shows neutron spectra of ^{9}Be(D, n)^{10}B reaction at 0.9–1.1 MeV deuteron energy.

^{9}Be(p, n)^{9}B and ^{7}Li(p, n)^{7}Be are endothermic reactions with threshold energy 2.06 MeV and 1.88 MeV, respectively. These reactions have very good similarities in neutron energy and angular distribution except ^{7}Li(p, n)^{7}Be reaction produces slightly higher neutron yield with the same incident proton energy. The multi-body break-up processes produce neutron continuum from lower energy region with thick target at higher incident energy. Figure 2.8 shows neutron spectrum measured at different proton energy on Be target with cyclotron U-120 M at NPI, Rez, Czechia [18].

2.6 Accelerators for Neutron Productions

Many types of particle accelerators with different types of ion beams have been developed and used in applied and fundamental science research. More than 30,000 accelerators are in operation across the world with beam energy ranging from keV

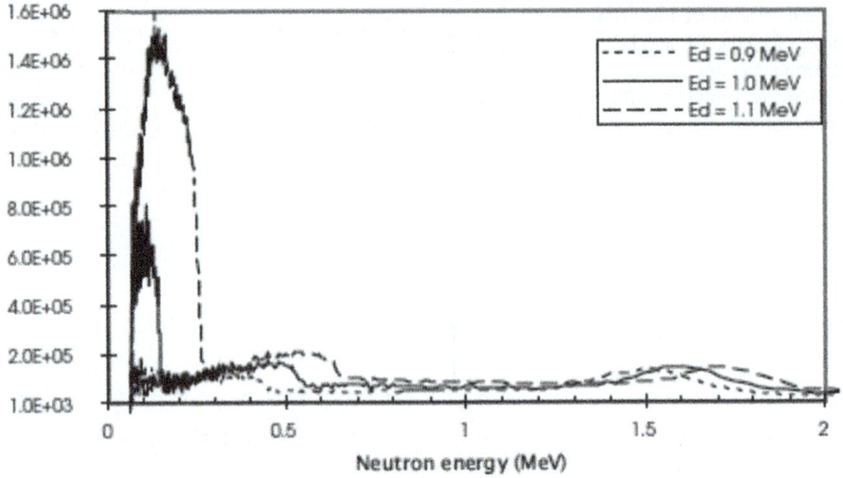

Fig. 2.7 Neutron Spectra of $^{9}Be(D, n)^{10}B$ reaction at 0.9–1.1 MeV deuteron energy [17]

Fig. 2.8 p-Be neutron spectrum measured at different proton energy with cyclotron U-120 M at NPI, Rez, Czechia (reproduced from [18] with Open Access)

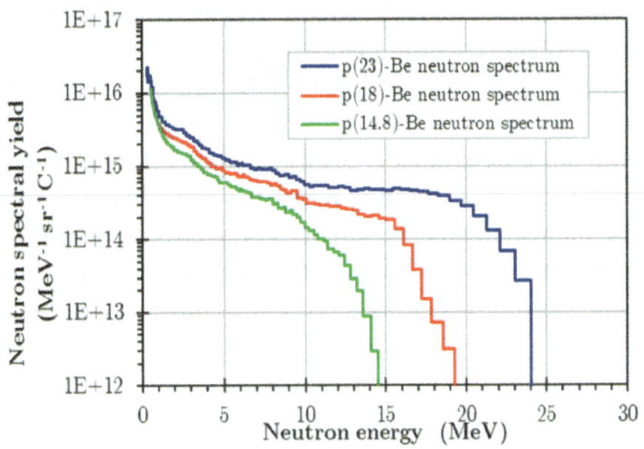

to TeV and many more are under development. Among these, sealed tubes, electrostatic accelerators, cyclotron, RFQ and linac, are commonly used accelerators for generating neutrons below 100 MeV. Working principle, design, construction, and experimental results of these accelerators are already very well described in many references. All neutron generators except sealed tubes are open-vacuum systems which require vacuum pumps to obtain high vacuum for its continuous operation.

Sealed tube neutron generators are high vacuum sealed tubes, produce neutrons using D-D and D-T reactions. These are mainly low energy (< 200 keV) single-gap DC accelerators. A photograph of very compact pulsed neutron tube used in IGN-1 and IGN-1-M developed by USSR for geochemistry and geology is shown in Fig. 2.9 [19]. Sealed neutron generators are simplest in design and most compact among all types of neutron generators. This technology has been developed since many years

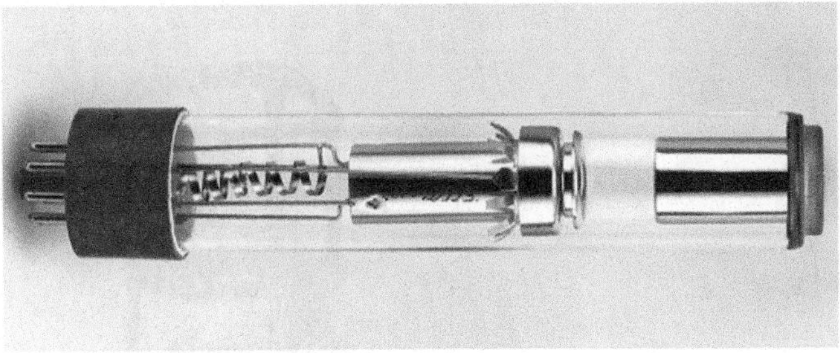

Fig. 2.9 Photograph of very compact pulsed neutron tube used in IGN-1 and IGN-1-M developed by USSR

and now it is available commercially. List of few manufacturing companies are given below [20].

1. Sodern
2. VNIIA
3. Adelphi Technology
4. Gradel (NSD Fusion)
5. Thermo Fisher Scientific.

These companies offer a range of DD and DT sealed neutron generators producing neutron yield from 10^6 to 10^{11} n/s. Sealed tube has overcome the constraints of handling of Tritium. Their small size and low power requirement make them most suitable for field applications. The main disadvantage of sealed tube is that it has a limited lifetime, couple of hundred to maximum few thousand hours. This has limited their use for neutron radiography applications. However, sealed tubes have been extensively used for many applications such as oil well logging, mineral mining, explosive and SNM detection, radioactive waste management, basic nuclear research, etc.

Cockcroft-Walton (CW), Van de Graaff, and Pelletron are commonly used electrostatic accelerators for neutron production. The Cockcroft-Walton (CW) generator is also known as cascade generator, is mostly used accelerator in low energy range as it is less expensive, small in size, easy to install and operate compared to other types of accelerators, and can be used for few tens of mA ion beam current. Technical limitation due to high voltage break down has restricted its use upto 2 MV accelerators. CW accelerator-based DD and DT neutron generators are still a basic tool for nuclear research in many laboratories. Figure 2.10 shows a CW accelerator-based neutron generator facility installed at BARC, India [21]. Phoenix, USA offers cascade neutron generator with neutron yield 10^{11} n/s in DD and 10^{13} n/s in DT which is suitable for neutron radiography [22].

Van de Graaff (VDG) generator is proven technology to achieve much higher DC voltage than Cockcroft-Walton generator. Pelletron generator is an improved version of Van de Graaff which provides better voltage stability at higher voltage and current. Few hundreds of Van de Graaff and Pelletron accelerators ranging from 1 to 25 MV

Fig. 2.10 Neutron generator facility installed at BARC, India

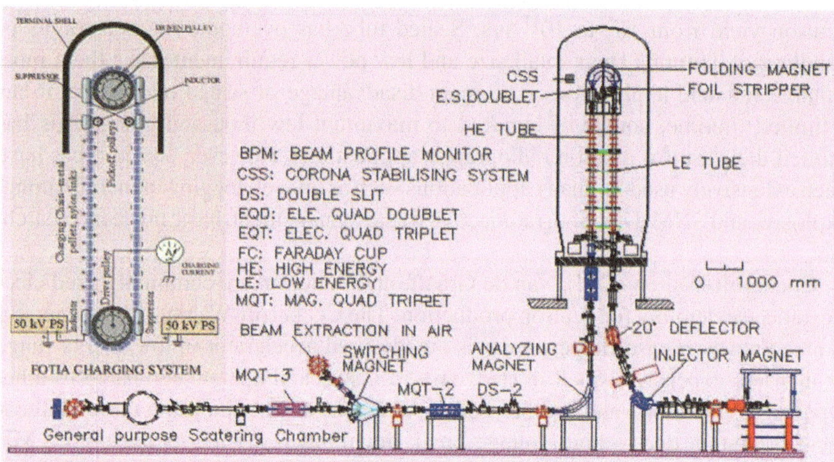

Fig. 2.11 6 MV Van de Graaff accelerator (FOTIA) installed at BARC

terminal voltage are operational around the world. They are used for positive and negative single and multi-charged ion beams. Figure 2.11 shows a 6 MV Van de Graaff accelerator (Folded Tandem Ion Accelerator—FOTIA) installed at BARC [23]. Low current capacity and large size of these accelerators have restricted the use for commercial applications. Another disadvantage of such accelerators is the lower neutron flux compared to reactors.

In contrast to electrostatic accelerators, cyclotron, linac, and RFQ uses RF field for accelerating the particles. For this reason they are also called RF accelerators. Cyclotron is a circular accelerator whereas linac and RFQ are linear accelerators. The main advantage of RFQ is that with acceleration it also focuses the beam. This allows its use for high-intensity beams but has limitation to accelerate upto 10 MeV. Cyclotron and linac are capable of accelerating charged particles more than few hundred MeV. The world's largest 540 MeV proton cyclotron at TRIUMF, Canada and 50 GeV electron linac at SLAC, USA are classic examples of high-energy accelerators [24, 25].

2.7 Photo Neutron Source

Another alternate to the isotope-based neutron source is the use of accelerator for production of high-energy photons. Linear accelerator, microtron, and betatron can be used to accelerate electron to high energy. The high energy electron produces bremsstrahlung photon upon bombardment on high-Z materials such as tungsten (W) or tantalum (Ta). Eshawarappa et al. [26] had discussed the bremsstrahlung spectrum from Ta target for different electron energy (8, 9, 10, 12 MeV). These high energy photons can produce a photoneutron via the reactions $^{9}Be(\gamma,n)^{8}Be$ and $^{2}H(\gamma,n)^{1}H$. The excess energy is shared as kinetic energy of the neutron and the residual nucleus [27]. Eshawarappa et al. [26] had also discussed the variation of neutron yield with beryllium thickness for different photon energy. Photo neutron source based on electron LINAC is also under operation at Hokkaido University, HUNS and Kyoto University Research Reactor Institute (KURRI)-LINAC facilities in Japan [28]. The electron energy at HUNS and KURRI are 35 and 40 MeV along with beam power of 1 and 6 kW respectively. Kim et al. [29] reported the production of photo neutron using 75 MeV LINAC with beam current of 40 mA at Pohang Accelerator Laboratory (PAL) in Korea. They had also reported that the neutron yield per kW beam power at the target is 2.0×10^{12} n/s.

2.8 Spallation Neutron Source

Spallation reaction using high-energy particle accelerator can also be used for production of neutron. In spallation reaction, a high energy charged particle (~1 GeV), such as proton or deuteron interacts with some high-Z target material such as mercury, lead, uranium, lead–bismuth eutectic. For particle energy comparable to its rest mass energy, the De-Broglie wavelength of the projectile is very small, and it may be considered as a classical particle. The projectile interacts with the nucleons, which in turn interact with other nucleons. As a result, there is an intra-nuclear cascade, and some nucleons are emitted from the nucleus. Subsequently, the emitted high-energy nucleons interact further with other nucleus in target material, and this is defined as inter-nuclear cascade. Finally, the residual nucleus is de-excited by evaporation of nucleons and fission. The spallation neutrons have a spectrum similar to fission with a peak around 3–4 MeV and high energy tail [30]. The total number of neutrons produced per incident particle depends on the projectile mass, energy, and target material. However, in the case of Pb target and 1 GeV proton, ~ 25 neutrons are produced per incident proton. The spallation has the highest yield for incident particle energy of 1–2 GeV. At higher energies, the projectile will lose its energy by neutral Pion production.

An intense spallation neutron source can also be used as an alternate to nuclear reactor for neutron imaging and other applications such as neutron diffraction, activation analysis, isotope production, etc. Furthermore, the spallation source has an advantage over nuclear reactor due to the pulsed nature of beam which is crucial for time-of-flight experiments. At present, several countries have spallation neutron sources, for example, SINQ in Paul Scherrer Institute (PSI), Switzerland with proton beam current of 1.4 mA and energy of 590 MeV, Los Alamos Neutron Science Center (LANSCE) with proton energy of 800 MeV, Oak Ridge National Laboratory (ORNL) with proton beam energy of 800–1000 MeV in the United States of America (USA), ISIS spallation neutron source with beam energy of 800 MeV in the United Kingdom (UK), Japan Proton Accelerator Research Complex (JPARC) in Japan with beam energy of 3 GeV and beam current of 90 μA and China Spallation Neutron Source (CSNS) with beam energy of 1.60 GeV. Details of these spallation neutron sources are presented in Table 2.3. India is also developing a high energy high current proton accelerator and beam target for spallation neutron source [31].

2.9 Design Consideration of Non-reactor Based Sources for Neutron Imaging

Design of thermal imaging beam line depends on the energy along with the yield of the neutron source, the required thermal neutron flux, and the L/D ratio for better image quality. For example, the design of isotopic source (Sb–Be) based neutron imaging assembly has been discussed by Fantidis et al. [32]. The neutron radiography system consists of the isotopic source encapsulated by beryllium for photo neutron production followed by 1 cm thick polyethylene layer to thermalize the 24 keV neutrons. The thermal source assembly is encapsulated by bismuth and borated polyethylene for photon and thermal neutron filtration respectively from the assembly. The schematic for such a system is shown in Fig. 2.12.

The design of Am-Be neutron source based thermal imaging systems have been discussed by Sinha et al. [33], Jafari et al. [34]. The thermal neutron assembly consists of isotopic source followed by polyethylene, lead, and borated polyethylene or cadmium. Several researches have discussed the optimized polyethylene

Table 2.3 Details of different spallation neutron sources

Spallation source	SINQ	LANSCE	ORNL	ISIS	JPARC	CSNS
Accelerator type	Cockcroft-Walton, Injector II (small cyclotron) (72 MeV), Ring cyclotron	Full energy LINAC, Accumulator	Full energy LINAC, Accumulator	LINAC (70 MeV), Synchrotron	LINAC (400 MeV), Synchrotron	LINAC (81 MeV), Synchrotron
Proton energy on the target (MeV)	590	800	1000	800	3000	1600
Beam current on the target (mA)	1.5	0.125	1.4	0.225	0.090	0.315
Beam repetition on the target (Hz)	Continuous Wave	20	60	50	25	25
Beam Power on the target (kW)	885	100	1400	180	300	504
Beam target	Lead–bismuth eutectic	Tungsten	Mercury	Tantalum-clad tungsten	Mercury	Tungsten
Status	Operational	Operational	Operational	Operational	Operational	Operational

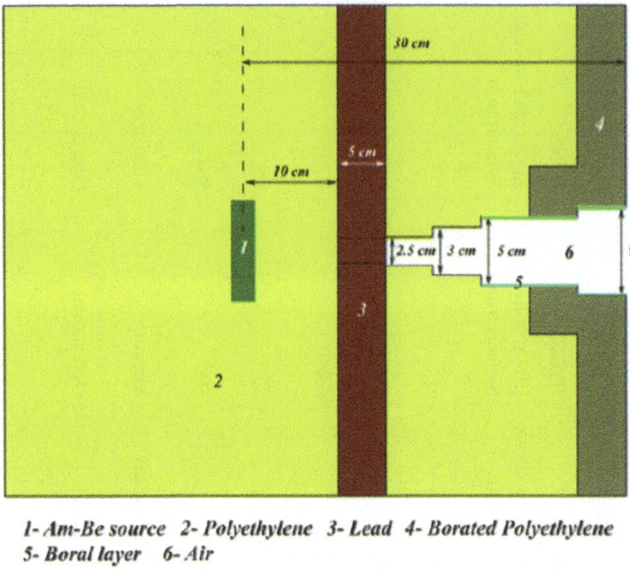

Fig. 2.12 Thermal neutron imaging system using Sb–Be neutron source (reproduced from [32] with permission from Elsevier)

thickness for thermalization of fast neutrons to thermal and it has been observed that the 15 cm polyethylene can slowdown fast neutrons to thermal energy. Lead and borated polyethylene or cadmium is used for the gamma and thermal neutron shielding respectively. The schematic for such a thermal neutron imaging system is shown in Fig. 2.13 [34].

The design of fusion reaction based neutron imaging source, such as D-D, D-T, etc. is similar to the Am-Be source-based system. The D-D and D-T produce mono-energetic neutron (2.45 and 14.1 MeV, respectively). Similar to the isotopic source

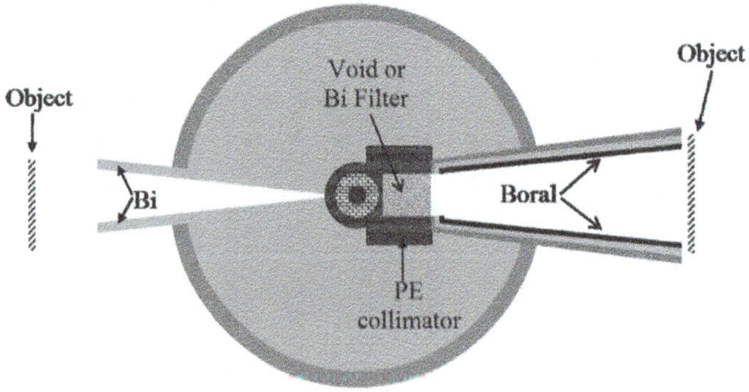

Fig. 2.13 Thermal neutron imaging system using Am-Be neutron source (reproduced from [34] with permission from Elsevier)

based imaging system, polyethylene, lead, and borated polyethylene are used for design of fusion reaction based thermal neutron assembly. The L/D design is as per the source strength and experimental requirement.

In case of spallation neutron source, for example in PSI spallation source, the high energy neutrons are slowed down by D_2O moderator at room temperature. The beam-line for thermal neutron imaging is directed towards the location of maximum thermal neutron flux inside D_2O moderator. Further, the design of thermal neutron imaging beam line is similar to the reactor-based imaging beam, which consists of beam filter, mechanical shutter, beam aperture, collimator, and flight tube upto the sample position. The thermal neutron at room temperature (300 K) has energy 0.025 eV and the neutron wavelength is approximately 2 Å. Furthermore, the absorption cross-section of different materials for cold neutron, having wavelength of approximately 10 Å, is higher compared to thermal neutron. Thus, liquid hydrogen or deuterium is used to further cool down the thermal neutron in different spallation neutron sources. In case of PSI spallation source, a liquid deuterium tank at temperature 25 K is placed inside the D_2O moderator tank for further cooldown of the thermal neutrons as shown in Fig. 2.14 [35]. This cold neutron assembly provides cold neutrons, which are extracted through guide tube for cold neutron imaging. Moreover, during the streaming of thermal and cold neutrons through beam line, there might be loss of neutron due to scattering and leakage. Thus, super mirror-coated guide tube along with vacuum are used to minimize the beam loss. Finally, the neutron beam has mixed energy spectrum, and time of flight (TOF) or beam monochromator are used for high-resolution energy selective neutron imaging. In case of pulsed source, for example, at LANSCE, ORNL, etc. the trigger pulse from the beam extractor is used to trigger the detector for TOF imaging. On the other hand, in continuous wave (CW) spallation source, for example, PSI spallation source, the mechanical single or dual beam chopper made of boral and cadmium are used to chop the beam at fixed

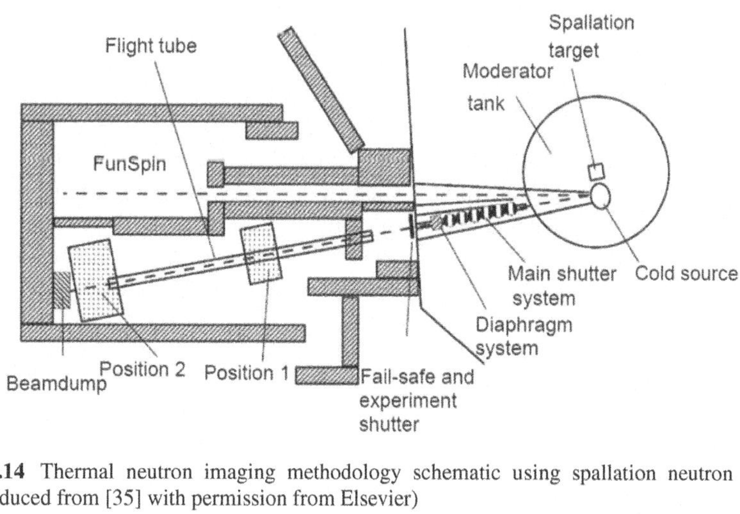

Fig. 2.14 Thermal neutron imaging methodology schematic using spallation neutron source (reproduced from [35] with permission from Elsevier)

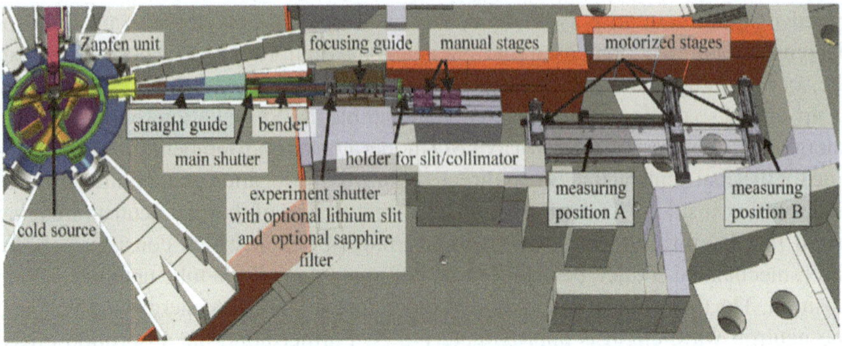

Fig. 2.15 Thermal neutron imaging system using spallation neutron source (reproduced from [36] with permission from Elsevier)

frequencies (25 and 60 Hz), which results in an energy resolution ($\Delta\lambda/\lambda$) of 1.5% and 4.5% respectively in TOF imaging. In case of monochromator-based imaging, a Double Crystal Monochromator (DCM) is used to select a narrow band of the beam energy from the white beam through Bragg scattering. The neutrons with a particular energy band, which is determined by the crystal mosaicity and the beam divergence, are reflected by the first crystal to the second crystal. The second crystal further reflects the beam in the direction of the initial beam. The schematic of such an imaging beam line using continuous-wave spallation source is shown in Fig. 2.15 [36]. Furthermore, neutron has both the up and down spin and the nuclear spin interacts with the magnetic field. Thus, neutron spin polariser is used for polarized neutron-based magnetic phase contrast imaging. Finally, the shielding of the neutron source and the imaging beam line is essential and erected by various materials, such as, lead, hematite concrete, borated polyethylene, cadmium, etc.

2.10 Summary

Over the years, nuclear reactors have been the major source of neutrons for neutron imaging facilities around the world. Recently, spallation neutron sources offering high neutron flux are also being used by researchers to explore further advancements in neutron imaging. However, non-reactor-based neutron sources such as isotopic sources or accelerator-based sources are also used for some limited applications notwithstanding the low neutron flux obtained from such sources. Due to the different characteristics of various neutron sources, the beam obtained from these sources needs to be suitably modified for use in neutron imaging. The advantages and limitations of various neutron sources and the modality of their use for neutron imaging have been discussed in this chapter.

References

1. Kallmann H (1948) Neutron radiography. Research 1(6):254–260
2. Peter O (1946) Neutronen-Durchleuchtung. Z Naturforsch 1(10):551–559
3. Thewlis J (1950) Neutron radiography. Br J Appl Phys 7:345–350
4. Hawkesworth MR (ed) (1973) Radiography with neutrons. British Nuclear Energy Society, London
5. Berger H (ed) (1976) Practical applications of neutron radiography and gaging, ASTM STP 586. American Society for Testing and Materials, West Conshohocken
6. Hawkesworth MR (1977) Neutron radiography issue. Atom Energy Rev 15(2):121–368
7. Berger H (1991) Early development of neutron radiography in the U.S.A. Mater Eval 49:1202–1214
8. Barton JP (1976) Neutron radiography—an overview. In: Practical applications of neutron radiography and gauging. ASTM STP 586. American Society of Testing and Materials, pp 5–19
9. Garrett DA, Berger H (1977) The technological development of neutron radiography. At Energy Rev 15(2):123–142
10. Hawkesworth MR (1977) Neutron radiography: equipment and methods. At Energy Rev 15(2):169–220
11. Glasstone S, Sesonske A (1994) Nuclear reactor engineering, vol 1. Chapman & Hall Inc., New York
12. di Novi RA (1962) Second symposium on physics and nondestructive testing, held at Argonne National Laboratory, October 3–5, 1961 (No. ANL-6515). Argonne National Laboratory
13. Chichester DL (2009) Production and applications of neutrons using particle accelerators. Idaho National Laboratory, Idaho, INL/EXT-09-17312
14. Zuo Y, Tang G, Guo Z, Guo J, Pei Y, Jianguo X, Wang H, Lu Y (2014) Neutron yields of thick Be target bombarded with low energy deuterons. Phys Procedia 60:220–227
15. Fantidis JG, Dimitrios BV, Constantinos P, Nick V (2012) Fast and thermal neutron radiographies based on a compact neutron generator. J Theor Appl Phys 6:20
16. Leung K-N, Reijonen J, Gicquel F, Hahto S, Lou TP (2004) Compact neutron generator development and applications. In: 16th World Conference on NDT, Montreal, Canada
17. IAEA-TECDOC-1153 (2000) Use of accelerators based neutron sources
18. Stefanik M, Simeckova E, Bem P, Majerle M, Novak J, Ansorge M, Mrazek J, Stursa J (2020) Neutron spectrum determination of p+Be reaction for 30 MeV protons using the multi-foil activation technique. In: EPJ web of conferences vol 239, p 17015
19. http://lampes-et-tubes.info/sp/sp179.php?l=e. Accessed date 03 June 2021
20. Valkovic V (2016) 14 MeV neutrons physics and applications. CRC Press, New York
21. Patel T, Bishnoi S, Adhikari PS, Sarkar PS, Gadkari SC (2018) D-D/D-T neutron generator facilities for basic and applied research. In: Proceedings of the DAE symposium on nuclear physics, vol 63, pp 1026–1027
22. https://phoenixwi.com/neutron-generators/high-flux-neutron-generator/. Accessed 03 June 2021
23. Santra S, Singh P (2002) Beam optics of the folded tandem ion accelerator at BARC. Pramana J Phys 59(1)
24. Dutto G, The TRIUMF 520 MeV cyclotron: recent and future developments. In: Proceedings of the 13th international conference on cyclotrons and their applications, Vancouver, BC, Canada, pp 138–142
25. Prescott CY (1994) The 50GeV Program at SLAC; SLAC-PUB-6428
26. Eshwarappa KM, Siddappa K, Kashyap Y, Sinha A, Sarkar PS, Godwal BK (2005) Estimation of photoneutron yield from beryllium target irradiated by variable energy microtron-based bremsstrahlung radiation. Nucl Instrum Methods Phys Res, Sect A 540(2–3):412–418
27. Eshwarappa KM, Sanjeev G, Siddappa K, Kashyap Y, Sinha A, Sarkar PS, Godwal BK (2007) Comparison of photoneutron yield from beryllium irradiated with bremsstrahlung radiation of different peak energy. Ann Nucl Energy 34(11):896–901

28. Kiyanagi Y (2018) Neutron imaging at compact accelerator-driven neutron sources in Japan. J Imaging 4(4):55
29. Kim GN, Kang HS, Choi JY, Cho MH, Ko IS, Namkung W, Chang JH (1999) Pulsed neutron source using 100-MeV electron linac at Pohang accelerator laboratory. In: Proceedings of the 1999 Particle Accelerator Conference (Cat. No. 99CH36366), vol 4, pp 2593–2595. IEEE
30. Kadi Y, Revol JP (2003) Design of an accelerator-driven system for the destruction of nuclear waste (No. INIS-XA-900)
31. Degweker SB, Bhagwat PV, Krishnagopal S, Sinha A (2017) Physics and technology for development of accelerator driven systems in India. Prog Nucl Energy 101:53–81
32. Fantidis JG, Nicolaou GE, Tsagas NF (2009) A transportable neutron radiography system based on a SbBe neutron source. Nucl Instrum Methods Phys Res, Sect A 606(3):806–810
33. Amar Sinha BDB, Panchal CG, Shyam A, Shrinivasan M, Joshi VM (1995) Exploratory studies on neutron radiography with a small neutron source using nuclear scintillation imaging technique. In: The 2nd international topical meeting on neutron radiography system design and characterization, 12–18 November 1995, The Shonan Village Center, Japan, pp 99–106
34. Jafari H, Feghhi SAH (2012) Design and simulation of neutron radiography system based on 241Am–Be source. Radiat Phys Chem 81(5):506–511
35. Kuehne G, Frei G, Lehmann E, Vontobel P (2005) CNR—the new beamline for cold neutron imaging at the Swiss spallation neutron source SINQ. Nucl Instrum Methods Phys Res, Sect A 542(1–3):264–270
36. Morgano M, Peetermans S, Lehmann EH, Panzner T, Filges U (2014) Neutron imaging options at the BOA beamline at Paul Scherrer Institut. Nucl Instrum Methods Phys Res, Sect A 754:46–56

Chapter 3
Neutron Optics and Detectors

P. S. Sarkar and Yogesh S. Kashyap

Introduction

Despite recent progress in the development of advanced and improved neutron sources, neutron-based experiments continue to be severely constrained by available neutron fluxes at various neutron facilities. Currently, the majority of neutron sources are either reactor-based (with power ranging from 1 to 100 MW) or accelerator-based, employing the spallation process (beam power 1 MW). Typically, the peak neutron flux at the core is in the range of 10^{13} to 10^{15} n/cm^2/s, with this range being limited mostly by target cooling or heat removal concerns. As a result, the neutron flux at the sample location typically varies between 10^6 to 10^9 n/cm^2/s. In comparison, X-ray flux at the sample locations in the contemporary synchrotron radiation-based X-ray sources varies between 10^8 to 10^{13} photons/s/mm^2/mrad^2/0.1% bandwidth. This means that neutron sources have a brightness several orders of magnitude lower than X-ray sources, and their phase space density is 1000–10,000 times smaller than X-ray phase space density. Therefore, in absence of extremely high brightness neutron sources, the advancement in neutron optics and the related instruments is of vital importance for efficient use of current generation of neutron sources [1].

Broadly speaking, one can classify neutron optics into the following major classes, based on their interaction with the material medium as shown in Fig. 3.1.

The Neutron optics is based on assigning a complex refractive index to the materials as $n = 1 - \delta - i\beta$. Neutron interactions with most materials are quite weak, resulting in a refractive index that is extremely close to unity. As a result, neutrons only reflect from surfaces at grazing angles, which are usually only a few degrees. Consequently, fabrication of efficient neutron optical components like lenses and

P. S. Sarkar (✉) · Y. S. Kashyap
Technical Physics Division, Bhabha Atomic Research Centre, Mumbai 400085, India
e-mail: pss@barc.gov.in

Y. S. Kashyap
e-mail: yskashyap@barc.gov.in

D. K. Aswal et al. (eds.), *Neutron Imaging*,
https://doi.org/10.1007/978-981-16-6273-7_3

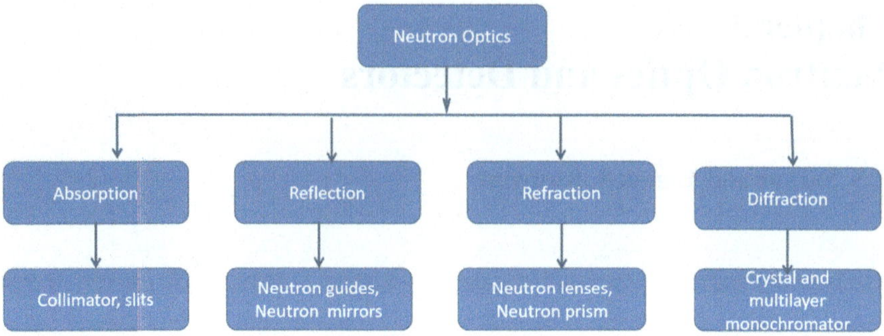

Fig. 3.1 Showing different types of neutron optics and the basic interaction mechanism

mirrors, which are often employed in visible-light optics, is quite challenging. We will briefly discuss a few of the optical elements relevant for neutron imaging.

3.1 Absorption-Based Optics

3.1.1 Neutron Collimators

Usually, the moderated neutrons from the reactor core are transported through the reactor shielding via hollow beam tubes or neutron guides. These extracted neutrons have the Maxwellian velocity distribution with average energy dependent on moderator temperature and an angular spread defined by the external collimation imposed on the beam. As the volume of the thermalized neutron in the moderator is enormous compared to the sample dimensions, one needs to have an efficient means to deliver the neutron beam of well-defined size and divergence for carrying out any measurement. A collimator is a passive device that filters a beam of neutrons so that only those travelling in a defined solid angle or directions are transmitted to the sample position. Usually, the type of collimators can be classified into three types namely, parallel collimator, diverging collimator and converging type collimator. An illustration with a cut-away view of a typical divergent collimator is shown in Fig. 3.2. Neutron radiography collimators typically included the following components:

(a) **Illuminator**—The role of the illuminator is to create a homogeneous neutron source. Normally, it is a 10–15 cm thick graphite or polythene block positioned at the beginning of beam tube to boost the thermal neutron flux entering the collimator.
(b) **Beam filters**—These filters are mostly used to remove or filter gamma rays and fast neutrons from the beam. Because of its low neutron attenuation and strong gamma-ray attenuation characteristics, Bismuth (Bi) is widely employed for gamma-ray filtering. Furthermore, to improve thermal neutron transmission, Bi is frequently utilized in a single-crystal structure. Similarly, single-crystal

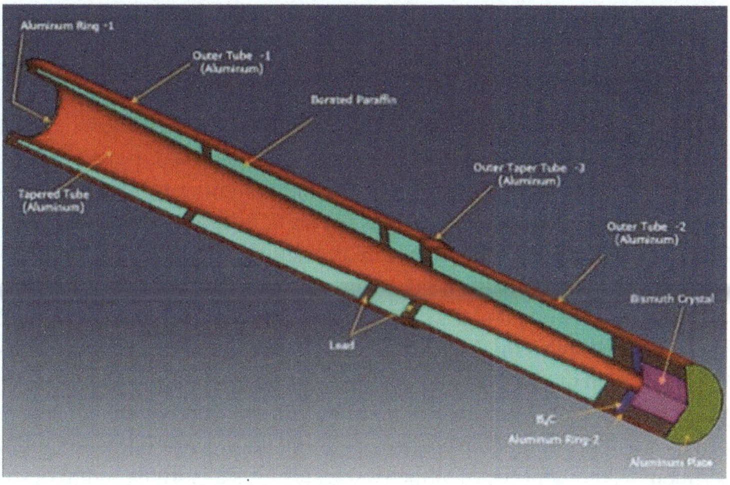

Fig. 3.2 Example of a divergent neutron collimator

Sapphire (Al_2O_3) and/or Silica (SiO_2) are often used to reduce the high energy neutron content travelling to the experimental hutch.

(c) **Aperture**—The aperture defines the pinhole in the pinhole camera analogy, which is frequently used to characterize collimator design. The pinhole size determines the maximum flux at the sample position and point spread function of the neutron imaging instruments. As the thermal neutrons should not be allowed to enter the beam tube other than through the pinhole, it's constructed of materials such as Gd, Cd, which have very high thermal neutron absorption cross-section.

(d) **Gamma shielding/cleanup plates**—Pb-based gamma shielding is commonly used, however heavy concrete or other materials may also be used. Moreover, to prevent streaming, gamma shielding is installed at every step change in beam-tube dimension. Similarly, in order to prevent capture gammas and scattered neutrons from reaching the experimental hutch, boron-based material in the form of plates is often employed in conjunction with lead sheet.

(e) **Filling gas**—It's worth noting that neutron scattering from nitrogen in the air decreases the neutron intensity at the sample by 5% per metre. These scattered neutrons not only degrade the sharpness of the neutron image or increase the background, but also interact with the collimator walls resulting in emissions of secondary gamma rays, that require further shielding. As a result, collimators are frequently either evacuated or filled with Helium.

3.1.2 Velocity Selector

A velocity selector is a mechanical device that allows only neutrons of certain wavelength or velocity to get transmitted through. The velocity selectors can be of the solid drum type with helical slot or of the multidisk type. A disk-type velocity selector device works on the principle of chopping of the incoming neutron beam twice in succession, at two different points separated by suitable flight path. Usually, the disks are displaced by certain angle (θ), therefore the neutrons transmitted by the first disk will be stopped by the second one if at rest. The second disk will transmit those neutrons whose velocity is equal to the distance between the two disks divided by the time corresponding to the rotation of disk by the angle (θ). Usually, more than two disks are used in modern devices and the selection of neutron energy is achieved by varying the speed of rotation. The number of disks and number of slots and its width is decided depending upon energy range and energy resolution requirement of the experiments. Another type of velocity selector uses absorbing radial blades, like multiple peripheral slits, usually set at certain angle to the incoming beam. When the device is stationary the neutrons will get completely absorbed and will get transmitted only if the device is rotating. Photograph of a typical mechanical velocity selector is shown in Fig. 3.3. Velocity selector performs a coarse monochromatization, and typical values are about 10–30% of the mean wavelength. The transmission of neutron through a velocity selector is determined by its geometrical parameters. The transmitted wavelength ($\lambda = K/\omega$) is inversely proportional to the rotational speed (ω), where K is a constant. The selector's geometry and the tilt angle between the selector axis and the neutron beam determine the constant (K). Another useful device based on the absorption optics is the Fermi chopper first proposed by Fermi in the 1940s, which is nothing but a rotating shutter consisting of neutron-absorbing and neutron transmitting materials placed into a very fast revolving cylinder. It can be used to obtain pulsed monochromatic beam of neutrons at the accelerator-based neutron sources or in combination of monochromator crystals at the reactor-based neutron sources.

Fig. 3.3 Example of mechanical velocity selector (reproduced from [2] © 2004, with permission from Elsevier)

3.1.3 ^{3}He Spin Filter for Neutrons

Polarized neutrons are widely used in neutron scattering and polarized neutron imaging experiments. Usually, neutrons with a specific spin orientation are chosen using suitable neutron optics, and their orientation after passing through the magnetic field region or sample is recorded on the neutron detector, which is positioned immediately after the neutron analyzer. The spin of scattered, transmitted, or reflected neutron after passing through the sample is analyzed to generate the neutron depolarization data, which in turn provides information about magnetic properties of the sample. The basic idea to obtain the polarized neutron is through spin dependence of magnetic scattering [3], nuclear scattering [4] or resonant neutron capture [5]. Absoprtion based polarized neutron spin filters are based on the spin-dependent absorption of neutrons by ^{3}He, and can produce a polarized neutron beam over a wide range of neutron wavelengths. The resonant absorption of the neutrons depends on compound nuclear spin states, $I + 1/2$ and $I - 1/2$, where I is nuclear spin before the neutron capture. The spin orientations of the nuclear and neutron spins affect the absorption of neutrons by ^{3}He; if the nuclear and neutron spins are anti-parallel, the absorption is extremely high. If the spins are parallel, however, there is far less absorption. The transmission of spin-parallel neutrons and spin anti-parallel neutron beam through polarized ^{3}He cylindrical cell can be simply expressed as $T_{\pm} = \exp\{-(1 \mp P)n\sigma l)$ where P is the ^{3}He nuclear polarization, n the number density of ^{3}He atoms, σ the absorption cross-section for unpolarized neutrons (σ_0 [barn] $\approx 3000 \cdot \lambda$ [Å]) and l is the length of the spin filter cell. Polarized ^{3}He is typically produced through metastability-exchange optical pumping (MEOP) [6] or spin-exchange optical pumping (SEOP) [7]. Figure 3.4 shows the Photograph of a typical ^{3}He cell was fabricated at J-PARC.

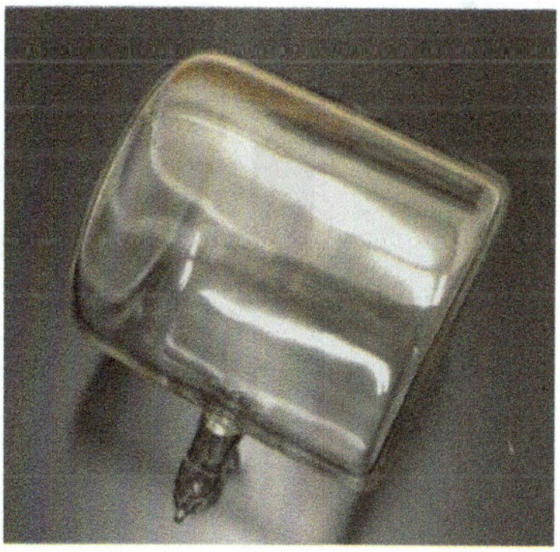

Fig. 3.4 ^{3}He cell fabricated at J-PARC (reproduced from [8] © 2020, with permission from Elsevier)

3.2 Reflective Optics

3.2.1 Neutron Mirrors

The neutron mirrors usually operate on the principle of total external reflection in exact analogy with the conventional light optics. However, in case of neutrons these mirrors work mostly in the grazing incidence mode due to small refractive index ($n \sim 10^{-6}$) of most of the materials for neutrons and they reflect the neutrons with wavelength greater than critical wavelength. As the refractive index depends upon the wavelength of the neutrons (λ^2), the mirrors are therefore most suitable for longer wavelength neutrons such as thermal or cold neutrons. Usually, neutron mirrors are used as focusing devices to effectively utilize the incoming neutron by efficiently converging them onto samples or detectors, thus improving flux density. The neutron focusing optics includes parabolic mirrors, elliptic mirrors, Kirkpatrick–Baez mirrors, Wolter mirrors, etc. [9]. Generally, neutron-focusing mirrors use supermirror coating, to reflect slow neutrons over a wide spectral range. Another completely different approach as suggested by Kumakhov [10] is to use thousands of glass capillaries and guide these neutrons using multiple reflections and thereby focusing the beam to a spot with a diameter less than 1 mm. Another recent development is the demonstration of novel neutron focusing optics based using Wolter optics suitable for use in neutron imaging and small-angle scattering experiments. The axisymmetric grazing-incidence focusing mirrors using identical pairs of confocal coaxial Hyperboloid and Paraboloid can improve the resolution of conventional neutron imaging camera to achieve sub-micron level resolutions [11]. Its schematics is shown in Fig. 3.5.

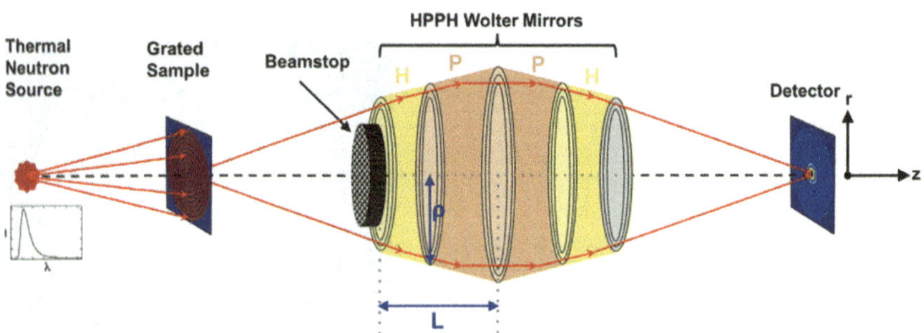

Fig. 3.5 Schematics of the neutron microscope optimized for thermal neutrons (reproduced from [11] © 2017 with permission from Elsevier)

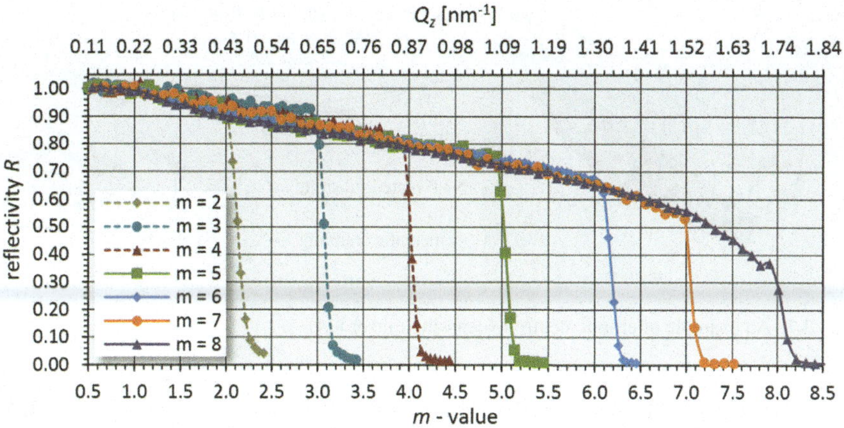

Fig. 3.6 Neutron reflectivity of a supermirror on a refined aluminium substrate (diamonds) and float glass (squares), with Ni/Ti supermirror [13]

3.2.2 Neutron Supermirror

These multilayer mirrors are constructed from alternating layers of two materials with different neutron scattering lengths. Alternating layers of materials such as nickel and titanium are deposited on a glass substrate to create these mirrors. Supermirrors, thus are a kind of multi-layer neutron mirror, with variable layer thicknesses. The advantage of this variable lattice spacing is that a neutron beam travelling through these layers encounters variable one-dimensional optical potentials. As a result, as the neutron beam transverses through these layers partial reflection and transmission of a neutron wave occur at every interface, resulting in the appearance of multiple waves within the system. By slowly varying the period of multilayer structures, the effective reflection angle can be larger than the total reflection angle. This idea was realized by Mezzi for extracting thermal neutrons with a sufficiently wide wavelength range. Ni/Ti is the materials system of choice for broadband neutron multilayer supermirrors due to the small absorption coefficient and large difference in the real part of the refractive index, with low values for the imaginary part.

One of the most important properties of a supermirror is its critical angle, θ_c^M the angle up to which it reflects and it is determined by the thickness of the thinnest layer. By convention it is measured in multiples (*m*) of the critical angle of natural nickel. The reflectivity of a mutilayer is generally proportional to the fourth power of its thickness and the square of the number of layers. As a result, the number of layers required to achieve the same reflectivity for a given bilayer thickness grows quadratically as the thickness decreases. The first neutron supermirror concept was proposed by Mezzi [12] inspired by earlier work with X-rays.

Hosobata et al. [14] have reported the development of precision elliptic neutron-focusing supermirrors. The evaluation experiments carried were carried out on BL16

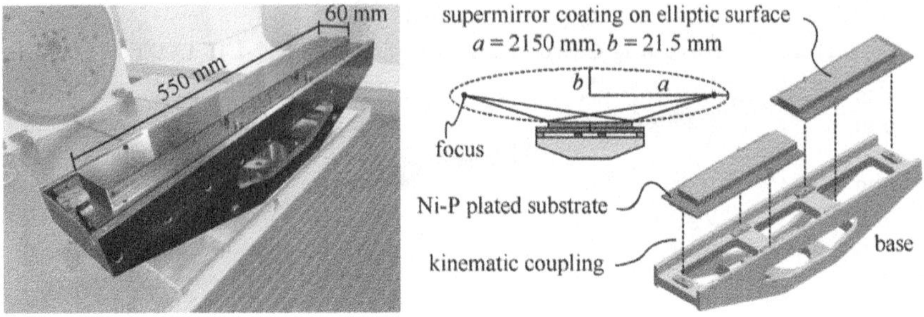

Fig. 3.7 An example of elliptic neutron-focussing mirror [14]

at the Materials and Life Science Facility, J-PARC, Japan and showed 68–90% reflectivity at a critical angle. The photograph of the developed supermirror is shown in Fig. 3.7.

The Polarizing multilayer supermirrors were reported by Lynn et al. [15]. The refractive index of neutron for magnetic materials can be expressed as

$$n^2 = 1 - \lambda^2\left(\frac{Nb_c}{\pi} \pm \frac{2m\mu_n B}{h^2}\right)$$

where N is particle density, b_c is average bound coherent scattering length, m and μ_n are mass and magnetic moment of neutron, h is Plancks constant and B is saturation induction of the magnetic material. For ferromagnetic materials, one can simplify the above expression as

$$n^2 = 1 - \frac{\lambda^2 N}{\pi}(b_c \pm b_m)$$

where b_m can be considered as effective magnetic scattering length. This shows that neutron with two different spins directions will interact very differently and for $b_c = b_m$, the reflected beam is completely polarized. Polarizing supermirrors make use of the fact that the scattering lengths of the two spin components in ferromagnetic materials are extremely different. Thus the idea of index matching, in addition to the property of two distinct scattering lengths, is employed in polarizing mirrors. A thin film of ferromagnetic material is deposited on a substrate whose refractive index matches that of the film, for only one of the two spin components of incoming beam of neutron. Under this condition, there will be no optical density difference between the ferromagnetic layer and substrate layer, and hence one of the spin component will get transmitted to the substrate layer. Therefore one of the spin component is then removed from the beam by absorption or by incoherent scattering, while the other spin component undergoes total external reflection at the magnetic layer, and hence polarized neutron beam can be obtained. The layer sequence is calculated based on the contrast of two materials for the one spin component, after selecting

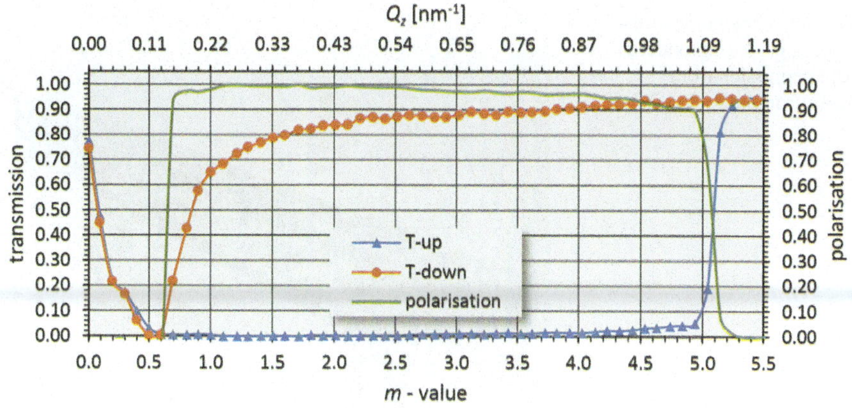

Fig. 3.8 Spin-dependent transmission and polarization of a polarizing supermirror Fe/Si with $m = 5$ coated on both sides of a Si wafer with a thickness $t = 0.3$ mm [13]

two materials with the same scattering length density. In most cases, polarizations of up to 98% are possible, while the polarized beam intensity of 30–40% of incident neutron beam intensity (Fig. 3.8). Historically, the material pairs Fe–Ag and Co–Ti were used to create the first polarizing mirrors. Fe–SiN_x and $Fe_{89}Co_{11}$–Si or Co–Ti, FeCo–TiZr, and $Fe_{50}Co_{48}V_2$–TiNx [16–18] are two of the most frequent supermirror combinations utilized nowadays.

3.2.3 Neutron Guide

The neutron guides work in the same way that of optical fibres: neutrons are transported through the neutron guide tube by total external reflection; only those neutrons which enter the guide with an angle of incidence that is smaller than the critical angle (depending upon their wavelength) will pass through the guide by repeated total reflections. For a non-magnetic material, the relation between critical angle (θ_c) and coherent scattering length (b_c) is given by:

$$\theta_c = \lambda\sqrt{\frac{Nb_c}{\pi}}$$

where λ is neutron wavelength and N is the N is the number of atoms per unit volume. The value of critical angle for representative material like for Ni is ~ 0.1° λ(Å). The guide tube usually has a little curvature to ensure that only slow neutrons with a wavelength higher than a certain value are guided towards exit of the guide tube. Fast neutrons or neutron with shorter wavelengths, as well as any gammas in the incident beam, pass through the guide tube wall and are usually absorbed in the shielding that surrounds it. This permits a very clean and highly collimated beam of slow neutrons

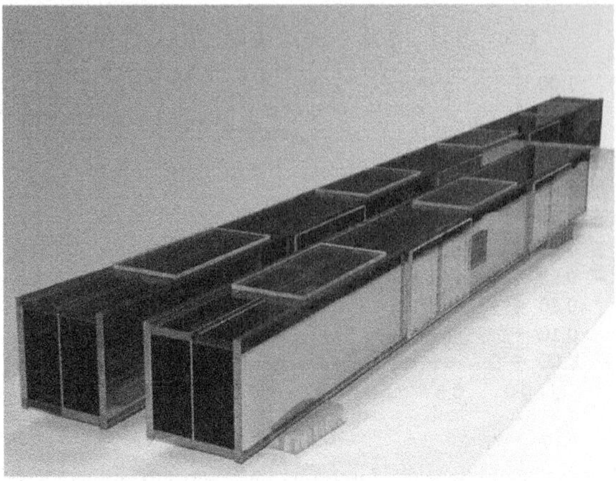

Fig. 3.9 Photograph of the neutron guide (reproduced with permission from Mirrortron Ltd.)

to be delivered from the original neutron source to the far away experimental station. As a result, using a neutron guide allows for extremely low background radiation, which is ideal for high precision neutron scattering and imaging experiments. Another advantage of the guide is that it can deliver neutrons to several instruments at the same time in a low-background environment. Moreover, the flux at the end of a neutron guide being proportional θ_c^2, therefore one can significantly improve the flux by using supermirror coatings such as Ni–Ti. Supermirrors consisting of 2600 Ni–Ti leading to refection angles more than $4\theta_c^{Ni}$ and reflectivity greater than 80% have already been achieved [13, 19]. Similarly, neutron guide made out of ferromagnetic material magnetized to saturation, it is possible neutron arriving at the exit of the guide tube is polarized. Such polarized neutron guides have been demonstrated with Fe–Co and Ti–Gd deposited on basic guide material. Photograph of a typical neutron guide is shown in Fig. 3.9.

3.3 Refractive Optics

3.3.1 Neutron Lenses and Prisms

A well-known phenomena in geometric optics is the refraction of a light at an interface with two different refractive indices. Neutron beams refract in the same way at the interfaces of different materials; but, as previously stated, neutrons have a very low refractive index for most materials, thus this effect requires a large number of interfaces for any practical uses. Therefore we need an array of refractive elements to focus the neutrons in a meaningful way. The first refractive lens was made using an array of 30 biconcave lenses from MgF_2 [20]. The advantage of the refractive lenses is that they can be placed directly in the beam path and are easy to align. Figure 3.10 shows a photograph of a 6- and 28-element array of symmetric spherical

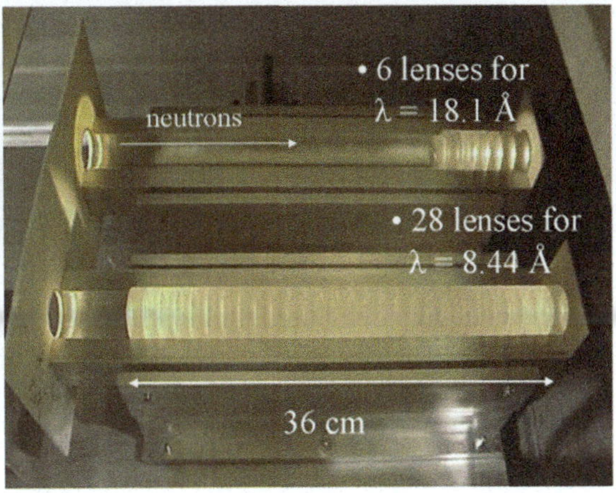

Fig. 3.10 Photograph of the 6/28 element array of 25-mm-diameter symmetric spherical MgF_2 biconcave lenses [24]

MgF2 biconcave lenses with a diameter of 25 mm. It may be noted that neutron transmission through these lenses decrease diagrammatically for larger beam size or larger neutron beam path, due to the lens shape effect. As a result, to address this issue, Fresnel lenses made of vitrous silica and MgF_2 single crystals have also been developed [21]. Similarly, compound refractive prism for cold neutrons made of MgF_2 single crystal have also been fabricated and with 50 single-sided lens element assembly, transmission of 0.83 with a focal length of 5 m for 1.14 nm neutrons was reported [21]. The focusing and polarization of neutrons by magnetic induction has been explored in the development of a magnetic compound refractive lens composed of 99 pairs of cylindrical permanent magnets in order to increase the focused intensity of a polarized white beam of cold neutrons. [22]. Recently, imaging of objects with very cold neutrons, using novel, permanent magnet ($Nd_2Fe_{14}B$) compound refractive lenses (MCRL) with a large 25 mm bore diameter has also been reported [23].

In order to overcome the absorption of the neutron in the convention refractive optics and diffuse scattering in the reflective optics, magnetic compound refractive lenses and prisms using have also been developed. The magnetic compound refractive lens consists of voids and magnetic field regions and it provides focused spin polarized neutrons. However, a magnetic refractive lens can focus only neutrons with spin parallel to the magnetic field, while neutrons with anti-parallel spins diverge, and therefore only 50% of neutron flux is available for the experiment, if the input beam is unpolarized. Shimizu et al. [25] used permanent magnets to demonstrate the feasibility of such a lens towards attaining intensity gain of 35 at 14.4 Å neutron wavelength, and further superconducting sextupole magnet (SSM) with Nb/Ti superconducting coils have been successfully demonstrated with cold neutrons. Development of lenses based on a permanent sextupole magnet with continuous field modulation via a rotating double-ring structure, called modulating permanent-magnet sextupole (mod-PMSx) [26–29] have also been reported. The feasibility of mod-PMSx for high resolution imaging through magnification has also been proven [29].

Fig. 3.11 Multiple single crystal based neutron monochromator (reproduced with permission from Mirrortron Ltd.)

3.4 Diffractive Optics

3.4.1 Crystal Monochromator

The monochromator is one of the most important optical components in almost all the reactor-based neutron instruments. A neutron monochromator is a device made up of a single or double crystal that receives polychromatic neutron beams and provides a specific wavelength at sample locations based on Bragg's rule. The wavelength range accepted by a monochromator is determined by its crystal structure and mosaicity. A higher mosaicity increases the integrated flux of neutrons that reach the sample, however, it reduces the wavelength resolution of the instruments. As shown in Table 3.1, most commonly used materials as monochromator crystals include pyrolytic graphite (PG), silicon (Si), copper (Cu), beryllium (Be), and Heusler crystals. The monochromator crystal is usually selected depending upon the incident energy range

Table 3.1 Examples of some materials used as single crystal monochromator

S. no.	Crystal	d-spacing (A)
1	PG (002)	3.35
2	Be (002)	1.79
3	Cu (220)	1.28
4	Ge (111)	3.27
5	Si(220)	1.92
6	Si(111)	3.13
7	Heusler (111)	3.44

required for the experiment and the desired energy resolution. The double-crystal monochromator with highly oriented pyrolytic graphite HOPG (002) single crystals has been demonstrated to be a useful device for energy-resolved imaging and Bragg-edge imaging applications. The benefit of this configuration is that the outgoing monochromatic neutron beam remains parallel to initial incoming neutron beam even when the wavelength of neutron is varied, simplifying the experimental setup.

On the other hand, neutron diffraction experiments require focussed neutron beams at the sample positions, and hence bent single crystal is used. Due to the size of neutron beams, very large single crystals are required for neutron monochromators, e.g. 10^5 mm^3. However, as the neutron beams are of relatively low intensity, and single crystals have too small band-pass ($\Delta\lambda/\lambda$), to accept large neutron beam, multiple single crystals are used, as illustrated in Fig. 3.11. This not only helps to increase the crystal size but allows a large degree of bending without approaching the breaking limits of the material. A more recent development is to produce very fine slices of single crystal, and to glue them together in near-perfect alignment, thus simulating on a rather large scale the effect of mosaic blocks.

3.4.2 *Neutron Filters*

Whan a neutron beam is passed through a system consisting of polycrystalline material, for $\lambda \leq 2d$ (d: interplanar spacing, λ: wavelength), the condition of Bragg scattering is always satisfied, leading to the removal of neutrons below this wavelength. On the other hand $\lambda \geq 2d$, and there would be no Bragg scattering and the beam will be transmitted. For a sufficiently large crystal, the effect will be the same as that sample is absorbing the neutrons with $\lambda \leq 2d$, while all the other wavelengths are transmitted and the crystal will act as a neutron wavelength filter. One can therefore arrange two such crystal materials with similar $2d$ spacing such that only a window of the wavelength is transmitted. Some useful polycrystalline materials are Be and BeO, which are usually used to select neutrons above 4 Å. On the other hand, some single crystals such as Bismuth, Sapphire, etc. are used as filters for spectral shaping in neutron imaging beamlines. Bismuth single crystal is usually employed in the reactor to block the incoming gamma rays and improve the neutron to gamma ratio. Sapphire (α-Al_2O_3) crystal exhibits some unique properties such as High Debye-Temperature (1047 K), high transmission of thermal neutrons, which help in improving the thermal neutron content at the experimental station by scattering out epithermal and fast neutrons. Figure 3.12a shows simulated total neutron cross-section per molecule of α-Al_2O_3 and Fig. 3.12b per atom of bismuth at 300 K, without the coherent elastic (Bragg) scattering part.

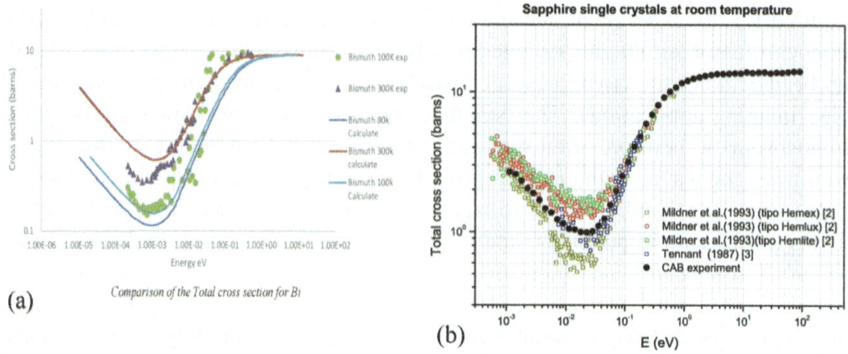

Fig. 3.12 Total neutron cross-section at 300 K for **a** Bismuth single crystal [30] (Al_2O_3) and **b** Sapphire single Crystal filter [31] (reproduced from [31] with permission from Copyright © 2015 Elsevier Ltd.)

3.5 Neutron Detection: Basic Principle and Methodology

Since neutrons have no charge, their interaction is through nuclear interaction and as they are not perturbed by the atomic electrons they can penetrate deeper into the materials. But this charge neutral status of neutrons makes their detection difficult. For detection of neutrons, they must interact and be absorbed by certain elements having high interaction (scattering/nuclear reaction) cross-section and the resulting particle(s) emitted from these reactions, deposit their energy and produce detectable emission(s). This forms the basis of neutron detection. As the cross section for neutron interactions in most materials strongly depends upon neutron energy, different techniques have been developed for neutron detection in different energy regions. The neutron energies can be roughly divided into two groups—slow neutrons which have energy below the cadmium cut off (0.5 eV) and fast neutrons with energy above this level. Based upon the methodology and application for neutron detection, the detectors can be subdivided into two groups: passive and active.

The interaction probability for thermal neutrons is high for elastic scattering such as $X(n, n)X$ and their detection is not possible owing low energy transfer via this mechanism. In case, the neutron energy is sufficient to impart some of its energy to the interacting nucleus, the recoil nucleus ionizes the surrounding medium which then produces some signal to be detected. Neutron-induced nuclear reactions are the mechanism by which charge particles are generated towards production of detectable signals in terms of scintillation or fluorescence. Possible outcome of neutron interaction with a target nucleus may be any or combination of various processes such as recoil nucleus, protons, charge particles, conversion electrons, gamma, fission fragment.

For thermal neutrons the materials having high absorption cross-section are ^{3}He (5333 b), ^{6}Li (940 b), ^{10}B (3835 b), Cd (^{113}Cd (12% abundance): 20743 b) and Gd

(^{157}Gd (15.7% abundance): 255011 b, ^{157}Gd: 60,791 b). He-3 and B-10 based gas-based detectors are in general used for monitoring and detection purposes though recent developments cite gas-based imaging detectors.

We here restrict ourselves to neutron detectors for imaging purposes only. For completeness, basic neutron detectors used for dose monitoring towards operational as well as personnel safety are mentioned in brief.

Neutron converters are the first most important part of neutron detection. They must have the following properties to qualify as suitable neutron converters. These are high interaction cross-section, creation of detectable signal in accordance with the used phosphor, reaction Q value must be high for good discrimination in mixed radiation fields and most importantly the material should be transparent to the emitted detectable photons. The materials which generate high mass charge particles on interacting with the neutrons are the most suited for imaging detectors because of two reasons; firstly their energy deposition will be more hence more light which in turn provide higher signal content and secondly owing to their less travel path localization of light signal takes place which translates to achieving higher spatial resolution.

Examples of various class of neutron imaging detectors are films, image plate, gas-based detectors, scintillator screens, single crystals and semiconductor detectors. Initial work on the development of area-based neutron detectors started with activation foils (Gadolinium, Dysprosium and Indium) and then transfer the image to a phosphor coupled with a photographic film. Development of neutron sensitive image plates (BaFBr(Eu):Gd_2O_3) on the lines of X-ray image plates were developed. Later on, neutron image scintillator screen having ^{6}LiF:ZnS(Ag)/Gd_2O_3 in Lucite were developed and made commercially available. Till date ^{6}LiF:ZnS(Ag) screens are the most widely used in combination with high resolution as well as cooled Charge Coupled Device (CCD) or CMOS cameras or semiconductor panels in various configurations to yield digital neutron images. In case the neutron flux falling onto the detector is low, image intensifiers are used to amplify the signal content but with an amplification in the noise content too. There has been constant development with newer materials or their combinations towards aiming high resolution, high image contrast and improved radiation hardness. Though thermal neutron radiography had been the mainstay technique for a long time, with the availability of high flux neutron sources like the spallation sources or high flux research reactors, the quest to look more deeper inside bulk materials with high resolution through high collimation factor without compromising on neutron numbers, carrying out dynamic imaging with high temporal resolution, high detection efficiency thereby decreasing experimentation time as well as increasing contrast has motivated the concerned scientific community towards development of better-improved neutron imaging detectors. Thrust was put on material research with an aim to develop new materials or their combination and even combining multiple detector technologies towards hybrid detector development providing information which were thought to be impossible to obtain a decade back. Moreover, with the advancement in computer technology, faster and robust data acquisition, archiving and analysis have prompted use of complex analysis methodologies in implementing imaging with such hybrid detectors. For example, the advanced highly efficient charge particle detectors developed for various particle accelerator programs were suitably modified by coupling with some neutron converter material such as ^{6}Li, ^{10}B or ^{157}Gd. They proved to be excellent high-resolution pixelated imaging detectors. Their throughput was also very high owing to their reduced read-out time. Medipix-2, nGEM and LiTA12 are some examples of such detectors. Special FPGA electronics and data acquisition including analysis techniques were also part of these developments. ^{3}He and BF_3 based gas detectors are already proven neutron detectors used for counting applications and dose measurements. A clever move by configuring these with multiwire proportional counter methodology and 2D read-out, produced gas-based hybrid detectors towards neutron imaging applications. Since these detectors are current based, appropriate mechanisms to incorporate high gain could be easily realized. μPIC, μNID are some newer detectors developed on this principle.

Application-specific imaging detector development is governed by various parameters. They are pixel size, dynamic range, frame rate, readout speed, overall noise (dark, white and readout), data readout mechanism, compatibility with existing protocols and finally the cost. All these factors are interlinked in some way or the other and an optimum solution is sought with all possibly achievable ones. Let's elaborate it a bit more. Since neutrons are of low interacting type, in the case of scintillator screens if the neutron converter thickness is increased (say 200–300 μm) to increase detection efficiency, light spread increases resulting loss in spatial resolution. On the contrary for very thin (say < 50 μm) scintillator screen, the achieved spatial resolution is high (~ screen thickness), but the light output is low which increases image acquisition time and this in turn increases photon detector noise. Detector throughput may not be that important for simple radiography but for tomography applications when a large number of projections are to be acquired, it is of great interest.

Various parameters linked to imaging detectors [32, 33] deciding their performance capability are linearity, response time, quantum efficiency, pixel size, gain, noise characteristics and dynamic range. The imaging system as a whole is qualified by parameters such as spatial resolution, Modulation Transfer Function (MTF), Detective Quantum Efficiency (DQE), Signal to Noise Ratio (SNR) and Noise Power Spectrum (NPS). The terms are discussed below.

Linearity: In radiation imaging detectors the signal output must be linear with the dose, equivalent to the number of radiation quanta, falling on it. For any detector, there is a linearity range defined by the manufacturer. The linearity is also dependent on the energy of the radiation quanta falling on it.

Response time: This is the time taken by the detector to generate response signal, after the interaction with the radiation, to be converted to measureable data. In the case of neutron imaging, the dependencies for response time are converter material property towards signal generation, methodology of coupling with photo-detector and its response characteristics.

(A) **Pixel size**: This is the smallest unit of the detector which generates signal proportionate to active volume/size and in combination with such large numbers of its replica forms the entire detector.

(B) **Quantum efficiency**: This term signifies the probability of primary ionization taking place as soon as it interacts with detector material.

(C) **Gain**: This parameter signifies the amplification introduced in the detector's signal generation module. This is helpful in case where sufficient signal does not exist to provide good quality images.

(D) **Noise**: The unwanted data associated with the actual signal decreasing its worthiness is called noise. The signal is represented by the number of incident particles and the noise associated with it the square root of the signal. The contributors towards noise are: (a) shot noise arising from the statistics of the signal generating process, (b) readout noise associated with the signal readout mechanism and (c) dark noise contributed by the background effects such as background radiation either same (scattered) or of other types. Shot noise decreases as the input radiation intensity increases, faster readout increases the noise and background or dark noise can be decreased by other cooling the detector and/or shielding to reduce such effects. The methodology to be implemented is different for different detectors.

(E) **Dynamic range**: This parameter designates the maximum number of discrete values the signal can be sampled to provide meaningful representations of variations present in the signal. In digital detectors, this is governed by the A/D converter and associated electronics. As, for example, an 8bit image contains 256 shades and a 16bit image contains 65,536 shades, each shade to shade variation brings in some contrast. So more the bit depth, more is the information and hence more discernible contrast embedded in the image.

(F) **Signal to Noise ratio (SNR)**: Denoted as S/N, this parameter signifies the signal level with respect to the noise. S/N must be high to obtain good quality data as well as good quality image. Though increasing exposure increases the signal strength but at the cost of increase in noise.

(G) **NPS**: This parameter determines the spectral density of the noise as a function of spatial frequency. Here spatial frequency is the inverse of the lengths, defined by the feature sizes looked into or imaged, in the spatial domain. Fine features in the spatial domain signifies high frequency component in the spatial frequency domain.

(H) **MTF**: This parameter describes how faithfully the detector as a unit translates the input signal to the image domain towards providing information being sought. MTF is described in terms of percentage contrast at a certain line pair per unit distance.

(I) **DQE**: In the case of imaging detectors, DQE [34, 35] is defined by the ratio of squared out SNR to the squared input SNR. With N number of radiation quanta falling on the detector and if N' numbers of such quanta are utilized to generate image information then the DQE of the detector is said to be N'/N. This measures the efficiency of the detector in terms of the degradation of the input SNR after being converted into image by the detector or the imaging system. For a detector with DQE = 1, all the input radiation has been translated into image information. DQE is dependent on detector construction, radiation exposure, spatial frequency and MTF.

Though the parameters such as DQE, NPS and MTF are evaluated in general for X-ray detectors as per the international standard IEC 62,220–1[36], the same is not strictly followed by all for neutron imaging detectors. It is worth mentioning that NR community has accepted MTF as the figure of merit as far as spatial resolution determination is concerned. But MTF alone is insufficient to determine image quality as it does not consider the noise components. In this regard, DQE which is deduced by taking into account both the noise and contrast information is the metric that can provide the ability of the detector towards fine feature detection. Masalovish [37] has suggested formula for evaluating DQE of a neutron image plate and optimize the exposure time for obtaining desired image quality. Radoslaw Lewandowski, Lei Cao, Danyal Turkoglu [38] have provided methodologies towards evaluation of DQE at NIST 20 MW research reactor using ^{6}Li-ZnS scintillator screen coupled with a digital camera as imaging unit and beam purity indicator as the standard test sample.

3.6 Neutron Detectors for Imaging Applications

3.6.1 Film Based Detectors

Nitrocellulose and X-ray-based films have been in use for ages for gamma/X-ray imaging in basic and applied science applications. Neutrons have very little effect on these types of films. To use them for neutron imaging one needs to introduce a suitable neutron converter just before the film. Charge particles are generated through recoil mechanism in the converter and these charge particles expose the film which is later processed. Direct and transfer techniques using metal converter foils and track-etch technique using X-ray film and nitro-cellulose film respectively are in general used for neutron radiography purposes. Neutron-based imaging using films is a two-step process. In the first step, specific thin foils (thickness ~ 125 μm) are activated spatially as per the un-attenuated neutron intensity by the neutrons passing through a sample. In the next step, they are taken out from irradiation site, kept in contact with an imaging film to get exposed from the decay radiation (beta rays) from the foil. The film is then developed as carried out in auto-radiography technique. The film then can be scanned for processing and archiving in digital format. The active component of a film comprises of an emulsion of radiation-sensitive materials (silver halide crystals suspended in gelatin) coated onto a transparent base material. One of the most important characteristics of film is its sensitivity, termed as film speed. This parameter determines the amount of exposure required to produce an image. A film with high sensitivity or speed requires less exposure than a film having lower sensitivity or speed. Another important parameter of film is its optical density and it is the quantification of the amount of light penetrating the film. Interested readers may delve into the book, "The photographic process and Film Sensitivity" by Perry Sprawls. In case of thermal neutron imaging, Dysprosium (Dy) and in case of epithermal neutron imaging, Indium (In) foil is used as the primary neutron detector. Shaikh [39] has described the use of Indium foils in his hydrogen sensitive epithermal neutron (HYSEN) radiography work. The imaging was carried out by placing the object in between a neutron beam filter comprising of 1 mm Cd and In foils and the imager consisting of 250 μm Indium converter screen. The filter efficiently gets rid of neutrons below 1.49 eV. They are chosen for their high neutron cross-section with the neutrons of the specific energy range. The beta energies for Dy and In foils are 1.28 and 1 MeV (Mega electron Volt), respectively. They are also developed far away from the irradiated site in order to get rid of gamma background-related issues. Though the transfer technique is time-consuming, but provides high-quality radiographic images. The transfer of image from the activated foils can also be taken by using Image Plate and then reading out through image plate scanner. In this case, the film processing is not required and the scanned image is available in digital format. This type of imaging with neutron sensitive image plate is known as neutron computed radiography or nCR. Detailed discussion on image plate is carried out in the next section.

3.6.2 Neutron Image Plate (NIP)

X-ray image plates are extensively used in X-ray diffraction and related studies in basic sciences and also for projection imaging in medical imaging techniques known as computed radiography. Combining thermal neutron convertors (B, Li, Gd) with these devices can make them employable towards neutron imaging applications providing resolution ~ 0.1–0.2 mm. Comparison between film and IP are provided in Table 3.3.

Definition: IPs are flexible film-like imager devices having specific phosphors to trap radiation energy and store for deciphering the data using laser-based scanning

Table 3.3 Comparison of some salient features of film and IP

	Film	Image plate
Sensitivity	High	Ultra high (1–2 order higher than film)
Dynamic range	10^2	10^4–10^5
Linearity	Good	Good
Intrinsic spatial resolution	High	Less than film
Reusability	No	Yes
Ease of use	No, elaborate developing mechanism involved	Yes, irradiate—scan—erase—use

method. The phosphors have special properties that utilize photo stimulated stimulation phenomena (PSL) which is different from phosphoresce and fluorescence. When these phosphorous are irradiated by radiation, say UV, X-ray, charge particles, neutron they produce colour centers where electrons get trapped. In other terms, we can say that proportional to the variation in un-attenuated neutron radiation, from the sample, is stored in the plate. When a PSL excitation light suitable for the absorption spectra of the colour center is irradiated, the trapped electron are liberated, producing luminescence and collected by a photo detector. The scanned data is registered in spatially resolved locations in a 2D format and an image is generated. The pitch by which the image plate is scanned, provides the resolution of the image, though it is not the actual resolution of the image plate. The finer the pitch, the higher the scanning time and larger is the size of the image. After scanning the IP can be erased of any background or trapped electron through UV irradiation and made ready to be used again.

Since the IPs are also sensitive to Beta, X-ray, gamma and alpha, for neutron imaging applications, the plates are positioned inside specially designed cassette having a lead sheet for gamma shielding. An important point to note is that the timing gap between the end of exposure and the start of readout should be minimized to get rid of fading phenomena. After exposure to radiation, the electrons are locked in metastable state. They are then stimulated by red light during the scanning process and they emit photostimulable luminescence. But due to thermal influence, some of them can de-excite and hence cause fading. Two types of scanners are available, flat IP type and curved IP type, though the mechanism of scanning and data readout is the same.

The basic construction of an IP for thermal neutron imaging is composed of the two admixtures coated on flexible support. The admixtures are of fine particles of (BaF(Br/I):Eu) and suitable neutron convertors such as nat Gd_2O_3 or nat LiF [40]. This mixture of photostimulated luminescence material forms the intrinsic part of the neutron IP. When using Gd_2O_3 as neutron converter, the stoichiometric ratio of Ba and Gd atoms is 1:1 in the substrate. Depending upon the requirement of detecting efficiency and spatial resolution, the intrinsic part thickness is decided. Their size can be as large as 40 cm * 46 cm. The main advantages of NIP's are short exposure time and broad dynamic range. The scanned image can have 16 bit data range which can help in revealing very low contrasts. The main disadvantages with NIP's are that they are not suitable for real-time imaging as well as tomographic applications and time-consuming processes. As an example, high-resolution scanning (~50 μm) of a 20 cm $\times$ 20 cm NIP can take 30 min. These types of NIP's are also used for cold neutron radiography and tomography applications.

FLA-9000 Starion by Fujifilm Life Science USA [41], as shown in Fig. 3.13, is commercially available Fujifilm's flagship image scanner. The FLA-9000 is capable of radioisotopic, fluorescent, chemiluminescent, and digitized imaging applications. It is a multipurpose image scanner designed to perform a variety of proteomic research applications. The scanner can accommodate up to four of a possible five lasers (473, 532, 635, 685, and 785 nm) at once. Three different models support a variety of your imaging needs:

Fig. 3.13 Photograph of FLA-9000 Starion by Fujifilm Life Science USA (downloaded from https://www.selectscience.net/products/fla-9000-starion/?prodID=84313#tab-2, accessed date 11/05/2021)

- IP Model: ideal for phosphor imaging
- RGB Model: ideal for multi-spectral 2D gel imaging
- IR + IP Model: ideal for conducting near-infrared and radioisotope imaging.

Commercially available neutron image plates (C-NIP) having mixture of $BaFX:Eu^{2+}$ ($X = Br, I$) and Gd_2O_3 are an excellent choice but they are also sensitive to gamma radiation which are always present in neutron-based experiments. New developments in this direction towards reducing gamma sensitivity has been carried out M Schlapp and co-workers of Germany, initial idea provided by Masalovich and Laffe [42], by fabricating ceramic image plates consisting of $KCl:Eu^{2+}$ and LiF. The drawback of this type of scintillator is the reduced absorption cross-section of LiF compared to Gd_2O_3 (940 barns for enriched ^6Li and 48,890 barns for ^{nat}Gd) for thermal neutrons and hence thicker version is required and this invariable results in loss in spatial resolution. This drawback can be overcome by structuring honeycomb-type cells in the ceramic and embedding in the image plate. Due to its design, they are named as Pix-NIP [43]. This novel technique allows read-out from particular illuminated cells during scanning procedure. Experimental results suggest that the Pix-NIP provides higher contrast transfer from original image to the detected signal compared to C-NIP. Typical resolution values at 20% MTF are 680 μm for C-NIP and 570 μm for Pix-NIP. The increase in resolution may be attributed to the confinement of light into individual pixels in the case of Pix-NIP. An important drawback of this Pix-NIP is that their resolution is dependent on the honeycomb structure. Moreover, NIP's based on $KCl:Eu^{2+}$ are still brittle and lots of developments are to be carried out. For fast neutron imaging copper and polyethylene sheets [44] are used as fast neutron converters. The charge particles generated in the converters cast image on the IP and later on read through scanners.

3.6.3 Scintillator-Based Detectors

Neutron detection is driven by the methodology of using composition of materials which upon interaction with neutrons provide charge particles or ions and these then deposit their energy to provide signals to be detected by suitable detectors. Since fast neutrons interact weakly with all materials in comparison to thermal neutrons, for detection of fast neutrons they need to be moderated to some extent to thermal or near thermal range to make their interaction viable and hence detectable.

3.6.3.1 Polycrystalline-Based Scintillators

(A) **Thermal neutron scintillators**: The development of scintillator-based thermal neutron detector was driven by the large interaction cross-section with 6Li, 10B and Gd. These sort of materials (in powder form) are mixed with ZnS(Ag) and fabricated as screens for area imaging purposes. A point to note is that for detection purposes the scintillator thickness must be high, within the limit that it does no stop its own emission to reach the photo detector, as one needs to stop maximum number of neutrons as spatial resolution issue is of no importance. But in case of imaging, spatial resolution matters and for this purpose there is a trade-off between the thickness of scintillator and the amount of radiation to be stopped or in other terms the detection efficiency. Out of these ^{6}LiF-ZnS(Ag) and Gd_2O_3S are the most used scintillator due to their high efficiency. The basic interaction of Li and thermal neutron is given by ^{6}Li $+ n \rightarrow {}^3H + \alpha$ (Q: 4.78 MeV). The energy of the triton and the alpha particles are $E_{3H} = 2.73$ MeV, $E_\alpha = 2.05$ MeV, respectively. Thickness of such scintillators ranges between 0.3 and 0.6 mm. For example, ^{6}LiF-ZnS(Ag) based scintilator screen has been developed by Eljen Technologies. The enrichment of ^{6}Li is ~ 95%. [35]. With the ^{6}LiF: ZnS mass ratio of 1:2 the theoretical and experimental thermal neutron detection efficiency reported is 36 and 32.4% respectively [45]. The light yield of its surface at the average value of the charge spectrum was approximately 8000 photons/neutron. They are very poor towards gamma sensitivity. Commercially, named as EJ-426, they are available with various options such as clear polyester sheet, aluminized mylar, pure aluminum, or highly reflective aluminum and chosen as per the application need. These scintillators have been under use for more than 4 decades and up-gradations (in terms of optimization of weight ratios of elements in the mixture) towards betterment in sensitivity, increase in light production, and image resolution. RC Tritec AG developed scintillators such as ^{6}LiF/ZnS:Cu, ^{6}LiF/ZnS:Ag and ^{6}LiF/Zn(Cd)S:Ag having thickness ranging from 50 to 400 μm are commercially available. Other than in sheet form, cerium activated lithium silicate glass scintillators are also used for thermal neutron radiography. In the upgraded Neutron Radiography Facility (INDLOVU) at the SAFARI-1 Research Reactor in South Africa scintillator materials such as 0.05 mm or 0.10 mm thick ZnS:Cu/6-LiF screens, a 0.01 mm thick (GADOX) Gd2O2S:Tb/6LiF: 80/20 screen and a 1.5 mm thick PP30 (30% ZnS:Cu) converter have been used depending upon the intended application [46].

Work in this direction are constantly progressing towards image quality improvement such as using micro-particles of ^{6}LiF and coating it with ZnS(Ag) [47]. Development of high-resolution high-efficiency micro-columnar LiI screens for thermal neutron imaging has been reported [48] by Radiation Monitoring Devices, Inc., Watertown, MA, USA. Feasibility studies on ^{6}Li loaded plastic scintillation films have been carried out by researchers [49] at the University of Tennessee, Knoxville, TN, USA. Screens based on ^{6}Li-enriched silicate glass cores with Ce activator have shown promising results for thermal neutron imaging. Commercial Li-glass scintillators for neutron detection and imaging are available from Scintacor [50].

Another detector known as scintillating fiber (SCIFI) based neutron imager has been reported by Michael Edward Moore [51] of University of Tennesse, Knoxville. Thermal neutrons interacting with ^{6}Li-doped optical multicore SCIFI produces charge particles which in turn produce primary and secondary electrons in the Li silicate glass matrix. These then interact with the Ce activator and provide enough scintillation to be guided to any suitable optical imaging detector through the microstructured waveguides.

GOS or $Gd_2O_3S(Tb)$ based scintillators also have proven a candidate for thermal neutron imaging. The guiding interactions are

^{155}Gd + n $\rightarrow$ ^{155}Gd + γ (0.09, 0.20, 0.30 keV) + conversion electrons; ($\sigma_{\text{th}}\ n =$ 61,000 b)
^{157}Gd + n $\rightarrow$ ^{158}Gd + γ (0.08, 0.18, 0.28 keV) + conversion electrons; ($\sigma_{\text{th}}\ n =$ 255,000 b)

The abundance of ^{157}Gd in natural Gd is 15.7%. The efficiency can be increased manyfold if this isotope can be enriched. With ~ 88% enrichment of $^{157}Gd_2O_3$, 3.8 and 3.6 factor of increase in absorption power and light yield with respect to its natural counterpart has been reported by researchers at Paul Scherer Institute (PSI), Switzerland [52]. RC Tritec AG [53] developed scintillator screen $^6LiF/Gd_2O_2S$:Tb provides 30–50% increase in brightness for the same resolution in comparison to Gd_2O_2S:Tb for neutron imaging. The emission is in the blue-green region and well suited for effective coupling with commercially available high-resolution CCD camera for imaging purposes.

Thermal neutron interaction with B-10 is as follows.

$$^{10}\mathrm{B} + \mathrm{n} \rightarrow ^7\mathrm{Li} + \alpha + \gamma(0.48\ \mathrm{MeV})(\mathrm{Q}: 2.31\ \mathrm{MeV})$$

The thermal neutron capture cross-section of B-10 is 4 times that of the Li-6. This certainly suggests that the scintillator screen containing B-10 will provide more detection efficiency than the Li-6 based ones. Though Gd has higher thermal neutron capture cross-section than B-10, it is sensitive to gamma and that is not the case with B-10. Also, the daughter products generated in case of thermal neutron interaction with boron have less energy and larger size than the daughter products with the case for Li and Gd. This has direct relation to the light spread on energy deposition; less energy and larger size means less travel in the base material and hence better spatial resolution. Because of these reasons, boron-10 based scintillators can provide higher performance than Li- or Gd-based scintillators for thermal neutron imaging even though they have lesser light output comparatively. RC Tritec AG developed B-10 based scintillators such as $^{10}B_2O_3$/ZnS:Cu, $^{10}B_2O_3$/ZnS:Ag and $^{10}B_2O_3$/Zn(Cd)S:Ag having thickness ranging from 50 to 400 μm are available. They are capable of providing high light output and high-resolution thermal neutron radiography images [53].

Study involving 25 different boron-based scintillators towards comparison with commercially available LiF and Gd-based scintillator screens and the parameters for the best image quality has been carried out by William Chuirazzi and his co-workers. It was found out that thinner boron-based scintillators could provide similar detection efficiency and better spatial resolution in comparison to ^{6}LiF-based scintillators. Interested readers may please go through the article [54].

Studies involving Dysprosium-based scintillator screens, fabricated by depositing mixtures of ZnS:Cu scintillator with Dy_2O_3 (in varied mixing ratios and grain sizes of ZnS:Cu) in Aluminum plate and Dysprosium foils have been reported. The isotope ^{164}Dy (28.18% of natural Dy) is useful in neutron imaging as its absorption cross-section is ~ 1000 b. Experiments at NEUTRA beamline (PSI) [55] suggests that the highest light output was provided by the screen having Dy foil as substrate and mixing ratio of 1:2 of Dy_2O_3: ZnS:Cu (grain size 9.67 μm). Though their spatial resolution was slightly poor than the standard ^{6}LiF:ZnS scintillator screen they are having much less sensitive to gamma radiation and have potential use in neutron imaging of irradiated fuels. The latent image created on the Dy foils by the neutrons crossing the sample can be read by a well-shielded sensitive CCD camera system kept at a distance from the actual neutron beam port.

Boron-rich Multi-Channel Plate (MCP) based imaging detectors [56, 57] have also been used. The neutron interacting with the boron coated on the inner walls of the MCP produce charge particles which then release secondary electrons. These are then multiplied in the MCP are read-out through specialized sensors and electronics. The best part of this type of detector is the high resolution provided by the MCP size (~10 μm) and the dedicated electronics for fast read-out and processing. But their drawback lies in their low efficiency (10–39% for thermal neutrons). Using neutron sensitive MCPs coupled to a Medipix2/Timepix active pixel sensor [58], feasibility experiments at both thermal as well as cold neutron beamlines at PSI established neutron imaging having spatial resolution ~ 55 μm. "Medipix2" [59] is a type of hybrid detector initially developed for charge particle detection with high temporal resolution for CERN experiments but various modifications have led to their use for applications including medical imaging, neutron imaging [60] etc. Basically, it is a semiconductor detector chip having pixelated matrix in the front and a common readout on the backside. Each pixel element has its own readout mechanism through individual preamplifier, discriminator and digital counter.

As an alternative to the abovementioned scintillator, (H_3 $^{10}BO_3$) in combination with ZnS(Ag) having mass ratio of 1:6 has shown potential in thermal neutron imaging applications [61]. It has been characterized at the d-Be neutron source facility at Peking University of China. The best performance with 0.35 mm thick scintillator was found out to be approximately 1.33 lp/mm (line pair per mm).

Another notable development in this field is the development of a thermal neutron imaging screen consisting of a pixelated microstructured CsI(Tl) sandwiched between two GdF_3 layers [62]. Thermal neutron interaction with Gd atoms produce conversion electrons which in turn produce scintillations in CsI(Tl). In this regard, aluminum-based substrates were found to be better than fibre-optic plate substrates in terms of light output, signal-to-noise ratio and contrast as per imaging tests at a thermal neutron port of the University of Massachusetts Lowell Research Reactor (UMLRR). One of the reasons for this can be attributed to the columnar structure of the CsI as well as the high absorption coefficient of Gd.

(B) **Fast neutron Scintillators**: Detection and imaging of fast neutron have always been a challenge. To achieve this, one need to slow down the fast neutron to energies where they interact through neutron-induced nuclear reaction to generate charge particles which then produce scintillations in accordance with the phosphor as detectable luminescent signals. As the number of collisions required to slow down the fast neutrons to make it compatible to (n, p) reaction, their thicknesses are more than the thermal neutron scintillators. For this very reason, the spatial resolution as well as contrast obtained in fast neutron radiography (FNR) is poor than Thermal Neutron radiography (TNR). The light signal can now be coupled with a CCD camera, amorphous Si panels or suitable semiconductor screens. Since the light output is very poor, when coupled with lens and CCD-based imaging components, the acquisition time goes very high. It becomes very time intensive in case of tomography experiments.

(C) **Plastic Scintillator (PS)**: PSs are compound formed by solid solutions formed by solid solution of organic scintillation materials in polymerized solvent. Owing to their physical properties, they can be easily molded and fabricated in any desired shape. These make them extremely useful towards development of organic scintillation-based detectors. Since these materials are of low Z, their gamma sensitivity is poor and effective toward charge particle detection with fast timing response. Fast neutron-induced (n, p) reaction generated protons directly or through further ionization deposit their energy in the matrix. Though the emission wavelength depends upon the organic activator present in the matrix, in general, the emission peaks around 400 nm.

As for example, Saint Gobain makes plastic scintillator (Trade name BC412, BC416) used for fast neutron imaging has light output and wavelength emission of anthtracene are (60% for BC412 and 38% for BC416) and 434 nm respectively. Saint-Gobain [63] (erstwhile Bicron), USA commercially also produces a scintillator in the trade name of BC 720 ZnS(Ag) for fast neutron detection. Using such pixelated scintillators and coupling to photo-detectors area imaging can be performed.

Thick transparent plastic scintillators (10–40 mm) have been successfully used for imaging with neutron (1–14 meV). High energy 10 meV neutron imaging has been successfully carried out at the LLNL, USA by imaging systems comprising of BC 408 and BC 400 scintillators (developed by Saint Gobain) and a compatible lens coupled CCD camera detector [64]. The schematic of a typical high-energy neutron imaging setup is shown in Fig. 3.14.

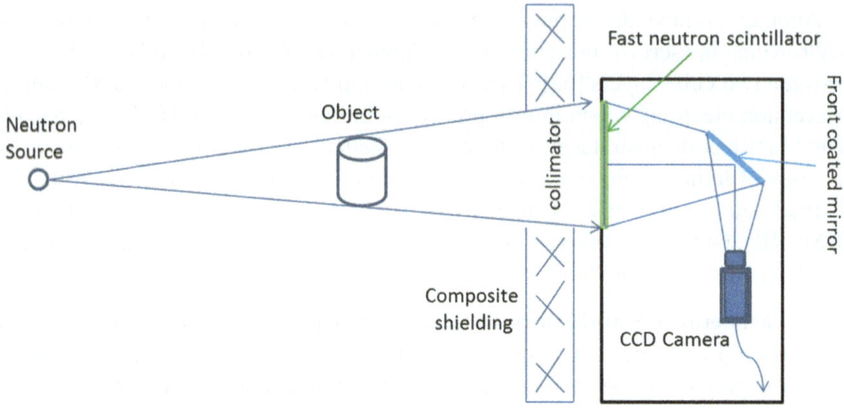

Fig. 3.14 Schematics of a typical high-energy neutron imaging setup

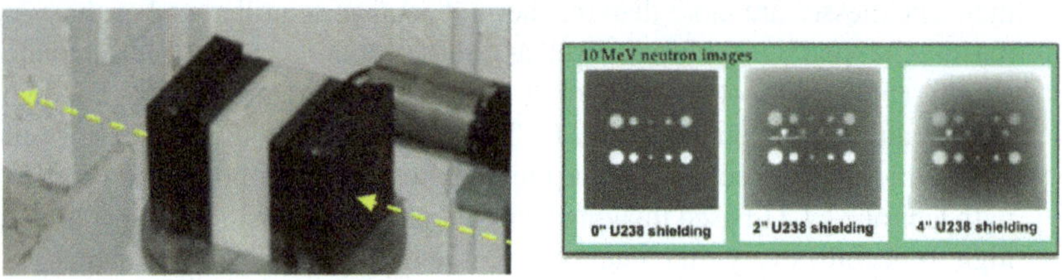

Fig. 3.15 (left) photograph of the test piece comprising of plastic piece sandwiched between DU with holes and (right (top) 10 MeV neutron radiographs with varying DU thickness (reproduced from Ref. [65] with permission from Rightslink by CCC)

Figure 3.15 shows the test sample, plastic piece sandwiched between depleted uranium (DU) blocks having holes of various diameters and its 10 MeV neutron radiograph at the Ohio University Accelerator Lab (OUAL) using an imaging system consisting of BC-408 scintillator, a turning mirror and a cooled CCD camera [65]. It is seen from the neutron radiographs could reproduce the holes clearly though the ability reduces with increase in DU thickness.

Researchers of Swiss Federal Institute of technology and PSI Switzerland has reported the development of a novel fast neutron detector system for tomography purpose using DD neutrons [66]. The primary component of the detector is aluminum mylar wrapped BC400 plastic and in the form of an array of 88 detectors. Each detector's dimension was 80 mm deep (In line with neutron path), 8 mm height and 5 mm wide (Perpendicular to the neutron path and parallel to the ground). Since individual detector represents a pixel, their response and output need to be normalized for imaging application. The achieved spatial resolution by the group was reported 2 mm, the scintillator output read by SIPMs connected to Multichannel DAQs and digitizer. The current in each channel is digitized to provide a grey value corresponding to a pixel in the 2D plane. The photograph of the detector assembly is shown in Fig. 3.16.

Feasibility experiments on fast neutron imaging using 14 MeV neutrons from DT neutron generator and 40 mm thick BC400 plastic scintillator and a EMCCD (Electron Multiplying CCD) camera have been demonstrated at BARC, India [67]. Fast neutron radiography images of step wedge samples of HDPE, MS and Pb with their intensity profile across the steps are shown in Fig. 3.17.

Fig. 3.16 Photograph of the source, rotating table and detector array (with scintillators) at the PSI, Switzerland (reproduced from Ref. [66] with permission from Rightslink by CCC)

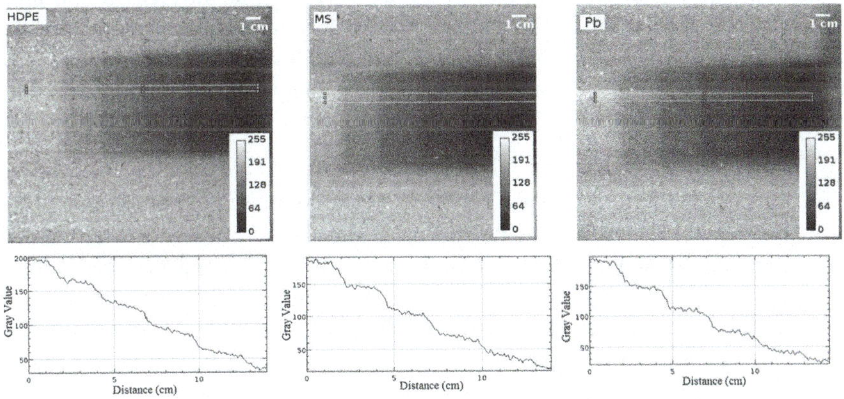

Fig. 3.17 Fast neutron radiography images of step wedge sample of (left) HDPE (middle) MS and (right) Pb with their intensity profile across the steps. The step wedge sample size is 50 mm × 100 mm × 20–100 mm [width (w) × height (h) × depth (d)] with 20 mm step height. The fast neutron imaging system at BARC, India has been upgraded by using a 3 mm PP-ZnS(Ag) fast neutron scintillator screen and an image intensified CCD camera (reproduced from Ref. [67] with permission from Rightslink by CCC)

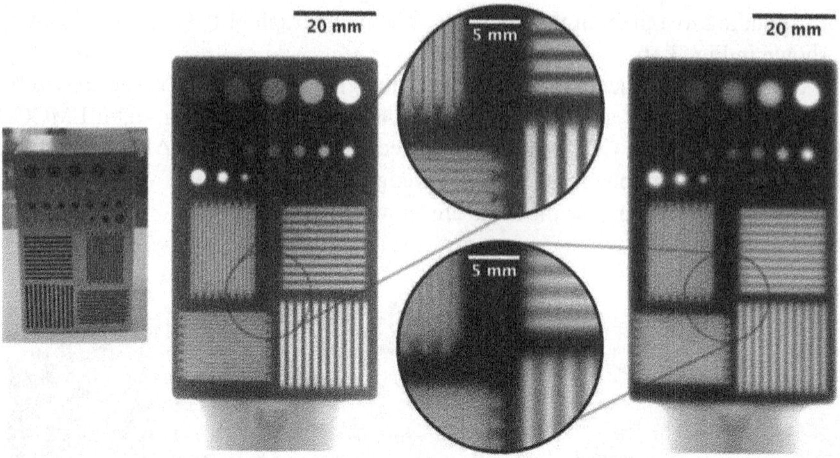

Fig. 3.18 Radiography of an iron block (shown on the right) having different diameter holes and slits along its thickness imaged with the 2 different scintillator types. Left with the 2 layer PE (3 mm)/ZnS:Ag (60 μm) and right with the standard 2.4 mm PP/ZnS:Cu (30%) scintillator at the NECTAR facility at FRM-II research reactor, MLZ, Garching (reproduced from Ref. [68] with permission from Rightslink by CCC)

Development of ZnS mixed in polypropylene base has been reported and also commercially available by RC Tritec AG (rctritec.com). This type of scintillator has been used for imaging with fast neutrons at the NECTAR facility located at FRM-II nuclear research reactor. Two variants tested with NECTAR facility are PP/ZnS (As) (Emission wavelength 450 nm) and PP/ZnS(Cu) (Emission wavelength 530 nm). They are available in thickness ranging from 1–5 mm. Figure 3.18 displays the radiograph of an iron block having different diameter holes and slits along its thickness imaged with the 2 different scintillator types, namely 2 layer PE (3 mm)/ZnS:Ag (60 μm) and 2.4 mm PP/ZnS:Cu (30%) at the NECTAR facility at FRM-II research reactor, MLZ.

(D) **Fibre Scintillator screen**: These screens are composed of polystyrene fibres arranged in 2D fashion on a PMMA base. The fibers are doped with fluorescent materials as per the radiation to be detected. At Lawrence Livermore National Laboratory, Livermore CA, the fast neutron imager consists of a 160-mm × 160-mm × 50-mm thick array of 250-μm round BCF-99-55 scintillating fiber, manufactured by Saint Gobain. In the lens coupled option the spatial resolution of the system is 5.5 lp/mm. This sort of system in combination [69] with a specially designed neutron pinhole aperture array was used at the National Ignition Facility, for imaging of size and shape of DT plasma emitting neutrons during the ignition phase. In order to increase the spatial resolution, some of the recommendations are as follows: development of liquid capillary array for use with liquid scintillators, high index fast liquid scintillator and improving quality of optical fiber arrays [70].

3.6.3.2 Single Crystal-Based Scintillators for Thermal Neutron Imaging

Rare earth based single crystal based scintillators are being used in various applications including healthcare imaging using X-rays for a long time. Gas detectors based on ^3He and ^{10}B provides very high detection efficiency owing to their high absorption cross-section. But the dwindling supply of ^3He gas has motivated researchers to look for promising candidates for neutron detection. Though inorganic scintillators are widely used for neutron imaging the light output spread in them, which depends upon the grain size of the converter material and substrate thickness, restricts their spatial resolution and contrast. Rare earth scintillators based on Ce-doped Li glass and Eu^{2+} doped LiI have shown neutron detection capabilities [71] but the former one has gamma sensitivity and the later one is highly hygroscopic. Crystal in solid form is supposed to increase detection efficiency because of their structure or in other terms higher atomic density of neutron capture atoms, if present, increases the interaction probability. Moreover, if it is single crystal, then the light spread will be minimal. Both these properties are the motivating factors for the development of single-crystal scintillators as a thin crystal (< 1 mm) can stop all thermal neutrons falling on it.

Some notable developments in this category of scintillators are discussed below:

A two-inch diameter and 2 mm thick wafer of Eu (2 mol %) doped $LiCaAlF_6$ (95% enriched ^6Li) developed by μPD (micro pulling down) technique in combination with a position-sensitive PMT has been reported for use in thermal neutron imaging providing spatial resolution better than 1 mm. Thermal neutron interacting with Li atoms produce α particles and these interactions with Eu^{2+} result in an emission peaking at 375 nm. Lots of research has gone into the understanding of the doping percentage, light emission dependence, etc. towards this development [72–75]. Experiments were carried out at the MUSASI (Multi-Purpose Thermal Neutron application and Science) beam port in JRR-3 (Japan Research Reactor-3). Test pieces made up of 1 mm thick cadmium plate having holes of diameter 1, 2, 3 and 5 mm and a complicated object made with 1 mm diameter cadmium wire were imaged (as shown in Fig. 3.19) with clarity. This crystal has also shown an order less scintillation under gamma radiations. Though the effective imaging area was small yet this work has proved the use of single-crystal neutron scintillators for thermal neutron imaging.

Ce-doped $Gd_3Ga_3Al_2O_{12}$ (GGAG:Ce) single crystal has been developed by Czochralski method at BARC, India and its study related to thermal neutron detection and imaging [76] has shown promising results. Gd atoms have the highest capture cross-section and a thin slice (< 1 mm) of GGAG:Ce can have almost 100% stopping power for thermal neutrons, which is directly related to the detection efficiency. Thermal neutron interaction with Gd atoms produces gamma and conversion electrons which are continuous upto 8.5 MeV. These excite Ce atoms (Ce^{3+}–5d-4f transitions) and the scintillations are recorded by a suitable photo detector. A 1 mm thin slice, cut from the ingot, was polished to a final thickness of 0.55 mm (Fig. 3.20a).

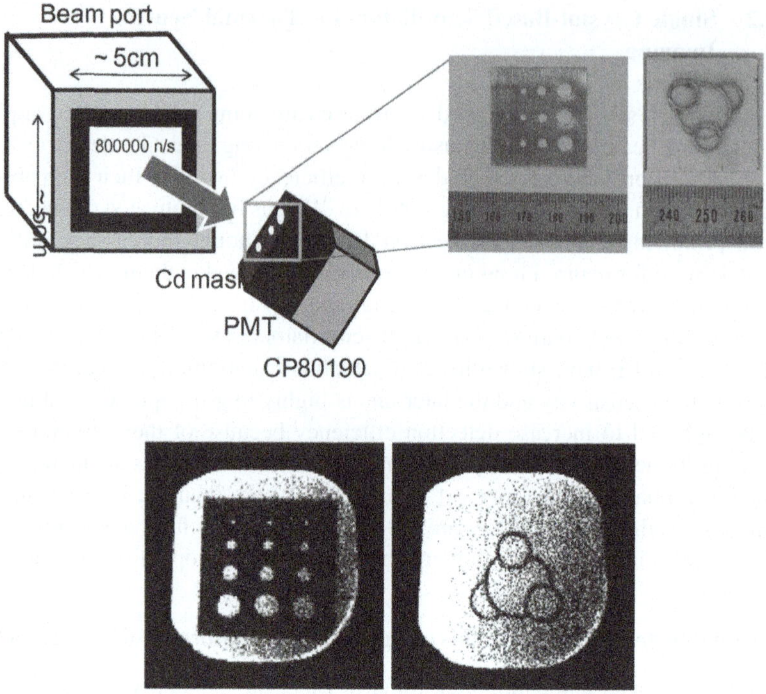

Fig. 3.19 (top) schematic drawing of the experimental setup and the test pieces imaged, (bottom) thermal neutron radiographs of the test pieces (reproduced from Ref. [73] with permission from Rightslink by CCC)

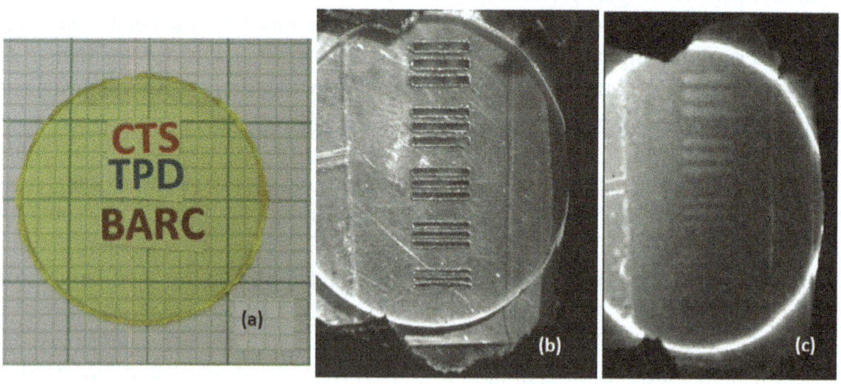

Fig. 3.20 **a** photograph of Ce-doped $Gd_3Ga_3Al_2O_{12}$ single crystal developed at BARC, India, **b** optical image showing the cadmium slits of varying dimension (0.5–4 mm) and **c** the thermal neutron image revealing the cadmium slits (reproduced from Ref. [76] with permission from Rightslink by CCC)

After characterization, its feasibility was tested at the thermal neutron imaging beamline at Dhruva reactor, BARC. Since its peak emission is around 550 nm, its coupling to CCD camera with lens was easily achieved. A portable neutron imaging system comprising of 0.55 mm thin GGAG: Ce crystal, a 90^0 turning mirror and off-the-shelf available Manta make CCD camera was developed. A mask made of Cd plate (1 mm thick having 0.5, 1, 2, 3 and 4 mm) was imaged. The optical and the thermal neutron image of the cadmium slit is shown in Fig. 3.20b, c respectively.

3.6.3.3 Semiconductor Based Detectors

(A) **Flat panel based detector**: In the same line of development of X-ray flat panel detector using amorphous silicon panels, coupling a neutron scintillator in place of X-ray sensitive screen, will give a neutron sensitive amorphous flat panel. As discussed above CCD camera based imaging systems using ^{6}LiF scintillator use lens coupling. This introduces some loss in the light collection which is already poor in intensity. This is not the case with flat panels as the scintillator screen is directly bonded with the panel [77] and hence they provide better light collection efficiency. They can be of large size but the cost increases sharply with size. The main drawback of these detectors is that in general their dynamic range is poor (14 bit) than CCD cameras (generally 16 bit), and 16 bit flat panels are very costly.

(B) **Semiconductor detector**: Complementary Metal Oxide Semiconductor (CMOS) devices gained attention over charge coupled devices (CCD) after the advancement of lithography technology as smaller pixel sizes could be realized incorporating the achievable fill factor. They have added advantage of low power consumption and fabrication of large panels using tiling mechanisms. With X-ray scintillating screens in their input side, they become X-ray imaging detectors. One such pixelated X-ray detector is PILATUS [78]. Researchers at the PSI have developed a neutron imager using PILATUS-II using Gd-157 as neutron converter. In the same line of development, using converter materials such as B-10, Li-6 and Gd incorporating on a new pixelated readout chip named EIGER (Enhanced Intensity and Greater Energy range) has been developed at PSI. EIGER, a hybrid detector with single photon counting capability, has been initially developed by PSI-SLS group for applications at Swiss Light Source. In Pilatus-II, 5 μm Gd layer was bump bonded using Indium to the chip. Initial results at the NEUTRA, PSI has provided encouraging results and improved its performance by removing issues such as edge smearing is being worked out. Also, the sensitivity of Gd with X-ray/gamma does pose some restrictions of their use for practical applications.

3.6.4 Gas-Based Detectors

Thermal neutron detection capability of Helium-3 and Boron-based gas detectors are well established. These gases have high absorption cross-section (He-3: 5840 barns; B-10: 3500 barns) for thermal neutrons. These detectors are in general cylindrical having diameter ~ cms and length ~ 10's of cms. Though they are routinely used for neutron dose and flux monitoring purposes.

He-3/BF_3 based neutron detectors:

These sort of gas-filled detectors were the first devices used for neutron detection and they work in the proportional region. They are in general housed in Aluminum or stainless steel cylinders with electrical connector at one end. Thermal neutrons entering the detector ionize the respective gas atoms, generate charge particles (positive ions and electrons) which then are collected and signal is generated. This happens only when designated positive potential is applied to the detector, otherwise the charge particles are lost due to recombination in the gas medium. The generated signal is processed further with nuclear instruments such as pre-amplifier, single channel analyzer, counter, etc. The interaction equations for both the gases with thermal neutrons are given below.

$$^{3}\mathrm{He} + \mathrm{n} \rightarrow ^{3}\mathrm{H} + \mathrm{p}(\mathrm{Q}\text{: } 0.764\,\mathrm{MeV})$$

The energies of the triton and proton are 0.191 meV and 0.573 meV respectively

$$^{10}\mathrm{B} + \mathrm{n} \rightarrow ^{7}\mathrm{Li} + \alpha + \gamma(0.48\,\mathrm{MeV})(\mathrm{Q}\text{: } 2.31\,\mathrm{MeV})$$

The energy of Li ranges from 0.84 to 1.01 meV and that of α ranges from 1.47 to 1.78 meV. In case where high detection efficiency is required, He-3 detector is preferred over BF_3 detectors as in the former case they can be operated at much higher pressures thereby enhancing gas multiplication and in turn more signal output. But in situation where gamma environment is very high, BF_3 are favoured over ^3He detectors. In cases where the neutron flux environment is very high, the internal walls are coated with activated charcoal which reduces the effect of output pulse reduction due to generation of electronegative gasses during irradiation. Some additive gas such as CO_2 or Ar are also put in these detectors to minimize γ-ray pile up by modifying the speed of electrons and thereby allowing use of fast processing electronics. The operating voltages for these gas detectors lie between 700 and 1000 V, off-course it also depends upon the construction and intended application. These detectors are routinely used in neutron facilities for area dose measurement, flux mapping etc. Since the active volume of interaction is ~ 10's of centimeters, they cannot provide spatially resolved information better than their size.

Gas-based detectors with oxide film coating of fissionable elements such as U235 or Pu239 on the inner surface of the cylinder can also be utilized for thermal neutron detection. They are called fission chambers. The thermal neutrons cause fission upon interacting with the fissile material and generates fission fragments which also include

(a) (b)

Fig. 3.21 **a** (left) Photograph of a 2D neutron detector based on multiwire chamber filled with He-3 developed at BNL and **b** complete detector set comprising of multiple segments installed on the goniometer on the Protein Crystallography Station (PCS) at Los Alamos (reproduced from Ref. [79] with permission from Rightslink by CCC)

charge particles. The rest of the process of signal generation is the same as discussed above. They are mostly used for reactor monitoring purposes.

Though the abovementioned detectors are highly efficient and robust in nature, they are directly not suited, in the present configuration, for area imaging applications.

In this direction, initial work towards development of area-imaging detector primarily on multi-wire chamber design but filled with He-3 gas mixture was carried out for studies of molecular biology and material science applications [79]. Figure 3.21 shows the photograph of the developed 2D detector with 5 cm × 5 cm sensitive area in (a) and neutron detector with 120° × 15° angular coverage on a radius of curvature of 70 cm in (b).

Researchers at BNL in collaboration with ANSTO has developed and tested a 1 m * 1 m (active area 0.96 m * 0.96 m), 3He based gas detector [80] where the generated signal is collected by discrete anode pads located in 2D separated by pitch. The signals are read by the low noise ASICs. The detector is kept inside a vacuum tank to reduce scattering of neutron and cooled to reduce noise in the electronics.

Development of the 2D neutron detection with an effective size of 200 mm*200 mm having ^{3}He gas and working on MWPC methodology has been reported [81]. Cathode and anode planes are composed of aluminum foil and gold plated tungsten wire respectively. The signals generated at specific (*x*, *y*) coordinates/positions are read out by electronic modules. The best positron resolution achieved using the detector is 1.18 mm.

BioRef is the standard versatile novel time-of-flight instrument developed at Helmholtz-Zentrum Berlin for applications [82, 83] on soft matter systems. It is a 30 cm × 30 cm active area 2D position-sensitive detector based on ^{3}He multi-wire chamber and delay line electronics. In comparison to CCD based imaging where the resolution is ~ 25 μm the spatial resolution of this detector is $2 \times 3\ mm^2$. But, because of ^3He its detection efficiency is superior for cold neutrons and also provides time resolution ~ μs, this instrument is best suited for TOF measurements. Experiment on strain measurements on a flat steel sample under tensile loading has been reported by Strobl et al. using this instrument.

In the detection and imaging with neutrons (40–400 meV), a proof of principle experiments was successfully carried out in 1998 by researchers of Lawrence national lab, USA at LOS Alamos Nuclear science centre (LANSCE) [84]. Imaging of 2.54 cm thick Lead (with drill holes) disk sandwiched between 5.08 cm thick uranium slab with a spatial resolution of 1 mm was demonstrated by using a position-sensitive multiwire detector in combination with a tungsten convertor.

Advanced Hybrid Detectors for Neutron Imaging is based on a combination of gas detectors with associated dedicated data acquisition. Though neutron imaging is an old and established technique, with the advent of improved sources such as accelerator-based Spallation Neutrons providing both high intensity and pulsed beams of neutrons newer techniques are being developed to look deeper into materials and processes with high contrast and temporal resolution. Advanced techniques such as energy-resolved neutron imaging, Bragg-edge imaging, resonance imaging have the potential in providing physical and chemical properties in a bulk in addition to its macroscopic 3D volumetric distribution. These are called counting-type imaging detectors. Initial development work [85] on such detector has been reported using Boron-10 (Purity $\geq$ 99%) coated gas electron multiplication foils stacked in a chamber and coupled to specific ASICs and FPGA . The prototype detector chamber has the following constructional details. Cathode plate for applying HV is made of aluminium, two numbers of B-10 coated GEM foils for neutron interaction, two normal GEM foils for gas gain towards detected signal and the read out boards (ASICs). The initial development though targeted for particle detection has been modified for neutron detection and imaging application. When only the boron coated aluminium is used and no B-10 coated GEM foil, the detector has shown potential application in neutron beam monitor. Using the complete detector i.e. incorporating B-10 coated GEM foils, experiment on energy selective neutron radiography using cold neutron beam from JPARC spallation source has been reported providing spatial resolution ~ 0.8 mm, temporal resolution ~ 10 ns and active imaging area of 10 cm × 10 cm. The experiments have revealed two-dimensional images of iron plates, which have gone through different bending conditions. Recent advances in this domain have resulted in development of various detectors such as μPIC, μNID, nGEM and the LiTA12.

μPIC (micro-pixel chamber) [86] is a micropattern gaseous detector having 400 μm pitch incorporating orthogonal anode and cathode strips for two-dimensional read-out. The chamber has Ar-C_2H_6-^3He (63:7:30) gas mixture at 2 atm. Demonstration experiments carried out using μPIC detector at Materials and Life Sciences experimental facility (JPARC) demonstrated its detection efficiency ~ 18% for thermal neutrons, spatial and temporal resolution of 100–120 μm and ~ 0.6 microsec respectively. This detector is developed for time-resolved neutron imaging applications.

μNID [87] has been developed using μPIC micropattern read-out plane coupled to a time projection chamber having an active volume of 100 × 100 × 25 mm^3. The detection is enabled by the gas mixture CF_4–iC_4H_{10}–^3He in the ratio of 45:5:50 kept at 2 atm pressure. Complex readout mechanism and analysis involving the μPIC in combination with the arrival time of the neutron along with the deposited energy (above some pre-decided threshold) provides spatial and temporal resolution ~ 100 μm and ~ 250 ns in the μNID. Further developments are being perused towards increasing both the resolutions.

The nGEM detector is a combination of a drift cathode in the time projection chamber and B-10 coated GEM (gas electron multiplier). The B-10 coating is ~ 1 μm. The detector is filled with a gas mixture of Ar-CO_2 (70:30) at 1 atm for its ionization by the charge particles emitted by thermal neutrons interacting with the boron. A pair of normal GEMs amplify this ionization and the signal is then read by FPGA-based data acquisition. The best part of the gas-based counter-type detectors is their very low sensitivity towards gamma rays.

The LiTA12 or ^{6}Li Time Analyzer, Model 2012 is made up of Ce activated Li-glass scintillator pixel (pixel size ~ 2.1 × 2.1 × 1 mm^3) detector coupled pixel-to-pixel with a multi-anode PMT. The scintillations produced by the charge particles generated in neutron interaction with Li are detected by the MA-PMT, converted to electrical signals and digitized by dedicated FPGA modules. Though the detection efficiency reported for this detector is 23%, the spatial resolution is poor (~3 mm) due to the limitation set by individual pixel size. A single scintillator plate when used in place of the Li-glass scintillator pixel assembly, provided improved spatial resolution of ~ 0.7 mm.

Summary

This chapter deals with the basics of various mechanisms of neutron optics employed towards spatial beam shaping, spectral shaping and intensity increase. These developments were driven by the need to carry out imaging for various applications. Once the beam is ready for imaging, one needs imaging detectors. Various detectors developed and are used till date are discussed with the applications intended for their use. With the development of high-intensity reactor or spallation sources, continuous as well as pulsed experiments with extra feature of wavelength dispersive imaging has been employed to dig deeper inside materials and processes related to them. Developments of advanced detectors and their use in these specialized experiments have also been discussed.

References

1. Klein AG, Werner SA (1998) Neutron optics. Rep Prog Phys 46(3):259–339
2. Rostaa L, Füzia J, Hományi L (2004) Neutron physical properties of a multi-blade neutron velocity selector. Physica B 350:e711–e716
3. Longpol I (1974) A long wavelength neutron polarization analysis instrument. In: Camphell SJ, Ahmed N, Hicks TJ, Ebdon FR, Wheeler DA (eds) J Phys E: Sci Instrum 7:195
4. Lushchikov WI et al (1970) J Nucl Phys 10:669
5. Williams WG (1988) Polarized neutrons.Clarendon Press, Oxford
6. Bouchiat M, Carver TR, Varnum CM (1960) Nuclear polarization in He^3 gas induced by optical pumping and dipolar exchange. Phys Rev Lett 5:373
7. Colegrove FD, Schearer LD, Walters GK (1963) Polarization of He3 gas by optical pumping. Phys Rev 132:2561
8. Okudaira T, Oku T, Ino T et al (2020) Development and application of a ^{3}He neutron spin filter at J-PARC. Nucl Inst Methods Phys Res A 977:164301
9. Ott F (2008) Focusing optics for neutrons. In: Erko A, Idir M, Krist T, Michette AG (eds) Modern developments in x-ray and neutron optics. Springer Series in optical science, vol 137. Springer, Berlin. https://doi.org/10.1007/978-3-540-74561-7_7
10. Kumakhov MA, Sharov VA (1992) A neutron lens. Nature 357:390
11. Rai DK, Abir M, Wu H, Khaykovich B, Moncton DE (2018) Focusing mirrors for enhanced neutron radiography with thermalneutrons and application for irradiated nuclear fuel. Nucl Inst Methods Phys Res A 879:141–146
12. Mezei F (1976) Novel polarized neutron devices: supermirror and spin component amplifier. Commun Phys (Lond) 1(3):81–85
13. Schanzer C, Scneider M, Boni P (2016) Neutron optics, towards application s for hot neutrons. J Phys: Conf Ser 746:012024. Accessed on 01 Sept 2021
14. Hosobata T, Yamada NL, Hino M, Yamagata Y, Kawai T, Yoshinaga H, Hori K, Takeda M, Takeda S, Morita S (2017) Development of precision elliptic neutron-focusing supermirror. Opt Express 25:20012–20024. https://doi.org/10.1364/OE.25.020012. Accessed on 01 Sept 2021
15. Lynn JW, Kjems JK, Passell L, Saxena AM, Schoenborn BP (1976) Iron–germanium multilayer neutron polarizing monochromators. J Appl Crystallogr 9:454–459
16. Hoghoj P, Anderson R, Siebrecht R, Graf W, Ben-Saidane B (1999) Neutron polarizing Fe/Si mirrors at ILL. Physica B 267–268:355
17. Schebetov F, Pleshanov NK, Pusenkov VM, Peskov BG, Shmelev GE, Kraan WH, Por PT, Rekveldt MTH, Mikhailova VE (1994) Construction and testing of a multichannel polariser for thermal neutrons. Nucl Instrum Methods B 94:575
18. Clemens D, Boni P, Friedli HP, Gottel R, Fermon C, Grimmer H, van Swygenhoven H, Archer J, Klose F, Krist Th, Mezei F, Thomas P (1995) Polarizing $Ti_{1-u}X_u/Fe_xCo_yV_z$ supermirrors. Physica B 942: 213–214
19. Boni P (1997) Supermirror-based beam devices. Physica B 234–236:1038
20. Eskildsen MR, Gammel PL, Isaacs ED, Detlefs C, Mortensen K, Bishop DJ (1998) Compound refractive optics for the imaging and focusing of low-energy neutrons. Nature 391:563
21. Adachia T, Ikedaa K, Okua T, Guoa J, Lina W, Ohmoria H, Morishimaa T, Shimizua HM, Sakaia K, Suzukic J, Littrelld KC, Loong C-K (2004) Possible application of compound Fresnel lens for neutron beam focusing. Physica B 350:e775–e778
22. Littrell KC, te Velthuis SGE, Felcher GP (2007) Magnetic compound refractive lens for focusing and polarizing cold neutron beams. Rev Sci Instrum 78:035101. https://doi.org/10.1063/1.2709844
23. Cremer JT, Filter H, Klepp J, et al. (2020) Focusing and imaging of cold neutrons with a permanent magnetic lens. Rev Sci Instrum 91:013704. https://doi.org/10.1063/1.5116759
24. https://www.ncnr.nist.gov/instruments/ng7sans/lens.html. Accessed on 27 Aug 2021
25. Shimizu HM et al (1999) Measurement of cold neutron-beam focusing effect of a permanent sextupole magnet. Nuclear Inst Methods Phys Res A 430(2–3):423–434

26. Iwashita Y, Tajima Y, Ichikawaa M, Nakamura S, Ino T, Muto S, Shimizu HM (2008) Variable permanent magnet sextupole lens for focusing of pulsed cold neutrons. Nucl Inst Methods Phys Res A 586(1):73–76
27. Yamada M et al (2009) Development of modulating permanent magnet sextupole lens for focusing of pulsed cold neutrons. Nucl Inst Methods Phys Res A 404(17):2646–2651
28. Iwashita Y et al (2010) Practical applications of permanent magnet multipoles. IEEE Trans Appl Superconduct 20(3):842–845
29. Yamada M et al (2015) Pulsed neutron-beam focusing by modulating a permanent-magnet sextupole lens. Progr Theor Exp Phys 2015(4):043G01
30. http://cis01.central.ucv.ro/pauc/vol/2019_29/5_45_49_2019.pdf. Accessed on 25 Aug 2021
31. Cantargi F, Granada JR, Mayer RE (2015) Thermal neutron scattering kernels for sapphire and silicon single crystals. Ann Nucl Energy 80:43–46
32. Seco J, Clasie B, Partridge M (2014) Review of the characteristics of radiation detectors for dosimetry and imaging. Phys Med Biol 59:R303
33. Lehmann EH et al (2004) Neutron imaging—detector options and practical results. Nucl Instrum Methods Phys Res A 531:228–237
34. https://radiopaedia.org/articles/detective-quantum-efficiency-1. Accessed on 31 May 2021
35. https://eljentechnology.com/products/neutron-detectors/ej-426. Accessed 14 May 2021
36. https://webstore.iec.ch/publication/21937. Accessed 30 May 2021
37. Masalovich S (2019) Analysis of the detective quantum efficiency of a neutron image plate detector. Nucl Instrum Methods Phys Res Sect A: Accelerat Spectrom Detect Associat Equip 930:151–155. ISSN 0168-9002. https://doi.org/10.1016/j.nima.2019.03.094
38. Lewandowski R, Cao L, Turkoglu D (2012) Noise evaluation of a digital neutron imaging device. Nucl Instrume Methods Phys Res Sect A Acceler Spectrom Detect Associat Equip 674:46–50. ISSN 0168-9002. https://doi.org/10.1016/j.nima.2012.01.025
39. Shaikh AM (2008) Pramana J Phys 71(4)–663–672
40. Nimura N (1994) An Imaging plate neutron detector. NIMA 349:521–525
41. https://www.selectscience.net/products/fla-9000-starion/?prodID=84313#tab-2. Accessed date 11 May 2021
42. Massalovitch S, Ioffe A, Kuessel E, Schlapp M, Brueckel T (2002) Appl Phys A 74:118
43. Schlapp M, Conrad H, von Seggern H (2004) Pixelated neutron image plates. J Phys D: Appl Phys 37:2607
44. Mikerov V, Bogolubov E, Samosyuk V, Verushkin S (2006) Fast neutron imaging with CCD detectors and imaging plates. In: International workshop on fast neutron detectors, University of Cape Town, South Africa, April 3–6, 2006
45. Wu C et al (2013) A study of ZnS(Ag)/(LiF)-Li-6 with different mass ratios. Radiat Meas 58:128–132
46. de Beer F et al (2020) Overview of the conceptual design of the upgraded neutron radiography facility (INDLOVU) at the SAFARI-1. In: Research reactor in South Africa, materials research proceedings, vol 15, pp 11–16. https://doi.org/10.21741/9781644900574-2
47. Osovizky A, Pritchard K, Ziegler J, Binkley E, Yehuda-Zada Y, Tsai P, Thompson A, Cooksey C, Siebein K, Hadad N, Jackson M, Hurlbut C, Ibberson R, Baltic GM, Majkrzak CF, Maliszewskyj NC. ^{6}LiF:ZnS(Ag) mixture optimization for a highly efficient ultrathin cold neutron detector. https://doi.org/10.1109/TNS.2018.2809567
48. Nagarkar VV, Tipnis SV, Gaysinskiy V, Klugerman Y, Squillante AR, Entine G (2001) Structured LiI scintillator for thermal neutron imaging. IEEE Trans Nucl Sci 48(6):2330–2334. https://doi.org/10.1109/23.983262
49. Sen I, Penumadu D, Williamson M, Miller LF, Green AD, Mabe AN (2011) Thermal neutron scintillator detectors based on poly (2-Vinylnaphthalene) composite films. IEEE Trans Nucl Sci 58(3):1386–1393. https://doi.org/10.1109/TNS.2011.2141149
50. https://scintacor.com/products/6-lithium-glass/. Accessed date: 20 May 2021
51. Moore ME, Trtik P, Lousteau J, Pugliese D, Brambilla G, Hayward JP (2019) Neutron imaging with Li-glass based multicore SCIntillating FIber (SCIFI). J Lightwave Technol 37(22):5699–5706

52. Trtik P, Lehmann EH (2015) Isotopically-enriched gadolinium-157 oxysulfide scintillator screens for the high-resolution neutron imaging. NIMA 788(11):67–70
53. https://www.rctritec.com/en/scintillators.html
54. Chuirazzi W, Craft A, Schillinger B, Cool S, Tengattini A (2020) Boron-based neutron scintillator screens for neutron imaging. J Imaging 6(11):124. https://doi.org/10.3390/jimaging6110124
55. Craft A, Grünzweig C, Morgano M, Chuirazzi W, Lehmann E (2020) Gamma discriminating scintillation screens for digital transfer method neutron imaging. Mater Res Proc 15:74–79. https://doi.org/10.21741/9781644900574-12
56. Feller WB, Gregory Downing R, White PL (2000) Neutron field imaging with microchannel plates. In: Hard x-ray, gamma-ray, and neutron detector physics II, vol 4141. International Society for Optics and Photonics
57. Tremsin AS, Bruce Feller W, Gregory Downing R (2005) Efficiency optimization of microchannel plate (MCP) neutron imaging detectors. I. Square channels with 10B doping. Nucl Instrum Methods Phys Res Sect A: Accelerat Spectrom Detect Associat Equip 539(1–2):278–311. ISSN 0168-9002. https://doi.org/10.1016/j.nima.2004.09.028
58. Tremsin AS, McPhate JB, Vallerga JV, Siegmund OHW, Hull JS, Feller WB, Lehmann E (2009) High-resolution neutron radiography with microchannel plates: proof-of-principle experiments at PSI. Nucl Instrum Methods Phys Res Sect A: Accelerat Spectrom Detect Associat Equip 605(1–2):103–106. ISSN 0168-9002. https://doi.org/10.1016/j.nima.2009.01.137
59. Medipix Collaboration. http://medipix.cern.ch/MEDIPIX
60. Jakubek J, Holy T, Lehmann E, Pospisil S, Uher J, Vacik J, Vavrik D (2006) Neutron imaging with Medipix-2 chip and a coated sensor. Nucl Instrum Methods Phys Res, Sect A 560(1):143–147
61. Wang Y, Han S, Hao L, He L, Wei G, Wu M, Wang H, Liu Y, Chen D (2013) New type of neutron image scintillator based on $H_3{}^{10}BO_3$/ZnS(Ag). Phys Procedia 43:216–222
62. Shestakova I, Tipnis SV, Gaysinskiy V, Antal JJ, Bobek L, Nagarkar VV (2005) A new sensor for thermal neutron imaging. IEEE Trans Nucl Sci 52(4):1109–1113. https://doi.org/10.1109/TNS.2005.852671
63. https://www.saint-gobain.com/en. Accessed date 16 May 2021
64. Hall J (2001) Science & Technology Report
65. Johnson MS, Anderson SG, Bleuel DL, Fitsos PJ, Gibson D, Hall JM, Marsh R, Rusnak B, Sain J (2017) Development of a high-brightness, Quasi-monoenergetic neutron source for neutron imaging. Phys Procedia 90:47–54
66. Adams R, Zboray R, Prasser H-M (2016) A novel fast-neutron tomography system based on a plastic scintillator array and a compact D-D neutron generator. Appl Radiat Isot 107:1–7
67. Bishnoi S, Sarkar PS, Thomas RG et al (2019) Preliminary experimentation of fast neutron radiography with D-T neutron generator at BARC. J Nondestruct Eval 38:13. https://doi.org/10.1007/s10921-018-0550-9
68. Lehman E et al (2021) Improvement in the spatial resolution for imaging with fast neutron. NIMA 988:164809
69. Merrill FE, Bower D, Buckles R, Clark DD, Danly CR, Drury OB, Dzenitis JM, Fatherley VE, Fittinghoff DN, Gallegos R, Grim GP, Guler N, Loomis EN, Lutz S, Malone RM, Martinson DD, Mares D, Morley DJ, Morgan GL, Oertel JA, Tregillis IL, Volegov PL, Weiss PB, Wilde CH, Wilson DC (2012) The neutron imaging diagnostic at NIF (invited). Rev Sci Instrum 83:10D317
70. Fittinghoff DN, Bower DE, Drury OB, Dzenitis JM, Hatarik R, Merrill FE, Grim GP, Wilde CH, Wilson DC, Landoas O, Caillaud T, Bourgade J, Buckles RA, Lee J, Weiss PB (2011) Performance improvements to the neutron imaging system at the national ignition facility, LLNL-TR-503351, 6 October 2011
71. Yanagida T, Kawaguchi N, Fujimoto Y, Fukuda K, Yokota Y, Yamazaki Y, Watanabe K, Pejchal J, Uritani A, Iguchi T, Yoshikawa A (2011) Basic study of europium doped LiCaAlF6 scintillator and its capability for thermal neutron imaging application. Opt Mater 33(8):1243–1247. ISSN 0925-3467. https://doi.org/10.1016/j.optmat.2011.02.016

72. Yoshikawa A, Yanagida T, Yokota Y, Kawaguchi N, Ishizu S, Fukuda K, Suyama T, Kim KJ, Pejchal J, Nikl M, Watanabe K, Miyake M, Baba M, Kamada K, Nucl IEEE (2009) Trans Sci 56:3796
73. Yanagida T, Yoshikawa A, Yokota Y, Maeo S, Kawaguchi N, Ishizu S, Fukuda K, Suyama T (2009) Opt Mater 32:311
74. Iwanowska J, Swiderski L, Moszynski M, Yanagida T, Yokota Y, Yoshikawa A, Fukuda K, Kawaguchi N, Ishizu S (2011) Nucl Instr Methods A 652:319
75. Yamazaki A, Watanabe K, Uritani A, Iguchi T, Kawaguchi N, Yanagida T, Fujimoto Y, Yokota Y, Kamada K, Fukuda K, Suyama T, Yoshikawa A (2011) Nucl Instr Methods A 652:435–438
76. Tyagi M, Sarkar PS, Singh AK, Kalyani, Patel T, Bishnoi S, Ray NK, Desai DG, Gadkari SC (2019) Development of neutron detector based on Gd3Ga3Al2O12:Ce single crystals. IEEE Trans Nucl Sci 66(4):724–728
77. Lehmann E, Vontobel P (2004) The use of amorphous silicon flat panels as detector in neutron imaging. Appl Radiat Isot 61(4):567–571
78. Lehmann EH, Vontobel P, Frei G, Brönnimann C (2004) Neutron imaging—detector options and practical results. Nucl Instrum Methods Phys Res, Sect A 531(1–2):228–237
79. Yu B, Harder JA, Mead JA, Radeka V, Schaknowski NA, Smith GC (2003) Neutron detector development at Brookhaven. Nucl Instrum Methods Phys Res Sect A: Accelerat Spectrom Detect Associat Equip 513(1–2):362–366
80. Schaknowski N, Fried J, Smith G, Mahler G, Yu B (2018) Design and construction of a 1m × 1m thermal neutron imager operating in ionization mode with Pad Readout. In: 2018 IEEE nuclear science symposium and medical imaging conference held at Sydney Australia. Brookhaven National Laboratory, Upton, NY, USA, 10–17 November 2018, BNL-211410-2019-PUCP. https://www.osti.gov/servlets/purl/1619254. Accessed date 10 May 2021
81. Tian LC, Chen YB, Tang B, Zhou JR, Qi HR, Liu RG, Zhang J, Yang GA, Xu H, Chen DF, Sun ZJ (2021) Study on the imaging ability of the 2D neutron detector based on MWPC. https://arxiv.org/pdf/1305.5011.pdf . Accessed date 9 May 2021
82. Strobl M, Steitz R, Kreuzer M, Rose M, Herrlich H, Mezei F, Grunze M, Dahint R (2011) BioRef: a versatile time-of-flightreflectometer for soft matter applications at Helmholtz-Zentrum Berlin. Rev Sci Instrum 82:055101. https://doi.org/10.1063/1.3581210
83. Strobl M, Woracek R, Kardjilov N, Hilger A, Wimpory R, Tremsin A, Wilpert T, Schulz C, Manke I, Penumadu D (2012) Time-of-flight neutron imaging for spatially resolved strain investigations. Nucl Inst Methods Phys Res A 680:27–34
84. UCRL-JC-124401, 11 Sept 1996
85. Uno S, Uchida T, Sekimoto M, Murakami T, Miyama K, Shojic M, Nakano E, Koike T (2012) Development of a two-dimensional gaseous detector for energy-selective neutron radiography. Phys Procedia 37(600–605):1875–3892
86. Parker JD, Harada M, Hattori K, Iwaki S, Kabuki S, Kishimoto Y, Kubo H, Kurosawa S, Matsuoka Y, Miuchi K, Mizumoto T, Nishimura H, Oku T, Sawano T, Shinohara T, Suzuki J, Takada A, Tanimori T, Ueno K (2013) Spatial resolution of a μPIC-based neutron imaging detector. Nucl Instrum Methods Phys Res Sect A: Accelerat Spectrom Detect Associat Equip 726:155–161. ISSN 0168-9002. https://doi.org/10.1016/j.nima.2013.06.001
87. Shinohara T, Kai T, Oikawa K, Nakatani T, Segawa M, Hiroi K, Yuhua S, Ooi M, Harada M, Iikura H, Hayashida H, Parker JD, Matsumoto Y, Kamiyama T, Sato H, Kiyanagi Y (2020) The energy-resolved neutron imaging system. RADEN Rev Sci Instrum 91:043302. https://doi.org/10.1063/1.5136034

Chapter 4
Major Neutron Source Facilities Across the Globe

Mayank Shukla, Nirmal Ray, and Tarun Patel

4.1 Introduction: Overview of Neutron Imaging Facilities—Past and Present

Neutron imaging has proven to be very useful in the areas of basic and applied research, industrial application, material research, non-destructive testing (NDT), space research, etc. as represented in Fig. 4.1. Kallmann and Kuhn in Germany [1] carried the first neutron radiography out way back in 1935, soon after the discovery of neutrons. They have used a Ra–Be radioisotope source and low neutron yield accelerator-based neutron source to get a radiograph with exposure of few hours using a vacuum cassette based radiographic converter-film system. For this work, they were awarded a joint US Patent entitled "Photographic Detection of Slowly Moving Neutrons" in January 1940 [2]. Thereafter, the neutron radiography work in Germany (upto World War-II) used intense accelerator-based source to record a neutron radiograph in few minutes [3]. Moreover, limitation of achieving high thermal neutron flux in accelerator-based sources coupled to a thermalizing assembly and realizing the potential of neutron imaging, researchers started looking for alternate intense neutron sources with stable beam started. The first reactor source for neutron imaging in practical applications was used by Thewlis and his co-worker Derbyshire in the mid-1950s, when they utilized a neutron beam with a flux of between ~10^9 n/cm^2 s from the BEPO reactor at Harwell in England [4]. Soon after that, several developments, using reactor sources, were carried out by many researchers during 1960s in neutron imaging. These include intense and collimated beam to produce good quality images

M. Shukla (✉) · N. Ray · T. Patel
Technical Physics Division, Bhabha Atomic Research Centre, Mumbai, India
e-mail: mayank@barc.gov.in

N. Ray
e-mail: nirmal@barc.gov.in

T. Patel
e-mail: pateltp@barc.gov.in

D. K. Aswal et al. (eds.), *Neutron Imaging*,
https://doi.org/10.1007/978-981-16-6273-7_4

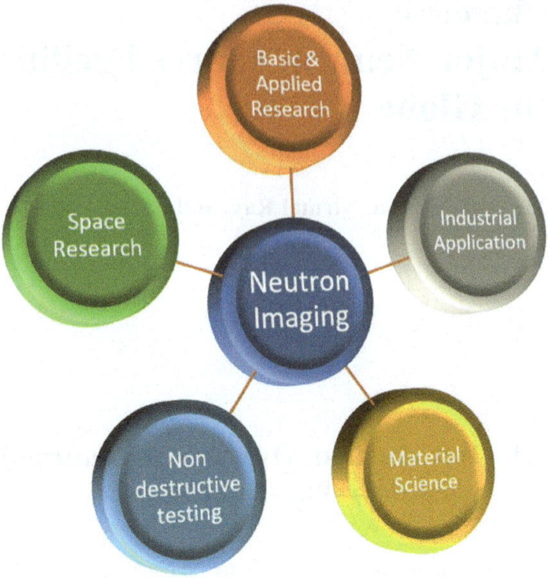

Fig. 4.1 Applications of neutron imaging

with high signal-to-noise ratio as compared to the earlier accelerator-based neutron source radiographs. Simultaneously, the improved detection methods adopted for NDT in neutron imaging provided qualitative as well as quantitative information of the object under examination. Reactor sources thus became the choice for neutron imaging, especially for industrial applications and nuclear fuels. Initially, in late sixties, the neutron radiography program was pursued at Argonne National Laboratory, United States, and at Melusine, Grenoble, France [5–9]. Thereafter, neutron imaging facilities were also developed in many other countries like Canada, Japan, South Africa, Germany, and Switzerland. Nearly 40 installations have come up for neutron radiography by the beginning of 1970 [10, 11]. These facilities were mainly involved in neutron imaging-based investigation of fresh and used nuclear fuel. At the same time, some of the facilities also started offering neutron radiography for commercial services that continue until today. General-Electric Vallecitos Center and the TRIGA type reactor at the Aerotest Operations Incorporated, both in the United States [7], were the first two reactors to offer such services. Similarly, neutron radiography as a service was also provided at the Fontenay-Aux-Roses in France and the Harwell NDT Center in England [7].

Neutron radiography mainly utilized attenuation property of thermal neutrons. However, fast neutron radiography investigations, using a 60-inches. cyclotron at two different neutron energies, at the University of California were also carried out [12]. At the same time, neutron radiography was also investigated using epithermal and cold neutron beam from 5 MW Herald reactor at the Atomic Weapons Research Establishment, England [13, 14].

Neutron imaging from 1970 onwards was focused on developing techniques and instrumentation for quantitative measurement rather than qualitative examination. Efforts were put to standardize neutron imaging in terms of reliability in finding defects and their characterization mainly for critical components such as those used in space programs. Thus, the relationship between standards and the acceptance of new NDT methods were recognized by researchers [15]. In 1969, Association of Neutron Radiographers (ANR) was formed to develop a personnel qualification standard under the authority of the American Society for Non-destructive Testing [16]. This helped in quantifying parts for aerospace industry.

First conference on neutron imaging with a theme of "Radiography with Neutrons," was held at the University of Birmingham in 1973 [17]. Upto this period neutron imaging facilities were geographically commissioned in the United States, in many European countries and in Japan. Researchers in these facilities were exploring real-time imaging methods and new detectors for investigation [18–20]. Efforts were being put in improving the beam quality such as collimation, cadmium ratio for high flux neutron beam along with achieving high spatial resolution. Neutron imaging revolutionized when neutron sensitive image intensifier tube coupled to a vidicon television camera was used as detector for real-time neutron imaging for the first time [19]. Subsequently, this detector system became the choice among the researchers for dynamic imaging through 1990s because of its commercial availability [18–20]. Besides this, high-speed neutron radiography was also accomplished using TRIGA reactor by utilizing a reactor pulse and a high-speed camera [21]. New imaging modalities such as computed axial tomography with neutrons were also demonstrated during this period [22, 23] though they were of poor resolution and image quality, compared to the present-day neutron computed tomography (NCT) [24, 25].

Between 1980 and 1990 the progress on neutron imaging facilities and related research did not take noticeable stride. This was due to development of new NDT techniques using portable and less expensive sources along with non-availability of adequate funds for carrying out research at university and national laboratory research reactors. However, since the start of mid-1990s there was a renewed interest in neutron imaging. This was due to the setting up of new improved facilities with upgraded power of reactor sources, setting up of new cold neutron and spallation neutron sources, availability of advanced imaging devices and sensors, availability of high-end computing stations, and image reconstruction software.

4.2 Global Facilities

Neutron imaging has come a long way over seven decades since mid-1950s. It has grown gradually from neutron radiography to advanced imaging such as tomography, phase contrast imaging, polarized neutron imaging, etc. Globally many facilities, built around reactors, spallation sources, have been setup catering the neutron imaging community either for in-house research work or for users as well [10, 11]. Some other neutron imaging facilities have also been reported using radioactive

In-House Usage : Algeria, Argentina, Australia, Austria, Bangladesh, Belgium, Brazil, Egypt, Germany, Hungary, Indonesia, Israel, Italy, Malaysia, Norway, Poland, Portugal, Romania, Slovenia, USA

User Facilities: Brazil, Germany, Hungary, Japan, Korea, South Africa, Switzerland, USA

Under Installation: China, Japan, Morocco, Vietnam, Greece, Thailand

Project : France, United Kingdom, India

Fig. 4.2 World map showing the neutron imaging facilities using reactor and spallation source [26]

sources, D-T fusion-based neutron sources, and photo neutron sources. A world map of neutron imaging installations is shown in Fig. 4.2. Geographically, the facilities have been widely spread across the globe from developed to developing nations. This section describes the neutron imaging facilities using reactor sources. Both thermal and cold neutron beam facilities are covered. Neutron imaging program in India started in mid-1990 using thermal neutrons from research reactors. A brief history along with the neutron imaging beamlines commissioned in India has been mentioned in a separate section.

4.3 Neutron Imaging at Reactor-Based Sources

Generally, neutron imaging facilities have a common setup as shown in Fig. 4.3. They consist of a collimator placed in the beam tube of the reactor port, a safety shutter, sample manipulator, detector system with a beam dump behind. Neutron beam from the reactor after filtering and collimation is extracted towards the sample position. The image is recorded in transmission mode using scintillator-based high-resolution imaging detector placed in proximity with the sample. A shutter placed in between the collimator and the sample manipulator, allows safe user access during beam operation. For tomographic investigation, the sample is rotated in steps to

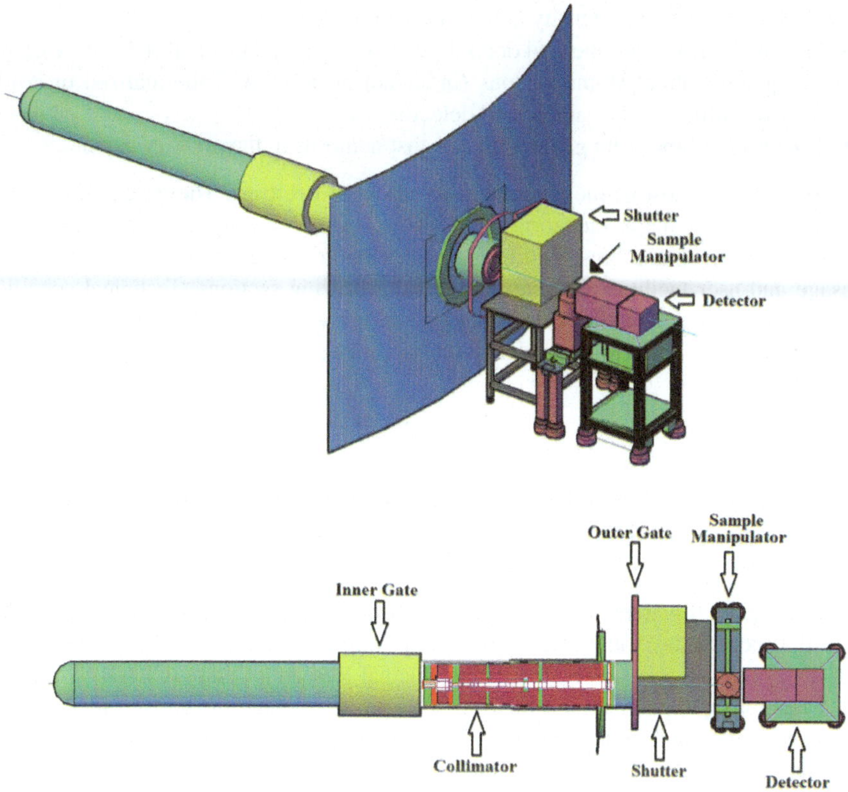

Fig. 4.3 Schematic of a typical neutron imaging setup consisting of collimator, safety shutter, sample manipulator and detector inside a shielding hutch

acquire multiple images and further used for reconstruction the volume image of the sample. The entire setup after the collimator is well shielded for neutron and gamma radiation using proper shielding materials such as high-density concrete, borated wax, high-density borated polythene, steel, and lead. A controlled access from outside the shielding helps user to safely conduct the experiment.

Depending upon the energy of the neutron beam (thermal or cold) extracted from the reactor, the imaging facility is categorized as thermal neutron imaging beamline or cold neutron imaging beamline. A good quality image, on one hand, can be acquired using resolution CCD-based detector system controlled by modern computers and post-processing of image using advanced image processing software. On the other hand, the beam quality also plays an important role in getting a quality image. A useful beam for neutron imaging application should have the following properties:

- Well collimated neutron beam with high *L*/*D*-ratio; *L* is the collimation length; *D* is the inlet aperture diameter
- Large beam size with flat-top beam profile suitable to the sample dimension

- High neutron beam intensity at the sample position
- Narrow energy band (thermal or cold), well-known spectral conditions, necessary for quantification. Homogeneous (in space) beam of well thermalized or cold neutrons fitting to the sample and detector size
- Low background from gamma rays or fast neutrons in the primary beam.

Worldwide, most of the beamlines use thermal neutrons. These beamlines can be put into four major categories namely in-house research, user, under installation, and project as represented in Fig. 4.2. Country-wise description of in-house usage and user facilities is described in the following sections. Technical specifications presented describes the reactor type along with the neutron imaging beamline parameters, detectors available at the facilities and research topics and services offered.

4.3.1 Algeria, Es-SALAM, Centre de Recherche Nucléaire de Birine (CRNB) 15 MW—In-House Usage

The Es-Salam research reactor is located in Birine, Algeria (Fig. 4.4). It is a 15 MW multipurpose heavy water reactor owned by Algeria's Atomic Energy Commission (COMENA) and operated by the Birine Nuclear Research Centre (CRNB) [10, 27]. The reactor, equipped with several irradiation positions and beam ports, was commissioned in 1992. It provides a high-quality thermal neutron flux that is used for experimental and training tool for nuclear techniques and reactor physics. One of the beam port is dedicated for neutron imaging activities.

Fig. 4.4 Es-SALAM reactor building, Algeria, Africa [27]

Technical Specifications

- Tank-type research reactor with a power of 15 MW
- Cooled and moderated by heavy water with a graphite reflector
- Six horizontal beam ports, including a thermal column and 45 vertical irradiation positions
- Maximum thermal neutron flux of 2.1×10^{14} cm^{-2} s^{-1}
- Radial Beam line alignment
- Thermal neutron energy spectrum
- Maximal beam intensity at sample position [n-cm^{-2} s^{-1}] ~2.25×10^{7}
- *L*/*D* collimation ratio ~300
- Effective Beam size at sample position ~240 mm
- Detectors—film, scintillator-based CCD camera system
- Radiography, Non-destructive testing
- Flow visualization and characterization
- Neutron Tomography.

4.3.2 Algeria, NUR, CRND, 1 MW—In-House Usage

The NUR research reactor, as shown in Fig. 4.5, is located near Algiers. It is an open pool-type reactor that was commissioned in March 1989 [10, 27]. With a 1 MW of power, the reactor is used for research, development, and training purposes. These include development of nuclear techniques, neutron activation analysis (NAA), Neutron reflectometry, Neutron radiography, small angle neutron scattering (SANS), production of radioisotopes, and radiopharmaceuticals and training of operators.

Fig. 4.5 Nur facility and reactor block, Algeria, Africa [27]

Technical Specifications

- Pool-type, 1 MW research reactor
- Materials testing reactor (MTR-LEU) plate-type fuel enriched to 20%
- Cooled and moderated by light water with a graphite reflector
- Maximum thermal neutron flux of ~5 × 10^{13} n-cm^{-2} s^{-1}
- Four radial and one tangential beam tubes, two vertical irradiation positions, two fast pneumatic transport systems, one hot cell, and one transfer cell
- Tangential Beam line alignment
- Thermal neutron energy spectrum
- Maximal beam intensity at sample position [n-cm^{-2} s^{-1}] ~6 × 10^6
- *L*/*D* collimation ratio ~113
- Effective Beam size at sample position ~340 mm dia
- Detectors—film based.

Applications

- Research topic—Non-destructive testing, Radiography.

4.3.3 Argentina, RA-6, 1 MW, In-House Usage

RA-6 is a multipurpose open pool type research reactor of 1 MW power [10]. It was commissioned in 1982 and is located in San Carlos de Bariloche, Rio Negro. It is used for research in physics and nuclear engineering. The neutron imaging facility is shown in Fig. 4.6.

Technical Specifications

- Open pool-type, 500 kW/1 MW research reactor
- Maximum thermal neutron flux of ~2 × 10^{13} cm^{-2} s^{-1}

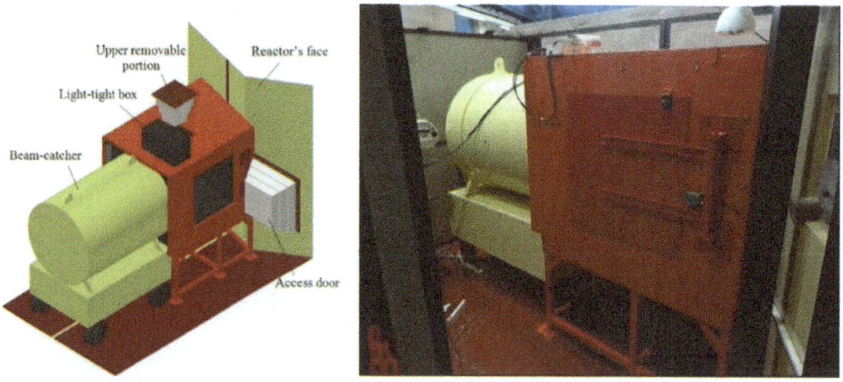

Fig. 4.6 Neutron irradiation and imaging facility at RA-6 reactor, Argentina [10]

- Radial Beam line alignment
- Thermal neutron energy spectrum
- Maximal beam intensity at sample position [n-cm^{-2} s^{-1}] ~2.54 × 10^6 (with the reactor operating at 500 kW)
- Filter Options—Sapphire
- *L*/*D* collimation ratio ~100
- Effective Beam size at sample position ~200 mm × 200 mm
- Detectors—Scintillator and CCD camera based.

Applications

- Research topic—Non-destructive testing, Radiography, Hydrogen technology, Cultural Heritage, Aerospace application.

4.3.4 *Australia, DINGO, OPAL 20 MW, User Facility*

The **Open-pool Australian light water reactor** (**OPAL**) is located at the Australian Nuclear Science and Technology Organization (ANSTO) Research Establishment New South Wales, Sydney (Fig. 4.7). It is 20 MW thermal reactor that uses low enriched fuel, heavy water as moderator, and light water as coolant. The reactor went critical on 12th August 2006. OPAL reactor has around nine states of beamlines using thermal and cold neutrons that are utilized in the areas of physical sciences, biology, chemistry, radio isotope production, residual stress measurement, neutron activation analysis, semiconductor material irradiation, and neutron radiography [28].

Technical Specifications

- Open pool-type, 20 MW research reactor
- Low enriched plate type fuel
- Maximum thermal neutron flux of ~4 × 10^{14} cm^{-2} s^{-1}
- Thermal neutron energy spectrum
- Maximal beam intensity at sample position [cm^{-2} s^{-1}] ~5.3 × 10^7 and 1.06 × 10^7 at two sample positions

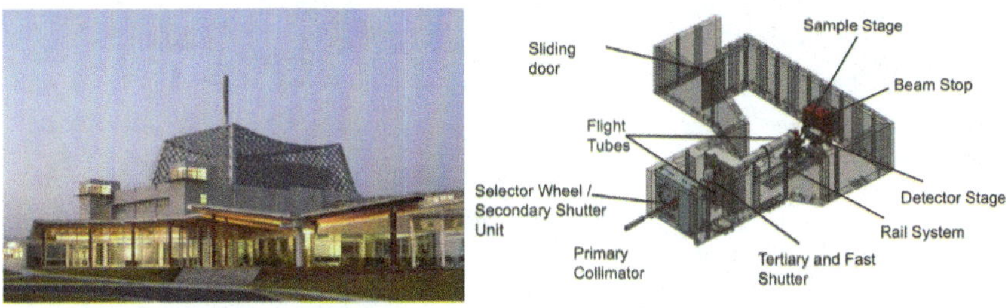

Fig. 4.7 OPAL research reactor facility [29] and DINGO neutron imaging setup, Australia [28]. Reproduced with permission from ELSEVIER

- *L/D* collimation ratio ~500 and 1000
- Effective Beam size at sample position ~200 mm × 200 mm
- Detectors—Scintillator-based CCD camera system
- Radiography, tomography, phase contrast imaging.

Applications

- Research topic—Non-destructive testing, Radiography, Fuel Cell research, material science, palaeontology and cultural heritage.

4.3.5 Austria, Atominstitut, Vienna, TRIGA II, 0.25 MW, In-House Usage

TRIGA Mark II research reactor facility, as shown in Fig. 4.8, is located at the Institute of the Technische Universität Wien (TU Wien, Atominstitut in Vienna). TRIGA is an acronym for **T**raining, **R**esearch, **I**sotope Production, **G**eneral **A**tomic. It is a swimming pool type reactor giving a continuous and pulsed thermal neutron beam of 250 kW and 250 MW respectively. The reactor is in operation since March 1962 and has been used for basic and applied academic research and teaching purposes. Since this reactor is close to the IAEA headquarters, it is also frequently used by IAEA staff for development and calibration of safeguards instruments [30].

Technical Specifications

- Swimming pool type research reactor, 250 kW in continuous mode and 250 MW in pulse mode (prompt pulse lifetime: 40 ms and maximum repetition frequency: 12/h)

Fig. 4.8 Atominstitut and reactor hall, Vienna, Austria [30]

- Maximum thermal neutron flux of ~1 × 10^{13} n-cm^{-2} s^{-1}, in continuous mode operation.
- Radial Beam line alignment
- Thermal neutron energy spectrum
- Maximal beam intensity at sample position [n-cm^{-2} s^{-1}] ~3 × 10^5 at NR-I facility and ~1.3 × 10^5 at NR-II facility
- *L*/*D* collimation ratio ~50 for NR-I facility and 130 for NR-II facility
- Effective Beam size at sample position ~400 mm diameter for NR-I facility and 90 mm diameter for NR-II facility
- Detectors—film, scintillator-based CCD camera system, Image plate
- Radiography, Tomography and timing sequence.

Applications

- Research topics—Absorber inhomogeneities, hydrogen detection
- Services for industry—Investigation of steels, borated materials, fuel cells, building materials
- Methodical developments—Systematic study of secondary effects in neutron transmission, beam hardening correction, NR and NT with weak beams.

4.3.6 Bangladesh, TRIGA MARK II, 3 MW, In-House Usage

TRIGA Mark-II type reactor is located in the capital city Dhaka, Bangladesh. It is a 3 MW research reactor operated under Bangladesh Atomic Energy Commission (BAEC). The reactor has been operating since September 14, 1986 as shown in Fig. 4.9. Utilization of this reactor includes radioisotope production (131 I, 99m Tc, 46 Sc), various R&D activities such as neutron activation, neutron scattering studies, neutron radiography and manpower training [10, 31, 32].

Fig. 4.9 Reactor hall and neutron radiography setup, Dhaka, Bangladesh [10, 31]

Technical Specifications

- Swimming pool type, BAEC TRIGA Research Reactor (BTRR) 3 MW
- Maximum thermal neutron flux of ~5.6×10^{13} cm^{-2} s^{-1}
- Tangential Beam line alignment
- Thermal neutron energy spectrum
- Maximal beam intensity at sample position (Tangential beam port) [n-cm^{-2} s^{-1}] ~1.32×10^7
- Neutron/gamma ratio (n-cm^{-2} per mGy)—2.23×10^6
- Cadmium Ratio—10.51
- Filter options—Bi
- *L*/*D* collimation ratio ~60
- Effective Beam size at sample position ~300 mm diameter
- Detectors—film based.

Applications

- Research topics—Non-destructive testing to study the internal defects and water absorption behaviour of different kinds of objects
- Services for industry—R&D collaboration with other industry/organization.

4.3.7 Belgium, BR-1, 4 MW, In-House Usage

The BR1, the first Belgian reactor, is an air-cooled reactor with natural metallic uranium as fuel and graphite as moderator (Fig. 4.10). It is located at SCK CEN Institute Mol, Belgium and is operational since 1956. It is a flexible instrument for fundamental research, training and education [10, 33]. The reactor is used for Neutron radiography, calibration and validation, NAA, Production of Neutron Transmutation Doped (NTD) Silicon.

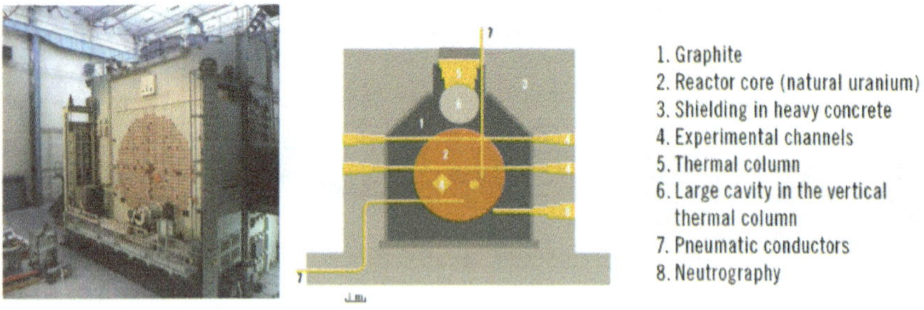

Fig. 4.10 BR-1 reactor, Belgium [33]

Technical Specifications

- Air-cooled natural metallic uranium fuel, graphite moderator research reactor BR-1
- Now operated at 0.7 MW/1 MW for short periods
- Maximum thermal neutron flux of ~2×10^{12} cm^{-2} s^{-1}
- Tangential Beam line alignment
- Thermal neutron energy spectrum
- Maximal beam intensity at sample position [cm^{-2} s^{-1}] ~4.2×10^5
- *L*/*D* collimation ratio ~75
- Effective Beam size at sample position ~200 mm dia
- Detectors—film based
- Radiography, NDT.

4.3.8 *Brazil, ARGONAUTA, 0.2 MW, In-House Usage*

The ARGONAUTA research reactor is located at the Nuclear Engineering Institute (IEN), university campus, Rio de Janeiro, Brazil. It is operated under Brazilian Nuclear Energy Commission located at the [10, 34]. "Argonaut" is an acronym for **ARGO**nne **N**uclear **A**ssembly for **U**niversity **T**raining and mainly refers to a class of small nuclear research and training reactors. The reactor uses U_3O_8-Al materials testing reactor (MTR) plate-type dispersion fuel type with 20% enriched uranium, giving 500 Wth of maximum power. Besides teaching and training in reactor physics and nuclear engineering, it is mainly used neutron radiography (Fig. 4.11) and neutron activation analysis.

Technical Specifications

- Type: Argonaut reactor Research Reactor
- Maximum thermal neutron flux of ~4.4×10^9 cm^{-2} s^{-1}
- Radial Beam line alignment

Fig. 4.11 ARGONAUTA research reactor and neutron radiography setup [10]

- Thermal neutron energy spectrum
- Maximal beam intensity at sample position [n-cm^{-2} s^{-1}] ~4.46 × 10^5
- *L*/*D* collimation ratio ~70
- Effective Beam size at sample position ~150 mm × 150 mm
- Detectors—film, scintillator-based CCD camera system, Image plate
- Radiography, Tomography.

4.3.9 Brazil, IAE-R1, 5 MW, User Facility

The Brazilian research reactor, IEA-R1, is located at Instituto de Pesquisas Energeticas e Nucleares (IPEN), São Paulo, Brazil. It was commissioned on September 16, 1957 and utilizes 20% enriched uranium fuel giving 5 MWth power [10, 35]. The reactor is widely used in areas of research such as physics, chemistry, biology, and engineering, as well as radioisotopes production for medical and other applications. The other service areas of IAE-R1 reactor utilization include neutron activation analysis, real-time neutron radiography and Tomography (as shown in Fig. 4.12), and neutron transmutation doping of silicon.

Technical Specifications

- Pool-type, IAE-R1, 5 MW research reactor
- 20% enriched uranium U_3O_8-Al and U_3Si_2-Al fuel
- Cooled and moderated by light water with graphite and beryllium reflectors
- Maximum thermal neutron flux of ~10^{13} cm^{-2} s^{-1}
- Eight radial and two tangential beam tubes
- Radial Beam line alignment
- Thermal/cold neutron energy spectrum
- Maximal beam intensity at sample position [n-cm^{-2} s^{-1}] ~5 × 10^6
- Cadmium ratio ~5
- Filter Options—Bi
- *L*/*D* collimation ratio ~90
- Effective Beam size at sample position ~130 mm diameter
- Detectors—film, scintillator-based CCD camera system

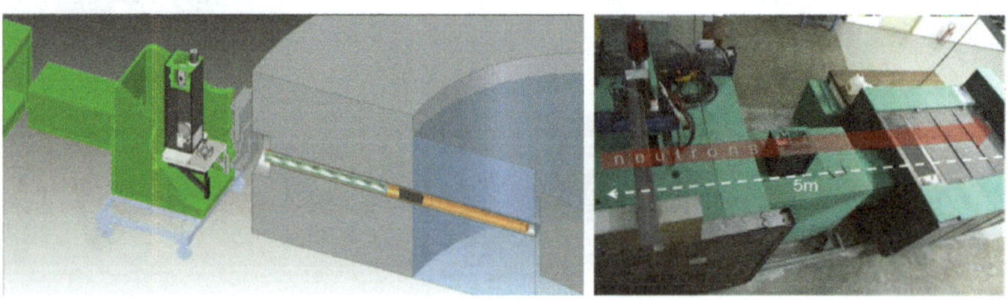

Fig. 4.12 Neutron imaging beamline, IAE-R1, Brazil [10, 35]

- Imaging Options—Radiography, Tomography and time sequence.

Applications

- Research Topics—Archeology, anthropology, engineering, aerospace, medicine.

4.3.10 Egypt, ETRR-2, 22 MW, In-House Usage

ETRR-2 (**Experimental Training Research Reactor-2**) reactor was supplied by the Argentine company, Investigacion Aplicada (INVAP). It is located at the Nuclear Research Center in the city of Inshas near Cairo, Egypt and is owned and operated by Egyptian Atomic Energy Authority (EAEA) (Fig. 4.13). The reactor achieved its criticality in 1997 and uses 20% enriched uranium fuel with light water moderator and beryllium reflector [10, 36]. ETRR-2 is widely used in areas of physics, materials/fuel testing, radioisotopes production, NAA, neutron radiography and teaching and training purpose.

Technical Specifications

- Pool-type, ETRR-2, 22 MW research reactor
- 20% enriched uranium fuel
- Maximum thermal neutron flux of ~1.2×10^{14} cm^{-2} s^{-1}
- Radial Beam line alignment
- Thermal neutron energy spectrum
- Maximal beam intensity at sample position [n-cm^{-2} s^{-1}] ~1.5×10^{7}

Fig. 4.13 ETRR-2 reactor facility, Egypt [27]

- Cadmium ratio ~10
- *L*/*D* collimation ratio ~117
- Effective Beam size at sample position ~220 mm diameter
- Detectors—film, scintillator-based CCD camera system
- Imaging Options—Radiography, Tomography and time sequence.

4.3.11 France, ORPHEE, 14 MW, User Facility

ORPHEE (or the Orpheus) nuclear reactor is located at the Leon Brillouin Laboratory (LLB), CEA-Saclay Centre, Paris, France (Fig. 4.14). LLB is a joint national laboratory between CNRS (French National Center for Scientific Research) and CEA (French Atomic Energy Commission). ORPHEE is a swimming pool-type light water-cooled and heavy water moderated reactor that uses highly enriched fuel. The reactor achieved its criticality in December 1980. This national facility has been aiming three missions (called TGIR in French) such as proper research, service and development, training and education [10, 37]. The large-scale facilities of the reactor are not only open to own research programs but also serves large user community from across the world.

Fig. 4.14 ORPHEE reactor, Saclay Center, France [38]

Technical Specifications

- Swimming Pool-type, ORPHEE, 14 MW research reactor
- MTR-type fuel with 93% U-235 enrichment
- Maximum thermal neutron flux of ~4.4 × 10^{13} cm^{-2} s^{-1}
- Tangential and *neutron guide tube** beamline alignment
- Thermal and *cold** neutron energy spectrum
- Maximal beam intensity at sample position [n-cm^{-2} s^{-1}] ~5 × 10^8
- *L*/*D* collimation ratio ~70
- Effective Beam size at sample position ~150 mm × 25 mm
- Detectors—film, image plate
- Imaging Options—Radiography.

** IMAGINE, ORPHEE is a cold neutron imaging beamline that is described in other section.*

4.3.12 France, Grenoble, ILL, NeXT, 58 MW, User Facility

The reactor is located at Institut Laue Langevin, Grenoble, France (Fig. 4.15). It is a high flux research reactor with highly enriched fuel, cooled and moderated with heavy water. The reactor achieved its criticality in July 1957. Multidisciplinary beamlines exist in this facility for wide areas of research in physics, chemistry, biology, and engineering along with the production of radioisotopes for medical and other applications. The reactor serves a wide user community across the world with their thermal and cold neutron sources. Neutron and X-ray Tomography (NeXT) beamline is a unique facility that hosts complimentary imaging setup: neutron and X-ray (as shown in Fig. 4.16) [39]. As per reports by Tengattini et al. the measured

Fig. 4.15 Institut Laue Langevin, Grenoble, France [39]

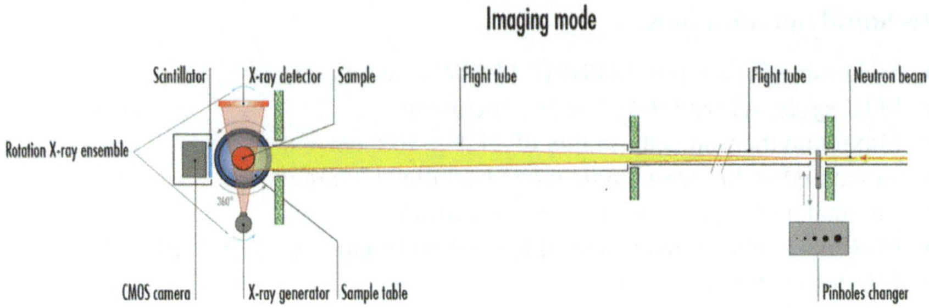

Fig. 4.16 NeXT beamline, ILL, Grenoble, France [40]. Reproduced with permission from ELSEVIER

flux at the end of the H521 guide is 1.4 × 10^{10} n-cm^{-2} s^{-1}, which makes this the highest neutron flux in the world [40].

Technical Specifications

- High flux research reactor, 58 MW research reactor
- Highly enriched fuel with 93% U-235
- Maximum thermal neutron flux of ~1.5 × 10^{15} cm^{-2} s^{-1}
- Tangential beamline alignment
- Thermal neutron energy spectrum
- Maximal beam intensity at sample position [n-cm^{-2} s^{-1}] ~3 × 10^8
- *L*/*D* collimation ratio ~330
- Effective Beam size at sample position ~30 mm diameter
- Detectors—film, scintillator-based CCD camera system
- Imaging Options—Neutron and X-ray Radiography and Tomography.

4.3.13 Hungary, DNR, 10 MW, User Facility

Budapest Research Reactor (BRR) is located at KFKI Science Campus in the capital city Budapest, Hungary. The reactor is a Soviet design VVR-type giving 10 MWth power. BRR is a tank-type reactor utilizing enriched uranium fuel, moderated and cooled by light water [10, 41]. It went critical for the first time on March 25, 1959. The reactor is operated by the Centre for Energy Research of the Hungarian Academy of Sciences. Many neutron-research related laboratories located in the KFKI campus participates in research activities under Budapest Neutron Centre (BNC) consortium formed in 1993 (Fig. 4.17). It is not only one of the leading research infrastructure in Hungary but also largest in Central-Europe. The facility has been widely used for neutron scattering investigations, neutron activation analysis and neutron imaging (Fig. 4.18).

Technical Specifications

Fig. 4.17 KFKI science campus and BRR hall, Budapest [42]

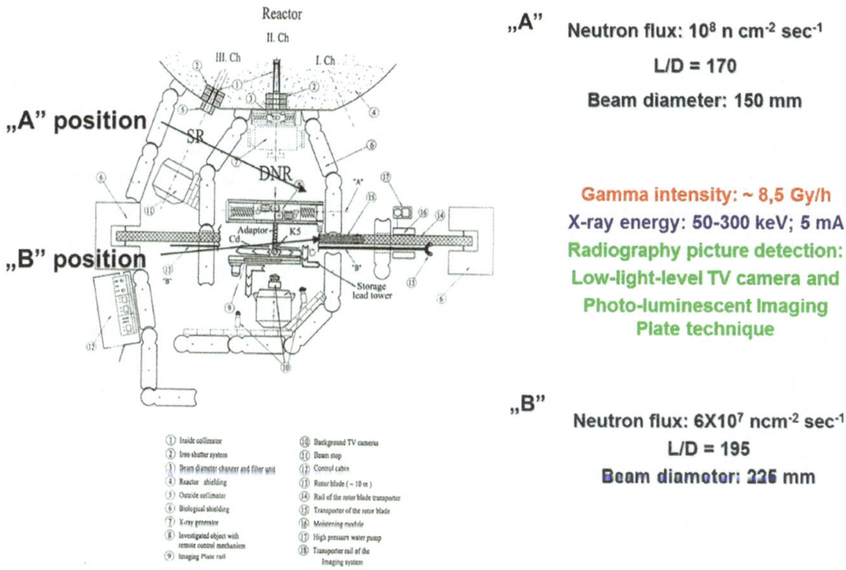

Fig. 4.18 Neutron, gamma and X-ray imaging setup at BRR [10]

- VVR Tank-type, DNR, 10 MW research reactor
- 20% lightly enriched Uranium VVR-M2 type fuel (LEU)
- Maximum thermal neutron flux of ~2.2×10^{14} cm^{-2} s^{-1}
- Radial beamline alignment
- Thermal neutron energy spectrum
- Maximal beam intensity at sample position [cm^{-2} s^{-1}] ~1.8×10^8
- Neutron/Gamma ratio ~7.82×10^7 (n-cm^{-2} per mGy)
- Variable inlet aperture—0.5–25 mm
- Variable *L*/*D* collimation ratio—170–600
- Variable effective beam size at sample position ~10–220 mm diameter
- Corresponding beam intensity—7×10^3–6×10^7 cm^{-2} s^{-1}

- Detectors—Scintillator-based CCD camera system, image plate, Low lux level TV camera with vidicon
- Imaging Options—Simultaneous Dynamic Neutron—and Gamma Radiography, time sequences, tomography, X-ray radiography.

Applications

- Research Topics—Study of the supercritical water, investigation of Fuel Cells in operation.
- Services for industry—Inspection of refrigerators, test of helicopter rotor blades.

4.3.14 Indonesia, GA Siwabessy MPR, 30 MW, In-House Usage

Reactor–Gerrit Augustinus Siwabessy is a multipurpose research reactor located in the Serpong, Tangerang, Indonesia. The reactor popularly known as BATAN is controlled and operated by BATAN, National Nuclear Energy Agency of Indonesia reactor (Fig. 4.19). It was designed and developed by Interatom GMBH from the Federal Republic of Germany. The reactor went critical in July 1987 and has played important role for nuclear industry development, radioisotope production, nuclear material science, reactor fuel element development, reactor safety, waste treatment, radio-metallurgy, nuclear-mechano laboratories, neutron imaging and teaching and training purpose [10].

Technical Specifications

- Swimming Pool-type, 30 MW MPR research reactor
- MTR-type enriched fuel
- Maximum thermal neutron flux of ~3×10^{14} n-cm^{-2} s^{-1}
- Tangential beamline alignment
- Thermal neutron energy spectrum
- Maximal beam intensity at sample position [n-cm^{-2} s^{-1}] ~10^7

Fig. 4.19 BATAN facility and neutron radiography beamline [10]

- Filter Options—Lead
- Neutron/Gamma ratio >10^5 (n-cm^{-2} per mGy)
- Cadmium ratio ~6.4
- *L*/*D* collimation ratio ~83
- Effective Beam size at sample position ~200 mm diameter
- Detectors—Film, scintillator-based CCD camera system
- Imaging Options—Radiography, tomography, time sequences.

Applications

Research Topics—Neutron radiography for examining automotive components and archaeological objects.

4.3.15 *Israel, IRR-1, 5 MW, In-House Usage*

The Israeli Research Reactor-1 (IRR-1) is located in Soreq Nuclear Research Center, Yavane, Israel. The reactor is a 5 MW pool-type light water reactor utilizing highly enriched uranium fuel. Construction on the IRR-1 began in January 1958 and the reactor reached criticality in June 1960. Major uses of this reactor include research and training in nuclear engineering, neutron radiography and tomography [10, 43], and diffraction, activation analysis, and changing colours of semi-precious and precious stones.

Technical Specifications

- Swimming Pool-type, 5 MW research reactor
- MTR-type fuel with 93% U-235 enrichment
- Maximum thermal neutron flux of ~7×10^{13} cm^{-2} s^{-1}
- Tangential beamline alignment
- Thermal neutron energy spectrum
- Maximal beam intensity at sample position [cm^{-2} s^{-1}] ~1×10^6
- *L*/*D* collimation ratio ~250
- Effective Beam size at sample position ~98 mm × 98 mm
- Detectors—Film, scintillator-based CCD camera system, Image plate, Amorphous Si flat panel detector
- Imaging Options—Radiography, tomography.

4.3.16 *Italy, TRIGA RC-1, 1 MW, In-House Usage*

The reactor is located at ENEA institute, Rome, Italy. It is a 1 MW TRIGA Mark-II research reactor that became critical in June 1960. The reactor is used for neutron activation, neutron radiography [10], training purpose.

Technical Specifications

- Swimming pool type, TRIGA RC-1, 1 MW
- Radial/Tangential Beam line alignment
- Thermal neutron energy spectrum
- Maximal beam intensity at sample position [cm^{-2} s^{-1}] ~2.2 × 10^5 (radial)
  ~2.5 × 10^6 (tangential)
- Inlet aperture—40 and 200 mm
- *L*/*D* collimation ratio ~108 and 116
- Effective beam size at sample position ~38 and 180 mm diameter
- Detectors—Film, scintillator-based CCD camera system
- Imaging Options—Radiography, Tomography, Time sequences.

4.3.17 Japan, JRR-3, 20 MW—User Facility

The reactor, which comes under Japan Atomic Energy Agency, is located in the Tokai, Naka-gun, Ibaraki, Japan. JRR3 achieved its criticality in 1962 (Fig. 4.20). However, it was modified in 1990 and renamed as JRR-3M. The reactor is a high-performance multipurpose research reactor with thermal power of 20 MW [10, 44]. This is a pool type light water moderated and cooled research reactor using low enriched aluminide fuels. The reactor has been utilized for nuclear research as well as industry. The areas of utilization of JRR-3 include neutron beam experiments, irradiation experiments for nuclear fuel and material, and production of radio isotopes and silicon semiconductors. Cold neutron beams are also available and utilized for research of life phenomena by analyzing the structure of polymer molecules.

Technical Specifications

- Swimming Pool-type, 20 MW multipurpose research reactor
- low enriched uranium aluminide fuel
- Maximum thermal neutron flux at core ~3 × 10^{14} n-cm^{-2} s^{-1}

Fig. 4.20 JRR-3 reactor building and neutron radiography facility, Japan [10, 44]

- Tangential beamline alignment
- Thermal neutron energy spectrum
- Maximal beam intensity at sample position [n-cm^{-2} s^{-1}] ~1.5 × 10^8
- Inlet aperture—39 and 48 mm
- *L*/*D* collimation ratio ~185 and 154
- Effective beam size at sample position ~255 and 305 mm diameter
- Detectors—Scintillator-based CCD camera system, Image plate, Neutron colour image intensifier
- Imaging Options—Radiography, tomography, time sequences.

Applications

- Research Topics—Fuel cell, Architecture, Fluid
- Services for industry—Fuel cell.

4.3.18 South Korea, HANARO, 30 MW—User Facility

The High Flux Advanced Neutron Application Reactor (HANARO) is located at Daejeon, Republic of Korea. It is a 30 MW multipurpose research reactor designed by Korea Atomic Energy Research Institute (KAERI) as a facility for research and development on the neutron science and its applications [10, 45]. The reactor reached its criticality in February 1995. It is an open pool type heavy water moderated and light water-cooled reactor that uses low enriched fuel. HANARO as a national as well user facility (as shown in Fig. 4.21) has played significant role in the area of neutron science, radioisotopes production, material testing, neutron transmutation doping (NTD), neutron activation analysis, and neutron radiography. The installation of a cold neutron source in 2010, has been also serving as a regional and international facility for neutron science.

Fig. 4.21 HANARO reactor building and hall, South Korea [46]

Technical Specifications

- Open pool-type, 30 MW multipurpose research reactor
- U_3Si, 19.75% enriched
- Maximum thermal neutron flux of ~4×10^{14} n-cm^{-2} s^{-1}
- Tangential beamline alignment
- Thermal neutron energy spectrum
- Maximal beam intensity at sample position [n-cm^{-2} s^{-1}] ~2×10^7
- *L*/*D* collimation ratio ~278
- Effective beam size at sample position ~350 mm × 450 mm
- Detectors—Film; Scintillator-based CCD camera system
- Imaging Options—Radiography, tomography.

Applications

- Research Topics—Fuel cell, Li-Ion Battery, Non-destructive testing for aircraft, Dynamic radiography for heat exchanger
- Services for industry—Fuel cell and Li-Ion Battery.

4.3.19 Malaysia, TRIGA II PUSPATI, 1 MW—In-House Usage

The PUSPATI reactor, owned by Malaysian Nuclear Agency, is located Selangor, Kajang in Malaysia [10, 47–49]. Also known as RTP, the reactor started its operation in 1982 and reached its first criticality in July 1982. It is a 1 MW TRIGA Mark-II pool type reactor that uses light water moderator and coolant, and graphite reflector (Fig. 4.22). The reactor is intended for medical, industrial, and agricultural radioisotope generation, as well as neutron radiography and small angle neutron scattering.

Fig. 4.22 TRIGA II PUSPATI 1 MW facility and reactor hall with neutron imaging beamline, Malaysia [47]

Technical Specifications

- Pool type, TRIGA II PUSPATI, 1 MW Research reactor
- Maximum thermal neutron flux of ~1×10^{13} n-cm^{-2} s^{-1}
- Radial Beam line alignment
- Fast and Thermal neutron energy spectrum in the ratio of 1:3
- Maximal beam intensity at sample position [n-cm^{-2} s^{-1}] ~1×10^5
- Neutron/gamma ratio—1.02×10^6 n-cm^{-2} mR^{-1}
- Cadmium Ratio—10.51
- Filter options—Bi
- *L*/*D* collimation ratio ~75
- Effective Beam size at sample position ~80 and 150 mm diameter
- Detectors—Film based; Scintillator-based CCD camera system
- Imaging Options—Radiography, tomography.

Applications

- Research topics—Non-metallic materials discontinuities assessment, Cultural heritage objects characterization.

4.3.20 Morocco, MA-R1, 2 MW—In-House Usage

The MA-R1 research reactor is located in Morocco's National Center for Nuclear Energy, Sciences, and Technology's Maâmora Nuclear Research Centre (CENM) in Rabat (CNESTEN) [10, 27]. As illustrated in Fig. 4.23, it is a 2 MW TRIGA Mark-II pool reactor with a light water moderator and coolant, as well as a graphite reflector. The reactor achieved first criticality in 2007 and has a wide range of capabilities,

Fig. 4.23 MA-R1, 2 MW building and reactor hall, Morocco [10, 27]

including radioisotope production for medical, industrial, and environmental applications, metallurgy, chemistry, and the use of nuclear analytical techniques such as NAA and non-destructive neutron radiography examination techniques.

Technical Specifications

- Pool type, TRIGA II MA-R1, 2 MW Research reactor
- Maximum thermal neutron flux of ~1.03×10^{13} n-cm^{-2} s^{-1}
- Tangential Beam line alignment
- Thermal neutron energy spectrum
- Maximal beam intensity at sample position [n-cm^{-2} s^{-1}] ~1×10^{6}
- Filter options—Bi, Sapphire
- *L*/*D* collimation ratio ~75
- Effective Beam size at sample position ~80 and 150 mm diameter
- Detectors—Film based; Scintillator-based CCD camera system
- Imaging Options—Radiography, tomography.

Applications

- Research topics—Non-metallic materials discontinuities assessment, Cultural heritage objects characterization.

4.3.21 Poland, Maria, 30 MW—In-House Usage

The Maria research reactor is named after Maria Skodowska-Curie and is located near Warsaw in Wierk-Otwock (Fig. 4.24). It is the sole reactor of Polish design, built by the Institute of Nuclear Research in 1970. In December 1974, the reactor reached criticality. The high flux research reactor is a pool-type reactor with a graphite reflector and pressurized channels that is water and beryllium moderated [10, 50, 51].

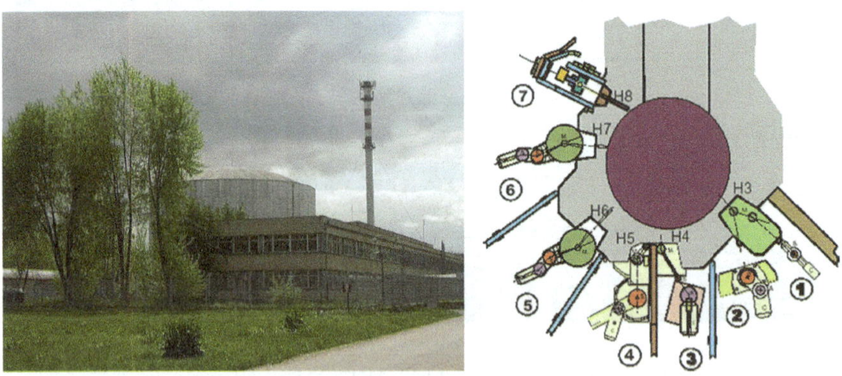

Fig. 4.24 Maria research reactor facility and beamlines arrangement in reactor hall: 1–6: neutron diffraction beamlines, 7: neutron and gamma radiography beamline [10, 50]

Maria is a multifunctional research tool with notable applications in radioisotope production, fuel and structural material testing for nuclear power engineering, neutron transmutation doping of silicon, neutron modification of materials, research in neutron and condensed matter physics, neutron therapy, neutron activation analysis, neutron radiography, and reactor physics and technology training.

Technical Specifications

- Pool type, 30 MW multifunctional research reactor
- U_3Si_2, 20% enriched
- Maximum thermal neutron flux of ~3 × 10^{14} n-cm^{-2} s^{-1}
- Radial Beam line alignment
- Thermal neutron energy spectrum
- Maximal beam intensity at sample position [n-cm^{-2} s^{-1}] ~1.5 × 10^7
- *L*/*D* collimation ratio ~170
- Effective Beam size at sample position ~180 mm diameter
- Detectors—Scintillator-based CCD camera system
- Imaging Options—Radiography, tomography, time sequences.

Applications

- Research topics—Water migration in porous media (imbibition, drying)
- Service for industry—Tomography of industrial samples.

4.3.22 Portugal, RPI, 2 MW—In-House Usage

The RPI reactor that is part of Instituto Superior Técnico, is situated in Lisbon, Portugal [10, 52, 53]. It is a 1 MW light water open pool type reactor constructed by AFM Atomics, US (Fig. 4.25). The reactor became critical in April 1961 and has facilities such as radioisotopes production, detector testing, fast neutron irradiation for electronic components, neutron tomography and training in the field of reactor physics and technology.

Technical Specifications

- Pool type, 1 MW research reactor
- MTR-type 20% enriched fuel
- Maximum thermal neutron flux of ~2.5 × 10^{13} n-cm^{-2} s^{-1}
- Radial Beam line alignment
- Thermal neutron energy spectrum
- Maximal beam intensity at sample position [n-cm^{-2} s^{-1}] ~2 × 10^5
- *L*/*D* collimation ratio ~40
- Effective Beam size at sample position ~60 mm diameter
- Detectors—Scintillator based CCD camera system
- Imaging Options—Radiography, tomography, real time.

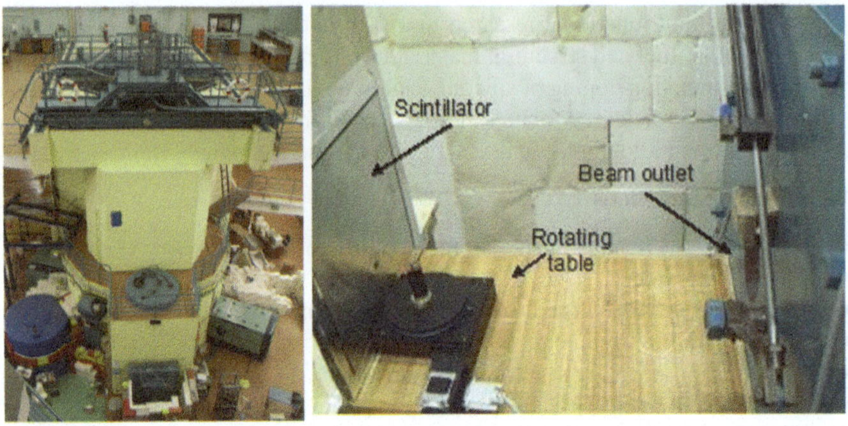

Fig. 4.25 RPI reactor hall and neutron tomography facility, Portugal [10, 52]

Applications

- Research topics—Archeological.

4.3.23 Romania, TRIGA 14 MW, INUS and INUM—In-House Usage

The reactor is located in the Institute of Nuclear Research (INR) in Pisteti, Romania, as seen in Fig. 4.26. At the INR, there are two high-intensity neutron sources: the 14 MW TRIGA research reactor and the TRIGA ACPR (Annular Core Pulsed Reactor). Both reactors are open-pool reactors. The reactors have been used for neutron activation studies, neutron scattering, neutron radiography, radioisotope generation, and material research [10].

Technical Specifications

- Pool type, TRIGA 14 MW and TRIGA ACPR
- U-ZrH fuel, 20% enriched uranium
- Maximum thermal neutron flux of ~1×10^{12} n-cm^{-2} s^{-1}
- Tangential Beam line alignment for INUS and under water for INUM
- Thermal neutron energy spectrum
- Maximal beam intensity at sample position [n-cm^{-2} s^{-1}] ~5.4×10^6 in continuous mode
- *L*/*D* collimation ratio ~95 for INUS and ~227 for INUM
- Effective Beam size at sample position ~290 mm diameter for INUS and ~385×100 mm rectangular for INUM
- Detectors—Film based; Scintillator based CCD camera system
- Imaging Options—Radiography, tomography.

Fig. 4.26 TRIGA 14 MW reactor facility, Romania [54]

Applications

Research topics—Archaeological, Fresh and spent nuclear fuel.

4.3.24 *Russian Federation, IBR-2 Pulsed Reactor, 2 MW Average—User Facility*

The IBR-2 reactor is a pulsed fast reactor located at Frank Laboratory of Neutron Physics, Dubna, Russia (Fig. 4.27). This reactor is part of one of the laboratories of the Joint Institute for Nuclear Research [10, 55, 56]. It is a 2 MW average power reactor that operates at 5 Hz repetition frequency with pulse duration of 350 μs of thermal neutrons using movable reflector as shown in Fig. 4.28. The reactor has been used for particle physics and condensed matter research, molecular biology, pharmacology, engineering diagnostics and neutron imaging. This reactor's pulsed operation is appealing not only for conventional neutron imaging application, but also for the development of contemporary energy-selective techniques for time-of-flight measurements.

Technical Specifications

- Pulsed Fast reactor, 2 MW average power (1850 MW pulsed)
- PuO_2 fuel
- Maximum thermal neutron flux of ~10^{13} n-cm^{-2} s^{-1} time averaged and 10^{16} n-cm^{-2} s^{-1} burst
- Radial Beam line alignment

Fig. 4.27 IBR-2 reactor building, Russia [57]

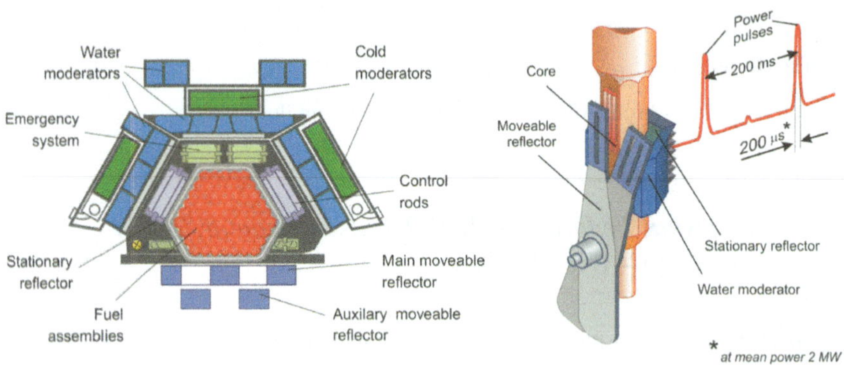

Fig. 4.28 Core of the IBR-2 reactor with a movable reflector [58]

- Thermal neutron energy spectrum
- Maximal beam intensity at sample position [n-cm^{-2} s^{-1}] ~5.6 × 10^6
- Filter options—Bi
- *L*/*D* collimation ratio ~200
- Effective Beam size at sample position ~200 mm × 200 mm
- Detectors—Scintillator based CCD camera system
- Imaging Options—Radiography, tomography, time sequence.

Applications

Research topics—Geology, Materials science, Planet science
Services for industry—non-destructive testing of various industrial components and products.

4.3.25 *Slovenia, Triga, Mark-II, 0.25 MW—In-House Usage*

The TRIGA Mark II research reactor is located at Jožef Stefan Institute (JSI) in Ljubljana, Slovenia. It achieved its first criticality on 31st May 1966 (Fig. 4.29). Since then, the reactor has played an essential part in Slovenia's development of nuclear technology and safety culture. It is one of the country's few modern technology facilities. The reactor has primarily been used for isotope production, neutron activation analysis, beam applications, neutron radiography, testing and development of a digital reactivity meter, verification of computer codes and nuclear data, primarily criticality calculations and neutron flux distribution studies [10, 59].

Technical Specifications

- Pool type, TRIGA 0.25 MW
- Thermal neutron energy spectrum
- Maximal beam intensity at sample position [n-cm^{-2} s^{-1}] ~4.5 × 10^5
- *L*/*D* collimation ratio ~65–80

Fig. 4.29 TRIGA 0.25 MW reactor hall, Slovenia [60]

- Effective Beam size at sample position ~120 mm diameter
- Detectors—Film based; Image Plate detector
- Imaging Options—Radiography.

4.3.26 South Africa, SAFARI-1, 20 MW—User Facility

South African Nuclear Energy Corporation (NECSA) owns and operates the **SAFARI-1** reactor at its site in Pelindaba, South Africa (Fig. 4.30). It is a 20 MW light water-cooled, beryllium reflected, pool-type high flux research reactor that went critical in March 1965. The reactor is primarily used for commercial production of medical and industrial isotopes, activation analyses, material modification (such as neutron transmutation doping of silicon for the semiconductor industry), and a variety of support services such as neutron radiography and neutron diffraction, which are of both industrial and academic interest [10, 27, 61].

Technical Specifications

- Pool type, 20 MW high flux research reactor
- MTR-type 20% enriched fuel
- Maximum thermal neutron flux of ~1×10^{14} n-cm^{-2} s^{-1}
- Radial Beam line alignment
- Thermal neutron energy spectrum
- Maximal beam intensity at sample position [n-cm^{-2} s^{-1}] ~1×10^{7} for *L*/*D*: 125
- Filter options—Bi
- Variable Inlet aperture—5–21 mm
- *L*/*D* collimation ratio ~525–125
- Effective Beam size at sample position ~130–300 mm diameter
- Detectors—Film based, Scintillator based CCD camera system
- Imaging Options—Radiography, tomography.

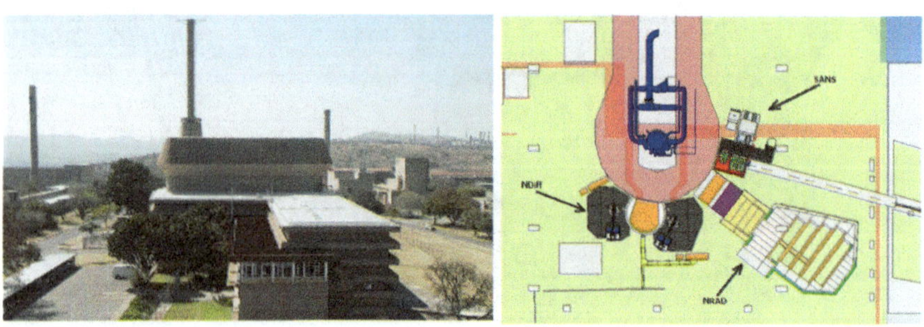

Fig. 4.30 SAFAR-1 reactor facility and beamlines layout, South Africa [10, 27, 61]

Applications

- Research topics—Archaeology, Paleontology, Porous media Civil Engineering Reverse Engineering Geosciences
- Service for industry—Geosciences, NDT.

4.3.27 *Thailand, TRIGA Mark III, 1 MW—User facility*

The Thai Research Reactor1/Modification 1 (TRR-1/M1) that went critical in November 1977, is located in Bangkok, Thailand (Fig. 4.31). It has been a key player in the establishment of the Office of Atomic Energy for Peace (OAEP) as well as nuclear applications in Thailand. The reactor has been used for radioisotope generation, neutron activation analysis, and neutron beam research, such as neutron scattering, prompt gamma analysis, neutron radiography, and training [10].

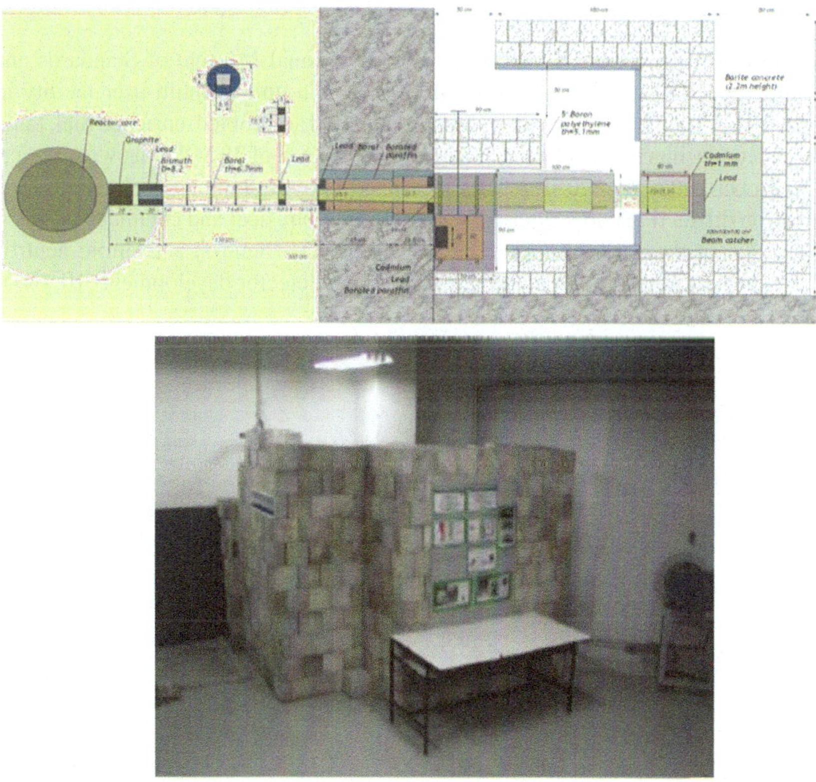

Fig. 4.31 Neutron radiography facility, Thailand [10]

Technical Specifications

- Pool type, TRIGA 1 MW, TRR-1/M1
- Maximum thermal neutron flux of ~3.1×10^{13} n-cm^{-2} s^{-1}
- Tangential Beam line alignment
- Thermal neutron energy spectrum
- Maximal beam intensity at sample position [n-cm^{-2} s^{-1}] ~1.1×10^6
- Filter Options—Pb, Bi, Graphite
- *L*/*D* collimation ratio ~50
- Effective Beam size at sample position ~200 mm × 200 mm
- Detectors—Film based; Scintillator based CCD camera system, Image plate
- Imaging Options—Radiography, tomography.

Applications

Research topics—Archaeological.

4.3.28 USA, NBSR-NIST, 20 MW—User Facility

Neutron Beam Split Core Reactor (NBSR) at National Institute of Standards and Technology (NIST) in Gaithersburg, MD, USA is a unique multi-user facility as shown in Fig. 4.32. It is heavy water (D_2O) moderated and cooled, enriched fuel, tank-type reactor designed to operate at a thermal power level of 20 MW, that went critical in December 1967. The NBSR reactor is utilized for wide-ranging applications such as neutron diffraction and scattering studies both in physical and chemical sciences, residual stress measurement, fuel cell research, NDT, neutron activation, neutron radiography, high-resolution neutron imaging, and detector development [10, 62].

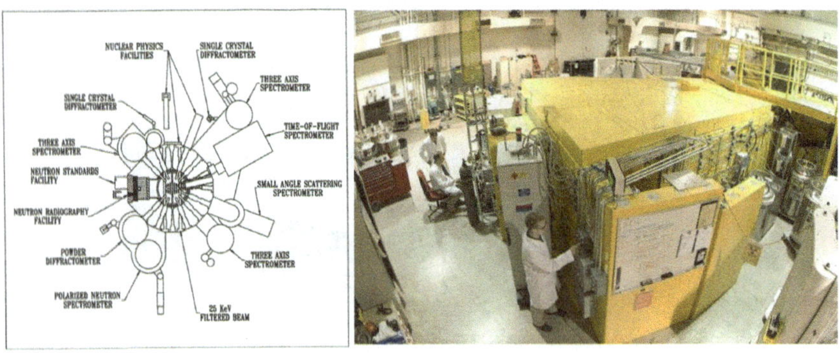

Fig. 4.32 Neutron beamlines and imaging facility at NBSR, USA [10, 62]

Technical Specifications

- Tank type, 20 MW reactor
- MTR-type enriched fuel
- Maximum thermal neutron flux of ~1×10^{14} n-cm^{-2} s^{-1}
- Radial Beam line alignment
- Thermal neutron energy spectrum
- Maximal thermal beam intensity at sample position [n-cm^{-2} s^{-1}] ~3×10^7
- Filter options—cooled Bi
- Cadmium ration >20
- Variable Inlet aperture—10, 15 and 20 mm option circular
  10 mm × 1 mm–20 mm × 2 mm rectangular
- *L*/*D* collimation ratio ~600–300 for circular
  ~6000 × 600–3000 × 300 for rectangular apertures
- Effective Beam size at sample position ~260 mm
- Detectors—Scintillator based CCD camera system, flat panel detector, microchannel plate
- Imaging Options—Radiography, tomography, phase contrast imaging, time sequence.

Applications

- Research topics—Neutron focusing optics, neutron detection, proton exchange membrane fuel cells, lithium batteries, two-phase flow, porous media, heat pipes, image analysis
- Service for industry—Free Collaborative beam time access and Proprietary access available via full cost recovery.

4.3.29 US, OSTR, TRIGA Mark II Research Reactor Oregon State, 1.10 MW—In-House Usage

The reactor is located in Corvallis, Oregon, at the Oregon State University Radiation Center. Oregon State University is one of the few educational institutions in the United States with a TRIGA research reactor. The Oregon State TRIGA Reactor (OSTR) is a water-cooled research reactor with a circular grid array of fuel components made of low-enriched uranium/zirconium-hydride. It can operate in both continuous and pulsed modes, with a continuous power output of 1.10 MW and a pulsed power output of 3000 MW [10]. OSTR's uses include student teaching, research, isotope synthesis, neutron irradiation, and neutron imaging, to name a few.

Technical Specifications

- Pool type, TRIGA 1.10 MW
- Tangential Beam line alignment
- Thermal neutron energy spectrum

- Maximal thermal neutron flux [n-cm^{-2} s^{-1}] ~1 × 10^{13}
- Maximal beam intensity at sample position [n-cm^{-2} s^{-1}] ~4.4 × 10^{6} at *L*/*D* of 101
- Filter Options—Bi
- *L*/*D* collimation ratio ~86 and 117
- Effective Beam size at sample position ~44 mm × 52 mm, 172 mm × 229 mm and 360 mm × 440 mm
- Detectors—Scintillator based CCD camera system, Image plate, Neutron sensitive microchannel plate
- Imaging Options—Radiography, tomography, time sequences.

Applications

- Research topics—Two-phased flow determination, Fuel Imaging
- Services for industry—MMC Material Boron Content.

4.3.30 USA, UC Davis, TRIGA Mark III, 1 MW

The reactor is located at University of California, Davis in the USA. It is a research reactor dedicated mainly for radiographic and irradiation studies. There are dedicated four beamlines neutron radiography [10].

Technical Specifications

- Pool type, TRIGA Mark III, 1 MW
- Four beams, All tangential Beam line alignment
- Thermal neutron energy spectrum
- Maximal beam intensity at sample position 4 stations [n-cm^{-2} s^{-1}] ~6 × 10^{6}, 7 × 10^{6}, 1.5 × 10^{7}, 1.5 × 10^{7}
- Filter Options—Bi
- *L*/*D* collimation ratio ~175, 200 and 300
- Effective Beam size at sample position ~500 mm diameter, 600 mm diameter and 450 mm square
- Detectors—Film based, Scintillator based CCD camera system, Image plate
- Imaging Options—Radiography, tomography.

Applications

- Research topics—PEM fuel cell design, Geology
- Services for industry—Turbine blades, pyrotechnical devices.

4.4 Cold Neutron Imaging Facilities

Cold neutron source-based imaging beamlines were setup at very few places across the world. The facilities are mainly located in developed countries. The use of cold neutrons in radiography enhances picture contrast and sensitivity for detecting tiny quantities of water and hydrogen-containing compounds in metal matrixes, for example. The cold neutron beam, on the other hand, may be easily changed using diffraction and neutron optical methods. This permits more advanced measurement techniques to be used in radiography and tomography investigations. These are mainly user-type facilities.

4.4.1 IMAGINE: ORPHEE, France—User Facility

IMAGINE is a cold neutron imaging station builds around ORPHEE reactor at the Laboratoire Léon Brillouin laboratory [10, 63]. As illustrated in Fig. 4.33, the station is intended for high-resolution neutron imaging and tomography. The facility is equipped with a variety of auxiliary equipment that allows for in-situ measurements as well as high-resolution tomographic measurements.

A highly curved guide (G3bis) that points at an H2 cold source provides the neutrons (20 K). The wavelength spectrum distribution begins at 3 Å because of the high curvature of the guide. The highest flux is at 4 Å, and the flux declines roughly in a $1/\lambda^3$ law above this wavelength. The high guide curvature also assures that the final beam used for experiments contains no fast neutrons or gammas coming from the reactor core is at 4 Å and above this wavelength the flux decreases roughly following a $1/\lambda^3$ law. The strong guide curvature also ensures that there are neither fast neutrons nor gammas originating from the reactor core in the final beam utilized for experimentation. Cold neutron flux at the sample position is 2×10^7 n/cm^2/s with a *L*/*D* ratio of ~200 and signal to noise ratio of ~2000. However, *L*/*D* ratio can also be increased to ~1000.

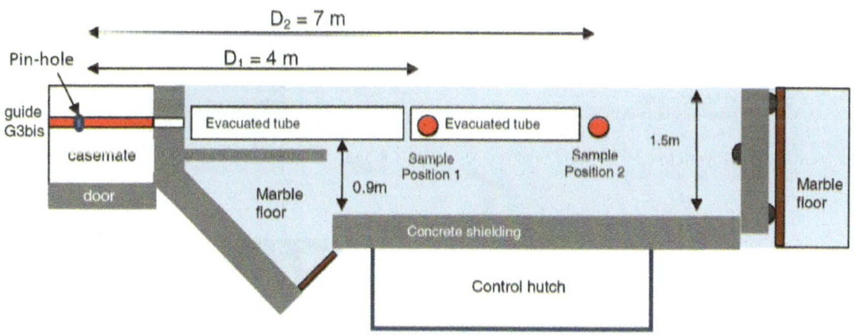

Fig. 4.33 Schematic: IMAGINE cold neutron imaging beamline, France [63]. Reproduced with permission from ELSEVIER

4.4.2 Germany, BER II, 10 MW (CONRAD and PONTO Instruments)—User Facility

The BER II, 10 MW research reactor located at Hahn-Meitner-Institut, Berlin was in service for 40 years before it was permanently shut down in 2019. However, it is important to describe this unique multi-disciplinary facility. Imaging at this facility was performed using cold neutron and polarized cold neutron beams (Fig. 4.34).

The new cold neutron radiography apparatus CONRAD, as depicted in Fig. 4.35, was a multifunctional facility for radiography and tomography utilizing cold neutrons at the BER II, 10 MW research reactor [10, 64, 65].

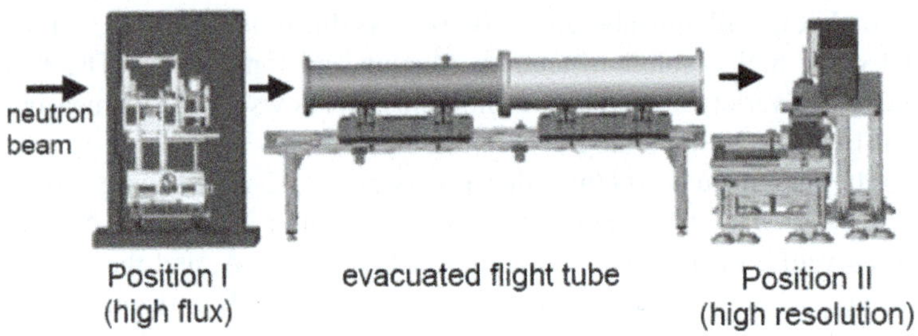

Fig. 4.34 The CONRAD imaging beamline, Germany [64]. Reproduced with permission from ELSEVIER

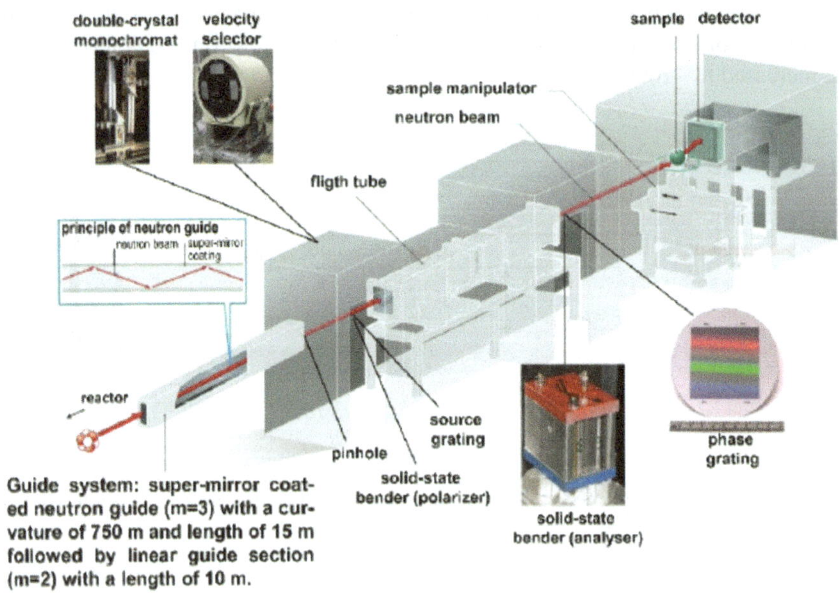

Fig. 4.35 The CONRAD-2 instrument setup [65]

The apparatus was constructed in 2005 and is placed at the end of a curved neutron guide that faces the BER-II research reactor's cold neutron source. The geometry yields a cold neutron beam with wavelengths ranging from 2 to 12 nm. For radiography and tomography studies, there are two measurement locations accessible. The first, which is at the end of the guide, was designed for in-situ investigations that demand a strong neutron flux. At this location, the available flux was ~10^9 n-cm^{-2} s^{-1}.

It was built in 2005 and was located at the end of a curved neutron guide, which faces the cold neutron source of the BER-II research reactor. The geometry provides a cold neutron beam with wavelengths between 2 and 12 Å. Two measuring positions are available for radiography and tomography investigations. The first one is placed at the end of the guide and it was optimized for in-situ experiments in which a high neutron flux is required. The available flux at this position was ~10^9 n-cm^{-2} s^{-1}. The second position has a pin-hole that allows for better beam collimation (L/D up to 1000) and higher image resolution in the 100 μm range, using CCD-based detector system.

The cold neutron source was changed, and the neutron guide system supporting the instruments in neutron guide hall was entirely rebuilt and upgraded, during the cold neutron instrumentation upgrade at BER-II from October 2010 to October 2012. The CONRAD instrument (now known as CONRAD-2) was moved to a new location in the facility with a 10-m collimation path [65]. The beam divergence was considerably reduced by replacing current neutron guides ($m = 1.2$) with new supermirror guides ($m = 2$). The device's adjustments increased the size of the usable beam and enhanced neutron transport efficiency. Furthermore, the curvature of the guide was improved by reducing its radius from $R = 3000$ m to $R = 750$ m in order to improve the distance between the shielding of neighbouring instruments and provide a more spacious experimental and user environment. Neutron tomography, energy-selective imaging, imaging with polarized neutrons, high-resolution radiography, and grating-interferometry were among the imaging modalities accessible at this facility.

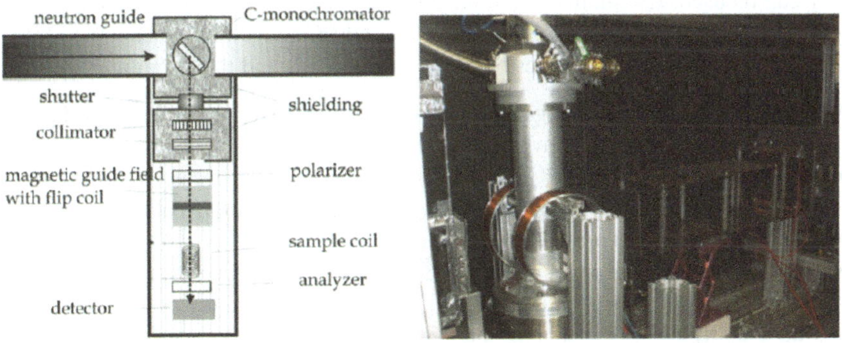

Fig. 4.36 The PONTO instrument, Germany [66]. Reproduced with permission from ELSEVIER

At the same facility, the PONTO equipment, as shown in Fig. 4.36, was dedicated to polarized cold neutron radiography and tomography [10, 66]. For a point-to-detector distance of 5 mm, the spatial resolution for un-polarized neutrons was less than 85 μm, and for polarized neutrons, it was less than 300 μm. These resolutions can be used to image magnetic fields in numerous lead samples below the critical temperature for the Meissner phase and magnetic flux pinning.

4.4.3 Germany, FRM-II (ANTARES), 20 MW—User Facility

The Heinz Maier-Leibnitz neutron source in Germany (Forschungsreaktor München II or FRM II) is one of the most powerful and advanced neutron sources in the world. It is named after physicist Heinz Maier-Leibnitz, who ran a highly successful research programme at the FRM I, its predecessor. It is run by the Technical University of Munich and is located on its Garching campus. FRM-II is a 20 MW reactor with a cold neutron source flux of 8×10^{14} n/cm^2/s at 3.5 Å.

The advanced neutron tomography and radiography experimental system (ANTARES) is installed at the FRM-II reactor's cold neutron beam port SR-4a, as shown in Fig. 4.37 [67]. At the beamport, the facility has a pinhole-based variable collimator that can be used for a variety of high-resolution and high flux imaging applications. In chambers 2 and 3, ANTARES offers two distinct detector positions, which can be selected based on sample size, beam size, neutron flux, and spatial resolution requirements. Built-in options such as a velocity selector, double crystal monochromator, interference gratings, and a Be-filter are also readily available for standard user operation on ANTARES. The *L*/*D* ratio can be set anywhere between 100 and 3500. At the sample position, the average cold neutron flux varies from 4×10^5 to 4×10^8 n-cm^{-2} s^{-1}. The ANTARES neutron imaging facility is designed to produce radiographs and computed tomography of materials. Because neutrons can penetrate metals (such as Fe, Al, and Pb) and have a high sensitivity for hydrogen, they can be used to visualize metal machine parts, as well as liquids, sealants, and polymers within them. Crack and void detection is done with liquid contrast agents. In this beamline, some examples of diverse approaches and their common uses are:

- **Standard neutron radiography**: Moisture in sandstone, Rubber gaskets in machine parts, aerospace pyrotechnical components, fuel cells
- **Computed tomography**: Geological samples, mineral phases, voids in carbon fiber structures (using contrast agents), machine parts, biological tissues
- **Continuous radioscopy**: Real-time radiography
- **Stroboscopic imaging**: Visualization of repetitive processes with high time resolution like oil distribution in running combustion engines
- **Phase contrast**: Edge enhancement in aluminium foams, interfaces of similar alloys
- **Energy/wavelength scan**: Scanning for Bragg edges, phase or material identification for welds investigation

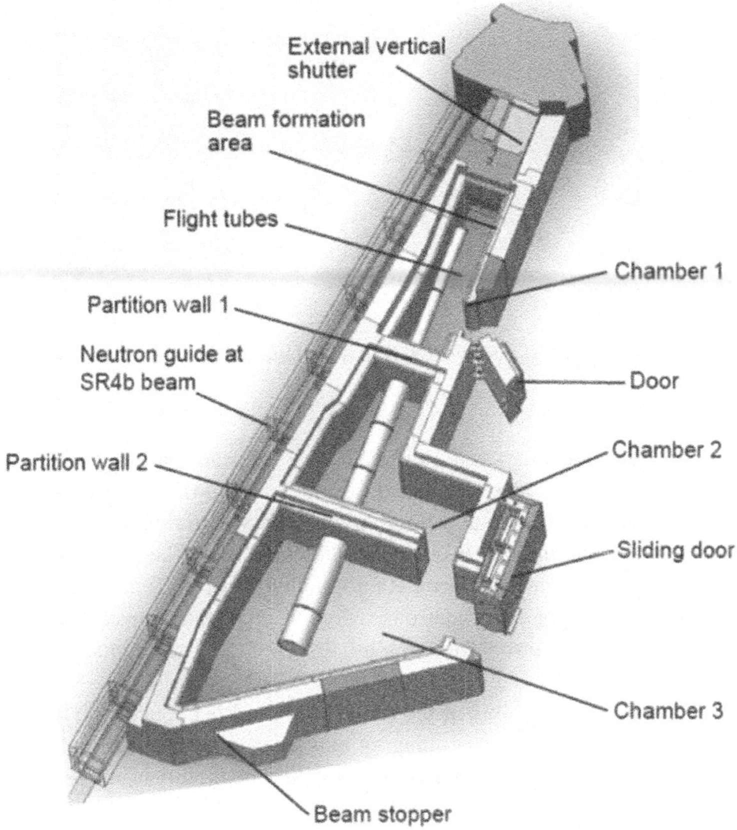

Fig. 4.37 ANTARES facility at FRM II, Garching, Germany [67]. Reproduced with permission from ELSEVIER

- **Polarized neutron imaging**: Metallurgical homogeneity of ferromagnetic materials, fundamental research on ferromagnetic phase transitions, visualization of magnetic field profiles
- **Neutron grating interferometry**: Measurement of the spatially resolved SANS or USANS signal of the sample. Detection of microstructures on length scales of 500 nm–10 μm, porous materials, magnetic and superconducting vortex lattice domains.

4.4.4 Hungary, NORMA, 10 MW Cold Neutron Source—User Facility

The 10-MW Budapest Research Reactor has a cold-neutron source (CNS) that has been operational since February 2001 [10]. It involves a 'direct-cooling' moderator

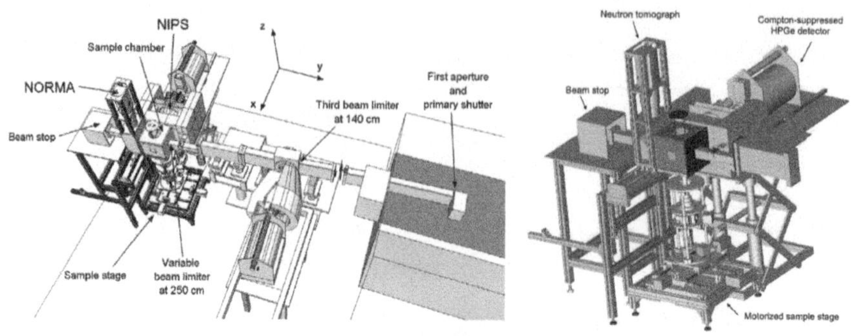

Fig. 4.38 NIPS and NORMA beamline at BPR [10, 69]. Reproduced with permission from ELSEVIER

assembly with the helium cryogenic system [68]. This beamline has two facilities, as shown in Fig. 4.38 [69], Neutron-Induced Prompt Gamma-ray Spectroscopy (NIPS) and imaging—Neutron Optics and Radiography for Material Analysis (NORMA). These can be used separately or in combination with one another. Both conventional PGNAA measurements and Neutron Radiography/Neutron Tomography imaging can be carried out using cold neutrons. More sophisticated investigations using Neutron radiography/Neutron tomography driven PGAI technique can also be performed.

The cold neutron source flux at the sample position is about ~2×10^7 n/cm^2/s. High-resolution and high-contrast imaging may be accomplished with an effective square beam size of 40 mm and an *L*/*D* ratio of 270. The facility is utilized for a wide range of research, including energy production and conversion, art, geology, and material science, object analysis and imaging, homogeneity of materials, coatings, and elemental characterization.

4.4.5 USA, ORNL, HFIR, 85 MW Cold Neutron Source—User Facility

The High Flux Isotope Reactor (HFIR), which runs at 85 MW, is the highest flux reactor-based source of neutrons for research (Fig. 4.39). It is located in Oak Ridge National Lab, USA, and produces one of the world's highest steady-state neutron fluxes [10, 70]. The high neutron flux, constant power density, and constant-length fuel cycles are used for neutron scattering studies of condensed matter. As shown in Fig. 4.40, this facility also performs neutron activation, neutron radiography [71, 72], and cold neutron tomography. The *L*/*D* ratio for neutron imaging may be set to 480 or 725, with an average cold neutron flux of 7.5×10^6 n-cm^2 s at the sample location. This beamline has been utilized in the field of energy storage, biomedical, materials science, engineering, geosciences, and cultural heritage.

Fig. 4.39 (HFIR), Oak Ridge National Lab, USA [70]

Fig. 4.40 Schematic view of neutron imaging facility at HFIR, ORNL [71]

4.5 New Facilities Using Research Reactors in Developing Countries

Radiographic imaging with cold neutrons is being proposed for the China Advanced Research Reactor (CARR). The China Institute of Atomic Energy's (CIAE) 60 MW CARR attained full capacity in March 2012. It's a tank-in-pool reactor using a D2O reflector. The expected optimal undisturbed thermal neutron flux of CARR is 8×10^{14} n-cm^2 s. However, the expected thermal neutron flux at sample for *L*/*D* of 293 and 585 are 2×10^8 n-cm^2 s and 5×10^7 n-cm^2 s respectively [73]. In the guiding hall, a cold neutron imaging facility will be installed. Its conceptual and practical designs are complete at this moment [74]. The cold neutron imaging facilities will be a flexible and cost-effective tool for non-destructive research in both fundamental science and industry.

4.6 India on Global Map

Neutron imaging facilities in India were built around research reactors. The program of imaging with thermal neutrons was started in mid-1990 initially with a low neutron flux beam at APSARA reactor, BARC and KAMINI reactor, IGCAR, Kalpakkam. Later on, the new and advanced imaging facilities were commissioned at CIRUS and DHRUVA reactor, BARC. The following sections describe the neutron imaging facilities commissioned in India.

4.6.1 Apsara Reactor Imaging Beamline

Apsara, India's first reactor, was developed at BARC in Mumbai in 1956 to undertake fundamental nuclear physics research (Fig. 4.41). It was light water-cooled and moderated swimming pool type 1 MW thermal reactor that attained criticality on August 4, 1956. The reactor was used for radio isotope production, fundamental nuclear research, shielding experiments, neutron activation analysis, neutron radiography, and neutron detector testing until it was permanently shut down in 2010. This reactor has now been upgraded to 2 MW Apsara-U facility in BARC.

Technical Specifications

- Swimming pool type, 1 MW research reactor
- Uranium—aluminium plate type highly enriched fuel
- Maximum thermal neutron flux of ~10^{13} n-cm^{-2} s^{-1}
- Radial beam line alignment

Fig. 4.41 APSARA research reactor facility

Fig. 4.42 Neutron imaging detector at Apsara facility

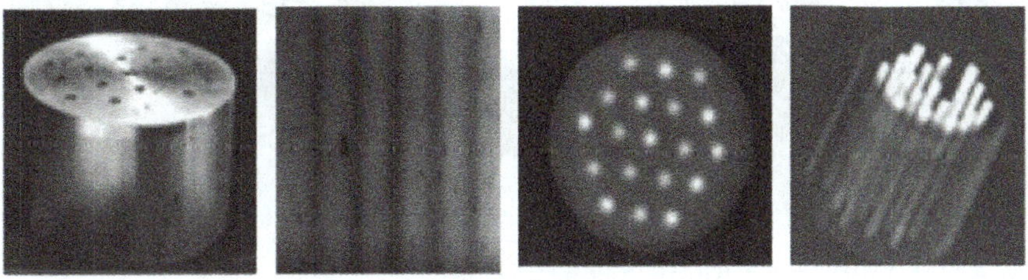

Fig. 4.43 Photograph of an object (an aluminium cylinder having 19 rods of SS and copper inside), its radiography image, 2D tomography and 3D tomography

- Thermal neutron energy spectrum
- Maximal beam intensity at sample position [n-cm^{-2} s^{-1}] ~1 × 10^6
- *L*/*D* collimation ratio ~90
- Effective Beam size at sample position ~150 mm × 150 mm square
- Detectors—Scintillator-based Intensified CCD camera system
- Imaging Options—Radiography, tomography.

This facility was mainly used for radiography, tomography and simulated two-phase flow experiments. Figure 4.42 shows the neutron imaging setup used [75]. An aluminium cylinder having 19 rods of SS and copper inside, its radiography image, 2D tomography and 3D tomography carried out at reactor power of ~400 kW is shown in Fig. 4.43.

4.6.2 *KAMINI Reactor Neutron Imaging Beamline*

KAMINI is a special purpose research reactor at Indira Gandhi Centre for Atomic Research (IGKAR), Kalpakkam [76]. This reactor was designed and built jointly by BARC and IGCAR. It is a 30 kW thermal reactor that uses metallic Uranium-233 as

Fig. 4.44 KAMINI reactor, IGCAR, Kalpakkam [76]

fuel, light water as moderator and coolant as shown in Fig. 4.44. The reactor achieved criticality on October 29, 1996. Since then, it has been used for neutron radiography, neutron activation analysis and radiation physics research.

Technical Specifications

- Open Tank type, 30 kW research reactor
- Enriched Metallic Uranium—233 fuel
- Maximum thermal neutron flux of ~10^{12} n-cm^{-2} s^{-1}
- Radial beam line alignment
- Thermal neutron energy spectrum
- Maximal beam intensity at sample position [n-cm^{-2} s^{-1}] ~1 × 10^7
- *L*/*D* collimation ratio ~160
- Effective Beam size at sample position ~70 mm × 210 mm square
- Detectors—Film based
- Imaging Options—Radiography.

4.6.3 Cirus Reactor Imaging Beamline

Cirus (Canada India Reactor Utility Services) was built in 1954 at BARC, Mumbai in collaboration with Canada as shown in Fig. 4.45. The 40 MW tank-type research

Fig. 4.45 CIRUS reactor facility

reactor used natural metallic uranium as fuel, heavy water as moderator and light water as coolant. It went critical on 10th July 1960 and was the second nuclear reactor to be built in India. The neutron beams produced by the reactor's core were used extensively in condensed matter research, material irradiation, fuel testing, neutron activation research, neutron radiography, and the production of radioisotopes. CIRUS reactor proved to be an excellent platform for training engineers and scientists and in understanding the intricacies of managing natural uranium, heavy water, reactor systems that eventually evolved into the Indian pressurized heavy water reactor programme. After 50 years of successful operation, it was permanently shut down in December 2010.

Technical Specifications

- Tank type, 40 MW research reactor
- Natural metallic Uranium
- Maximum thermal neutron flux of ~6.5×10^{13} n-cm^{-2} s^{-1}
- Radial beam line alignment
- Thermal neutron energy spectrum

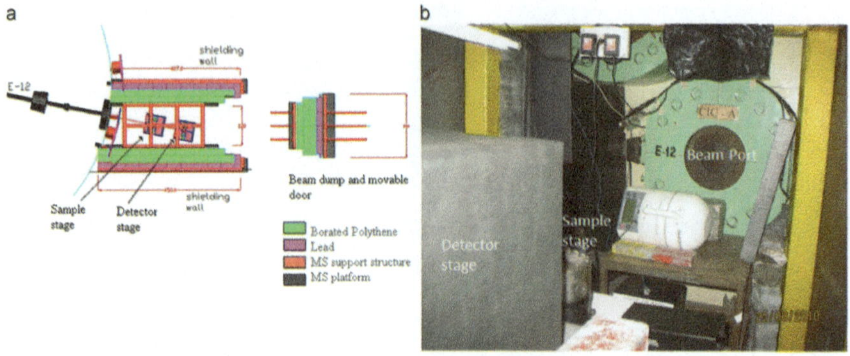

Fig. 4.46 **a** Schematic of experimental along with shielding arrangement. **b** Photograph of the set-up

- Maximal beam intensity at sample position [n-cm^{-2} s^{-1}] ~1 × 10^6
- Filter Options—Bi
- *L*/*D* collimation ratio ~90
- Cadmium ratio ~10
- Effective Beam size at sample position ~100 mm diameter
- Detectors—Scintillator-based CCD camera system, Image plate
- Imaging Options—Radiography, tomography, phase contrast imaging.

The neutron imaging beamline was commissioned at E-12 beamport of Cirus reactor as shown in Fig. 4.46. The shielding hutch and the motorized shielded door were fabricated using borated polythene and lead blocks [77]. This beamline was designed in such a way that conventional neutron radiography and tomography along with phase contrast imaging studies could be performed without any substantial modification of the experimental set-up. Radiography was carried out using main collimator permanently installed in the beamport. In order to carry out phase based neutron imaging a separate collimation insert was designed. Its dimensions were chosen in such a way that it can completely fit in the empty space of the preceding collimator.

4.6.4 Dhruva Reactor Imaging Beamline

The **Dhruva reactor** is indigenously made India's largest nuclear research reactor that is located in BARC, Mumbai as shown in Fig. 4.47. It is a 100 MW vertical tank-type thermal reactor that uses natural metallic uranium as fuel and heavy water as moderator and coolant. The reactor went critical on August 8, 1985. The building of Dhruva marked a turning point in India's indigenous nuclear technology development and use. Inside the reactor, many beamlines exist to satisfy the demands of a diverse multidisciplinary user group. It has been in extensive use for condensed

Fig. 4.47 Dhruva reactor facility

matter research, material irradiation, fuel testing, neutron activation analysis, neutron radiography and production of radioisotopes for application in the fields of medicine, agriculture and industry.

Scientists from BARC, other units of the Department of Atomic Energy (DAE), universities, and national labs collaborate on joint projects at Dhruva, which has been recognized as a National Facility for Neutron Beam Research to meet the requirements of the Indian scientific community. Many of the collaborations are supported by the University Grants Commission—DAE Consortium for Scientific Research (UGC-DAE-CSR), the Board of Research in Nuclear Sciences (BRNS) and other agencies.

A state of art neutron radiography beamline was commissioned at one of the radial beam port as shown in Fig. 4.48. This beamline is operational since 2016 [78]. The collimator has been designed using dual aluminium cone shaped housing

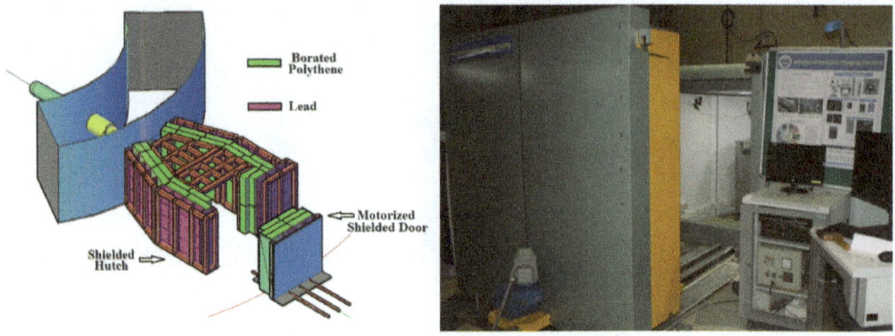

Fig. 4.48 Neutron imaging beamline at Dhruva reactor

filled with mixture of sand-B_4C powder for absorbing scattered neutrons and lead rings at different positions to absorb gamma radiation. A sapphire crystal as neutron filter followed by a bismuth crystal for gamma filtering has been used at the input of the collimator. Sand-B_4C powder, cadmium sheets, lead rings and Boral has been used as filler materials in the annular housing. The collimator has been designed in such a way that radiography-tomography or phase contrast imaging studies can be performed on the same beamline. Shielding hutch and a motorized shielding door of the beamline have been made using borated polythene and lead blocks. The beamline has been utilized for non-destructive evaluation of fuel pin, ceramic to metal joints, locomotive parts. It is also utilized for advanced studies like hydrogen diffusion in Ziraclloy, water retention in plants, lead melting.

Technical Specifications

- Vertical tank type, 100 MW research reactor
- Natural metallic Uranium
- Maximum thermal neutron flux of ~1.8×10^{14} n-cm^{-2} s^{-1}
- Radial beam line alignment
- Thermal neutron energy spectrum
- Maximal beam intensity at sample position [n-cm^{-2} s^{-1}] ~4×10^7
- Filter Options—Bi, Sapphire
- *L*/*D* collimation ratio ~160
- Cadmium ratio ~250
- Effective Beam size at sample position ~120 mm diameter
- Detectors—Scintillator-based CCD camera system, Image plate
- Imaging Options—Radiography, tomography, phase contrast imaging.

4.7 Radioisotope-Based Neutron Imaging Facilities

Radioisotope-based neutron sources (Am–Be, Pu–Be, Sb–Be, etc.) can be used for neutron imaging with suitable moderator and collimator geometry. In this context,

Fantidis et al. [79] had reported design consideration of neutron radiography system using Sb–Be neutron source. Authors had reported thermal neutron flux for different *L*/*D*, with and without Bismuth filter along with 1.85×10^{13} Bq ^{124}Sb source. Sinha et al. [80] had also reported neutron radiography using Pu–Be neutron source of strength 2×10^{7} n/s along with image intensifier and CCD camera. Sinha et al. had reported the thermal neutron flux of 70 n-cm^{-2} s^{-1} with *L*/*D* of 10 using thermalizing assembly of HDPe. However, the lower thermal flux with isotopic neutron source compared to reactor and spallation neutrons is the major issue for having good image quality.

4.8 Accelerator-Based Neutron Imaging Facilities

The accelerator-based neutron sources as discussed in Chap. 2 are used for neutron imaging, advanced research and commercial applications. The different accelerator-based neutron imaging facilities are discussed in the following sections.

4.8.1 Fusion Reaction-Based Neutron Imaging Facilities

Rapid advancement in the development of high energy and high current accelerators for high neutron yield production has increased their scope for neutron imaging. D-T, p-Be and D-Be are most common reactions used in these accelerators. List of few thermal and fast neutron imaging reported facilities are given in Tables 4.1 and 4.2 respectively [81–84].

4.8.2 Photo-Neutron Based Neutron Imaging Facilities

The photo neutron source is based on electron accelerator as discussed in Sect. 2.7 can be used for neutron imaging. The Hokkaido University neutron source (HUNS), powered by 45 MeV electron accelerator, is under operation since 1973, has three different neutron beamlines for experiment with fast neutron, thermal imaging and cold neutron imaging [84]. The *L*/*D* for both the imaging beam line is 60 and the thermal and cold neutron fluxes are 10^{3} n-cm^{-2} s^{-1} and 10^{4} n-cm^{-2} s^{-1} respectively. The cold neutron beam line is used for Bragg imaging, magnetic field imaging. The thermal beam line uses polyethylene moderator and it can couple 3.6 Qc super mirror. The beam line is used for resonance imaging and temperature measurement using time of flight method. The photograph of various beamlines at HUNS is shown in Fig. 4.49. Experiments such as Bragg-edge transmission imaging of welded iron samples, change of lattice plane distance with increasing of charge of a commercial

Table 4.1 Details of several thermal neutron imaging facility

Facility	Accelerator type	Accelerated particle	Max. energy of incident particle (MeV)	Target	Thermal neutron flux at sample (n-cm^{-2} s^{-1})	L/D or imaging distance from source	Applications
VSSC, Trivandrum, India	Sealed tube	Deuterium	0.16	Tritium	~10^4	~40	Pyrotechnic devices used in aerospace applications
CPHS, China	Linac	Proton	13	Beryllium	~1×10^6	5.4–9.4 m	Student training, scientific and industrial applications
PKUNIFTY, China	RFQ	Deuterium	2	Beryllium	5×10^5	25–200	Pedagogical training, scientific application, and technological development of neutron radiography
LENS, USA	Linac	Proton	13	Beryllium	~1×10^6		Structural studies of materials
PNL Neutron Facility, USA	Cascade	Deuterium	0.3	Deuterium	4.4×10^3	20	Inspection of munitions and other critical defence and aerospace components
RANS, Japan	Linac	Proton	7	Beryllium	10^4	21–781	Water penetration in a concrete block, Study for rusting of corrosion-resistant alloy under a paint film

(continued)

Table 4.1 (continued)

Facility	Accelerator type	Accelerated particle	Max. energy of incident particle (MeV)	Target	Thermal neutron flux at sample ($n\text{-}cm^{-2}\ s^{-1}$)	*L/D* or imaging distance from source	Applications
KUANS, Japan	Linac	Proton	3.5	Beryllium	1.2×10^3	20	Time-resolved neutron radiography, CT-imaging
SHI-ATEX Co. Ltd. Saijo, Japan	Cyclotron	Proton	18	Beryllium	2×10^5	44	Aerospace applications such as checking explosive devices, extraneous materials in casting and inspection of shield materials
Aomori Prefecture Quantum Science Center, Japan	Cyclotron	Proton	20	Beryllium	6.1×10^5	44	Plastic detection in metal materials and imaging of water in plants
DT neutron generator, China	Cascade	Deuterium	0.25	Tritium	$\sim 1 \times 10^4$	~25	Aviation and munitions components such as alloy blade, detonator
Peking University, China	Van de Graaff	Deuterium	4.5	Beryllium	5×10^3	20	Study for the design of future RFQ based NR facility and NDT for materials

VSSC Vikram Sarabhai Space Centre; *CPHS* Compact Pulsed Hadron Source; *PKUNIFTY* PeKing University Neutron Imaging FaciliTY; *LENS* Low Energy Neutron Source; *PNL* Phoenix Nuclear Labs; *RANS* Riken Accelerator-driven Compact Neutron Source; *KUANS* Kyoto University Accelerator-driven Neutron Source

Table 4.2 Details of some noteworthy fast neutron imaging facilities

Facility	Accelerator type	Accelerated particle	Max. energy (MeV)	Target	Neutron flux (n-cm^{-2} s^{-1})	*L/D* or imaging distance	Applications
LENS, USA	Linac	Proton	13	Beryllium	1.4×10^6	~100	Structural studies of materials
Necsa RFQ Facility, South Africa	Linac	Deuterium	4.6	Deuterium	1×10^4	5 m	Student training, scientific and industrial applications
PTB Accelerator Facility, Germany	Linac	Deuterium	13	Beryllium	$2–4 \times 10^5$	3 m	Pedagogical training, scientific application, and technological development of neutron radiography
DT neutron generator, China	Cockcroft-Waltron	Deuterium	0.25	Tritium	~2×10^4	200–300	Aviation and munitions components such as alloy blade, detonator
Peking University, China	Van de Graaff	Deuterium	4.5	Beryllium	1×10^5	330	Study for the design of future RFQ based NR facility and NDT for materials

(continued)

Table 4.2 (continued)

Facility	Accelerator type	Accelerated particle	Max. energy (MeV)	Target	Neutron flux (n-cm^{-2} s^{-1})	*L/D* or imaging distance	Applications
PIAF, Germany	Cyclotron	Deuterium	12	Beryllium	CW: 5×10^6 Pulsed: 5×10^5 @ 3 m	400	Fast-neutron resonance radiography (FNRR) and fast neutron tomography
Sames T400, Turkey	Cockcroft-Waltron	Deuterium	0.4	Tritium	1×10^5	100	Determination of thickness of iron in concrete materials

Fig. 4.49 Hokkaido University neutron source and different neutron beam lines [84]

Li battery, elemental distribution along with temperature measurements have been reported from this facility.

The KURRI-LINAC was installed in 1965 for nuclear physics experiments [84]. Presently, the neutron source is used for neutron imaging to observe the impurities and voids in nuclear fuel. The neutron beam has flight path of 12.7 m and the corresponding *L*/*D* is 80. The thermal neutron flux at the sample position is 2.36×10^4 n-cm^{-2} s^{-1}. The schematic of the imaging beam line is shown in Fig. 4.50.

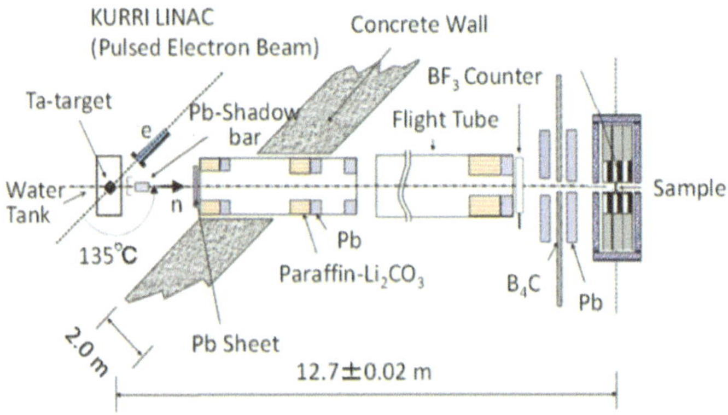

Fig. 4.50 Layout of the beam line of KURRI-LINAC [84]

4.9 Spallation Neutron-Based Neutron Imaging Facilities

High intense spallation neutron source is an alternate to reactor for neutron imaging and other applications. The advantages of using spallation source rather than nuclear reactor are the fewer nuclear safety concern, no need of fissile materials, chain reaction, and no issues related to minor actinide production. The high-intensity spallation neutron sources are also used for thermal and cold neutron imaging. The details of the neutron imaging facilities at different spallation source are as follows.

4.9.1 SINQ

SINQ is a continuous spallation neutron source providing a flux of about 10^{14} n-cm^{-2} s^{-1}. It is the first and only one of its kind in the world. Both thermal and cold neutrons are available making it a unique user facility for materials research and in the investigation of biological substances. It has three imaging beamlines, namely NEUTRA, ICON and BOA, having varied resolutions and dedicated for specific applications.

4.9.1.1 NEUTRA (An Acronym for NEUtron Transmission RAdiography)

NEUTRA [85] is the thermal neutron imaging beam line at SINQ. The initial portion of the beam line contains a convergent inner collimator tube, guiding neutrons to a fixed beam aperture of 2 cm in diameter followed by a divergent collimator. Bismuth has been used for gamma and high energy neutron filters. The flight tube of NEUTRA is directed towards the heavy water moderator tank of maximum thermal neutron density. The beamline has three different sample positions (at 3.82, 7.29 and 10.54 m) along beam direction with different beam size (15, 29 and 40 cm) for different sample dimensions. The thermal neutron fluxes (n/(cm^2 s mA)) at different sample positions are 1.6×10^7, 5×10^6 and 3×10^6 with *L*/*D* ratio of 200, 350 and 550, respectively. The beam line has a special setup (NEURAP) for imaging of highly radioactive samples. In addition to neutron imaging, the beam line is also equipped with an X-ray tube at sample position of 3.82 m to study the X-ray imaging, as a complimentary technique, with identical geometry. Various benchmark experiments such as in-situ study of crystal growth, moisture content estimation and their transport in building materials, dynamic imaging of two-phase flow, etc. have been reported from this facility. Depending upon the applications image plate detector, digital camera, intensifier systems have been used at NEUTRA.

4.9.1.2 ICON (An Acronym for Imaging with Cold Neutrons)

ICON [86] is the cold neutron imaging beam line with high image resolution. The beam line is directed towards the Deuterium liquid at temperature of −250 °C. The beam line has beam aperture of diameter 1, 10, 20, 40 and 80 mm. In addition, 20 mm beryllium aperture has been used to stop the fast neutrons. The beamline has two sample positions (6.86 and 12.08 m) with different field of view (15 and 30 cm). For most applications, the 20 mm aperture is used and cold neutron fluxes (n/(cm^2 s mA)) at the two positions are 1.3×10^7 and 3.9×10^6 with L/D of 343 and 604 respectively. The beam line provides advanced neutron imaging with an installation of velocity selector for the thermal neutrons. The control system provides the neutron wavelength in the range of 2.5–9 Å. The energy selector can be used for material study with very low content of light nuclide and Bragg-edge imaging in the cold energy range. A setup for neutron grating interferometry is also available for grating-based phase contrast and dark field imaging. Various imaging systems with varied sizes of scintillator screens coupled with high dynamic range digital cameras are used as per application. The highest resolution of 20.3 μm has been achieved using a 10 μm Gd-based scintillator.

4.9.1.3 BOA (An Acronym for Beamline for Neutron Optics and Other Approaches)

Beamline for neutron Optics and other Applications (BOA) [87] is the third imaging beam line at SINQ. BOA is also directed towards the liquid Deuterium moderator and is a cold neutron beam line. The beam line is equipped with neutron guide tube coated with super mirror, multi-channel polarizing bender unit, horizontal focusing unit, etc. The beam line has mechanical chopper with special detector for time-of-flight experiments. The beam line has also double crystal monochromator which is able to select only a narrow energy band from the white beam by using two crystals to reflect only one particular wavelength, through Bragg scattering. In the BOA beamline, the monochromator crystals are made of pyrolytic graphite with a mosaicity of 0.61°. The polarizing bender is used to polarize the spin of the neutron beam and the beam line has capability of polarized neutron imaging. The sample can be placed at two different locations (13.1 and 16.5 m) with motorized sample manipulator and the corresponding neutron flux are 2.71×10^7 and 4.42×10^6 n-cm^{-2} s^{-1}. Various benchmark experiments such as diffractive neutron imaging for single crystal, time of flight neutron imaging, polarized neutron imaging, etc. have been reported using the cold neutron beam line BOA.

4.9.2 LANSCE

LANCE is a pulsed beam based spallation neutron source. The accelerated proton beam is used to bombard on the spallation target at frequency of 20 Hz. Both the thermal and cold neutron beam are available for neutron imaging and different applications. The pulse beam of LANSCE is also suitable for time-of-flight experiments.

4.9.2.1 ERNI (Flight Path 5)

FP5 (ERNI) [88] provides neutrons for energy range of 1 meV–1 keV for different nuclear physics experiments and ~10 meV–100 eV for energy-resolved neutron imaging. It uses the thermal neutrons from the 1L target station. ERNI has two experimental locations; A cave in ER1 of the Lujan Center, having source-detector distances from ~6.5–11 m, and a station accessible by a silo for source-detector distances of 58–62 m. An evacuated 45 m long guide tube connects the two stations. Variable collimation allows beam spots from mm to several cm in the cave to 1 m in the silo. A 3000-channel time-of-flight imaging detector with 512 × 512 pixels over 28 × 28 mm^2 is available for energy-resolved neutron imaging. The ability to record ~3000 neutron radiographs, each at different neutron energy, allows imaging contrast across neutron energies from cold to epithermal. The beam line is used for imaging of isotopes having neutron absorption resonances between 1 and 100 eV, 3D isotope distribution non-destructively. The beamline is also used for characterization of nuclear fuels and actinides materials, energy-resolved neutron imaging to characterize materials opaque to thermal neutrons, optimization of crystal growth, biological study and fissile elements.

4.9.2.2 Asterix

Asterix [89] is a multipurpose beamline to study the structure of interfaces and cold neutron imaging. Asterix operates on the Lujan Center's cold neutron moderator (liquid hydrogen) and views an intense polychromatic neutron beam with a wavelength range of 4–13 Å through a 36 cm^2 neutron guide. Imaging techniques using cold neutrons (0.5–5.0 meV) can be performed on Asterix. The beam has variable aperture size from <1 mm^2 to 9 cm^2 and the aperture to detector distance ranges from 1.0 to 5.5 m. The beam line is used for phase contrast imaging. The beam line is also used for cold neutron radiography and is ideal for materials science studies of very small hydrogen concentrations in materials that are often difficult to study with other techniques.

4.9.3 ORNL

Oak Ridge National Laboratory (ORNL) has a pulsed beam based intense spallation neutron source. The neutron beam is being used for different material and biological studies. The pulse beam of 60 Hz is also suitable for time of flight neutron imaging.

VENUS

Oak Ridge National Laboratory started the development of a world-class neutron imaging beam line (VENUS) [90] in 2019 using Spallation Neutron Source (SNS). VENUS beam line to be commissioned in 2023. The epithermal to cold neutron will be used for versatile neutron imaging. The decoupled, poisoned para-H_2 moderator will be used to provide sharp neutron pulses across the cold to thermal neutron range. The source to detector distance is 25 m and the sample to detector distance is from few cm to meter. The VENUS is expected to provide a wavelength resolution of approximately 0.12%, enabling techniques such as Bragg edge imaging and texture mapping. The maximum field of view is 28×28 cm^2. The neutron fluxes are 1×10^8 n-cm^{-2} s^{-1} and 1×10^7 n-cm^{-2} s^{-1} in white beam and time-of-flight (TOF) mode respectively. The TOF will be used for energy-selective neutron imaging, hence making use of neutron scattering Bragg features for improved contrast and identification of phases in absorption image. The beam line will also be used for grain, 2D strain and porosity mapping in materials, research on energy storage materials, fracture propagation, magnetic properties and biological application.

4.9.4 ISIS

ISIS is also pulsed beam based spallation neutron source and used for neutron imaging and basic research related to materials.

IMAT

IMAT (Imaging and Materials Science and Engineering) [91] is a neutron imaging and diffraction beam line for a broad range of materials sciences. IMAT currently provides neutron radiography, neutron tomography, and energy-resolved neutron imaging. IMAT has liquid hydrogen as moderator for cold neutron beam. The beam has repetition rate of 10 Hz at target station 2. A supermirror straight neutron guide is used to transport neutrons to the aperture at 46 m from the moderator. The aperture size can be selected based on the imaging requirement (5, 10, 20, 40, 80 mm). The distance from the aperture to the sample position is 10 m. The maximum field of view is 200×200 mm^2. IMAT is used for imaging of diverse applications, for example, aerospace and transportation, civil engineering, power generation, fuel cell technology, archaeology, earth and bioscience.

4.9.5 JPARC

Japan Proton Accelerator Research Centre (JAPRC) has pulsed spallation neutron source and is being utilized for imaging and different research activities.

RADEN (BL22)

Energy-Resolved Neutron Imaging System (RADEN) [92] is the pulsed neutron imaging system for neutron radiography, tomography, energy-resolved neutron imaging experiments (Bragg-edge imaging), neutron resonance absorption imaging and polarized neutron imaging. The beam line has decoupled hydrogen moderator for cold neutron imaging. The incident wavelengths are $\lambda < 8.8$ Å ($L = 18$ m, 25 Hz), $\lambda < 6.8$ Å ($L = 23$ m, 25 Hz). RADEN has wavelength resolution of ~0.2% (minimum) and the beam size at the sample position is 30×30 cm^2. The neutron fluxes at the sample positions are 9.8×10^7 n/cm^2 s ($L/D = 180$), 5.8×10^7 n/cm^2 s ($L/D = 230$).

4.9.6 CSNS

China Spallation Neutron Source (CSNS) is pulsed neutron source and an imaging beamline using the pulsed beam is under construction.

ERNI (BL13)

An energy-resolved neutron imaging instrument (ERNI) [93] is under construction and the beam line is designed for basic research, evaluate the processing and manufacturing, operating performance of materials and devices/components in the fields of new energy, materials and high-end equipment manufacturing.

ERNI will also be used to analyze the 3D distribution of microstructure, defects, morphology and stress in materials and components. Its applications include the areas of new energy (i.e., renewable energy) materials and devices, new materials like advanced functional magnetic materials, engineering materials and structures in high-end equipment manufacturing, cultural heritage and archaeological research, plant physiology, geology, etc.

Summary

The present chapter has discussed major neutron source facilities across the world that are used for neutron imaging and applications. Broadly, two types of sources have been described: 1. Neutron imaging using reactor-based sources; 2. Neutron imaging using non-reactor-based sources. There are nearly forty reactor-based facilities across the world that carry out neutron imaging using thermal neutrons. These facilities are for either in-house usage or multidisciplinary user type. While conventional neutron radiography is carried out at almost all the facilities described, some of them also perform other advanced neutron imaging like neutron tomography, phase contrast imaging. Besides this, cold neutron facilities have also been described. Imaging with cold neutrons helps in enhancing picture contrast and sensitivity of detection. In the Indian context, neutron imaging facilities have been discussed separately.

The present chapter has also discussed radioisotopes, charge particle induced nuclear reactions and photo neutron source which are being used for imaging experiments. The radioisotope based neutron imaging assembly has limitation due to lower flux. The accelerator-based neutron source, such as, Fusion reaction, stripping reaction and photoneutron-based neutron imaging system has an advantage over isotopic source due to higher flux and can be switched On and Off according to the requirement. But, the flux of the above-mentioned source is also lower compared to reactor-based neutron imaging beamline. On the other hand, the spallation source has higher strength and the flux is comparable to the reactor. The spallation has advantage over reactor due to no issues related to minor actinide production. The present chapter has also discussed the different imaging beamline using spallation neutron source. The SINQ imaging beamlines utilize the continuous neutron beam for imaging application. For time of flight neutron imaging a mechanical beam chopper is used in BOA beamline. The rest of the spallation sources are based on pulsed beam and time flight experiment uses the pulsed beam. Some of the imaging beamline has liquid Deuterium or Hydrogen for cold neutron beam. The cold neutron beam is suitable for energy-resolved neutron imaging to characterize materials, which are opaque to thermal neutrons.

References

1. Kallmann H (1948) Neutron radiography. Research 1:254–260
2. Kallmann H, Kuhn E (1940) Photographic detection of slowly moving neutrons. United States patent 2,186,757, 9 Jan 1940
3. Fischer C (1994) The history of the first neutron radiographs in Berlin 1935–1944. In: Barton JP (ed) Neutron radiography (4). Proceedings of the fourth world conference, San Francisco, CA. Gordon and Breach Science Pub, pp 3–9
4. Thewlis J (1956) Neutron radiography. Br J Appl Phys 7:345–350
5. Barton JP (1976) Neutron radiography—an overview. In: Berger H (ed) Practical applications of neutron radiography and gaging: ASTM STP 586. American Society for Testing and Materials, pp 5–19
6. Berger H (1965) Neutron radiography: methods, capabilities, and applications. Elsevier Pub. Co., Amsterdam, New York
7. Berger H, Beck WN (1963) Neutron radiographic inspection of radioactive irradiated reactor fuel specimens. Nucl Sci Eng 15:411–414
8. Barton JP, Perves JP (1966) Underwater neutron radiography with conical collimator. Br J Non-Destruct Test 8:79–83
9. Barton JP (1967) Divergent beam collimator for neutron radiography. Mater Eval 25:45A
10. Neutron imaging facilities survey (2019) IAEA document on neutron imaging facilities worldwide. https://www.isnr.de/images/facilities/NI_Facilities_Database_v2019-02-04.pdf
11. Lehmann EH (2017) Neutron imaging facilities in a global context. J Imaging 3:52. https://doi.org/10.3390/jimaging3040052
12. Tochilin E (1965) Photographic detection of fast neutrons: application to neutron radiography. Phys Med Biol 10:477–490
13. Barton JP (1965) Neutron radiography using a crystal monochromator. J Sci Instrum 42:540

14. Barton JP (1965) Radiographic examination though steel using cold neutrons. Br J Appl Phys 16:1833–1841
15. Haskins J (1977) Standards for neutron radiography. In: Berger H (ed) Nondestructive testing standards—a review: ASTM STP 624. American Society for Testing and Materials, pp 108–114
16. Whittemore WL (1976) Personnel training and certification. In: Berger H (ed) Practical applications of neutron radiography and gaging: ASTM STP 586. American Society for Testing and Materials, pp 87–92
17. Hawkesworth MR (ed) (1975) Radiography with neutrons. In: Conference, University of Birmingham, 10–12 Sept 1973. British Nuclear Energy Society, London
18. Berger H (1966) Characteristics of a thermal neutron television imaging system. Mater Eval 24:475–481
19. Kawasaki S (1968) A thermal neutron television system using a high yield neutron generator. Nucl Instrum Methods 62:311–315
20. Hendry IC, Spowart AR, Robertson JA, Oliphant AJ (1969) The display of neutron radiography results by direct viewing of a scintillating plate. J Phys E Sci Instrum 2:191–192
21. Verat M, Rougeot H, Driard B (1983) Neutron image intensifier tubes. In: Barton JP, Von Der Hardt P (eds) Neutron radiography. Proceedings of the first world conference, San Diego, CA. D. Reidel Pub. Co., pp 601–607
22. Schlapper GA, Brugger RM, Seydel JE, Larsen GN (1977) Neutron tomography investigations at the Missouri University research reactor. Trans Am Nucl Soc 26:39
23. Matsumoto G, Krata S (1983) The neutron computed tomography. In: Barton JP, Von Der Hardt P (eds) Neutron radiography. In: Proceedings of the first world conference, San Diego, CA. D. Reidel Pub. Co., pp 899–906
24. Barton JP, Von Der Hardt P (eds) (1983) Neutron radiography. In: Proceedings of the first world conference, San Diego, CA. D. Reidel Pub. Co.
25. Barton JP, Farny G, Person J. Röttger H (eds) (1987) Neutron radiography. In: Proceedings of the second world conference, Paris. D. Reidel Pub. Co.
26. World map showing nuclear imaging facilities created by user on Google map and the picture is taken from https://www.google.com/maps/d/viewer?mid=1Y5wmdZWLf39qmflgNUQ-xUCKapg&ll=17.778172920233665%2C10.310944250000034&z=2, Oct 2021
27. Research reactors in Africa. IAEA document, Oct 2021. https://www.iaea.org/sites/default/files/18/09/research-reactors-in-africa.pdf
28. Garbe U, Randalla T, Hughesa C, Davidsona G, Pangelisa S, Kennedya SJ (2015) A new neutron radiography/tomography/imaging station DINGO at OPAL. Phys Procedia 69:27–32
29. OPAL reactor facility picture taken from https://nuclear.australianmap.net/lucas-heights/, Oct 2021
30. The TRIGA Mark-II reactor of the Atominstitut, Vienna, Austria. https://www.rertr.anl.gov/meeting_announcements/2014/Atominstitut.pdf, Oct 2021
31. Hossain SM (2021) Current status and perspectives of nuclear reactor based research in Bangladesh. http://www naweb.iaea.org/napc/physics/meetings/TM34779/Presenattions-PDF/Bangladesh-Hossain.pdf, Oct 2021
32. Islam MN, Alam MK, Zaman MA, Ahsan MH, Molla NI (2000) Application of neutron radiography to building industries. Indian J Pure Appl Phys 38:348–354
33. BR1 history brochure. https://science.sckcen.be/-/media/Files/Public/Publications/BR1_historybrochure.pdf?la=en&hash=1608BFA8CD0521AB0F2F562D94729F75F9063A31, Oct 2021
34. Ferreira FJO, Silva AX, Crispim VR (2010) Electronic imaging system for neutron radiography at a low power research reactor. Radiat Meas 45(7):806–809
35. Genezini FA, Fernando AJ, Marra Neto A, Aoki PE, Calvo WAP (2017) Industrial applications of the IAE-R1 research reactor. In: 1st international conference on applications of radiation science and technology. https://media.superevent.com/documents/20170427/b263e310f2aa0ad7d58128aee6668928/w.-a.-p.-calvo.pdf, Oct 2021
36. ETRR-2 facilities, Cairo, Egypt. https://www.nti.org/learn/facilities/364/, Oct 2021

37. Alba-Simionesco C, Menelle A, Visticot J-P (2011) The Laboratoire Léon Brillouin and the Orphée reactor: the French national neutron facility. Neutron News 22(4):10–14
38. ORPHEE reactor image taken from https://iramis.cea.fr/en/Phocea/Vie_des_labos/Ast/alltec.php?id_ast=110, Oct 2021
39. Nuclear reactors. http://pd.chem.ucl.ac.uk/pdnn/inst3/reactors.htm, Oct 2021
40. Tengattini A, Lenoir N, Andò E, Giroud B, Atkins D, Beaucour J, Viggiani G (2020) NeXT-Grenoble, the neutron and X-ray tomograph in Grenoble. Nucl Instrum Methods Phys Res A 968:163939, pp 1–11
41. Rosta L, Baranyai R (2011) Budapest research reactor—20 years of international user operation. Neutron News 22(3):31–36
42. KFKI science campus and Budapest research reactor hall pictures taken from https://www.bnc.hu/?q=BRR, Oct 2021
43. https://www.researchgate.net/publication/350314122_Development_of_a_CCD_based_thermal_neutron_imaging_detector_for_the_Israeli_Research_Reactor_IRR-1_at_Soreq_NRC, Oct 2021
44. Japan research reactor. https://www.jaea.go.jp/english/04/ntokai/kasokuki/kasokuki_01.html, Oct 2021
45. Choo KN, Cho MS, Yang SW, Park SJ (2014) Contribution of HANARO irradiation technologies to national nuclear R&D. Nucl Eng Technol 46(4):501–512
46. Hanaro pictures taken from https://businessdocbox.com/docview/88/115786217/#page=1, Oct 2021
47. Puspati reactor pictures taken from https://www.massa.net.my/agensi-nuklear-malaysia-nuklear-malaysia/, Oct 2021
48. https://www-pub.iaea.org/MTCD/Publications/PDF/SupplementaryMaterials/D482/AsiaAndPacific.pdf, Oct 2021
49. Awais M, Ramli AS, Mohamed AA, Idris FM, Harun MM, Azman A, Zin MRM (2016) Determination of neutron thermal and epithermal at reactor TRIGA PUSPATI thermal column beam port exit using neutron spectrometry technique. J Nucl Relat Technol 13(2):15–23
50. Maria reactor pictures taken from https://www.ncbj.gov.pl/en/o-nas/maria-research-reactor, Oct 2021
51. Krzysztoszek G (2016) The characteristics and irradiation capabilities of MARIA research reactor in NCBJ Świerk. EPJ Web Conf 115:01004
52. RPI reactor pictures taken from https://slideplayer.com/slide/9064195/, Oct 2021
53. Ramalho AJG, Marques JG, Cardeiria FM. The Portuguese research reactor: a tool for the next century. https://www-pub.iaea.org/MTCD/publications/PDF/csp_004c/PDFfiles/005.pdf, Oct 2021
54. Institute of Nuclear Research, Pisteti, Romania picture taken from https://stresa.jrc.ec.europa.eu/facilities/fipred-eq, Oct 2021
55. Shvetsov VN (2017) Neutron sources at the Frank Laboratory of neutron physics of the joint institute for nuclear research. Quant Beam Sci 1:6
56. Kichanov S, Saprykina I, Kozlenko D, Nazarov K, Lukin E, Rutkauskas A, Savenko B (2018) Studies of ancient Russian cultural objects using the neutron tomography method. J Imaging 4:25, pp 1–9
57. IBR-2 reactor building taken from http://flnph.jinr.ru/en/, Oct 2021
58. Core of the IBR-2 reactor with a movable reflector picture taken from open access https://www.mdpi.com/2412-382X/1/1/6, Oct 2021
59. Annual report, Jožef Stefan Institute (JSI), Ljubljana, Slovenia, 2010 from http://citeseerx.ist.psu.edu/viewdoc/download;jsessionid=5605BBE5670AD6ED59170703909E4346?doi=10.1.1.229.2787&rep=rep1&type=pdf, Oct 2021
60. TRIGA 0.25MW reactor hall, Slovenia picture taken from http://www.animma.com/index-2019.php/scientific-fields/, Oct 2021
61. De Beer FC, Franklyn CB, Venter AM, Nothnagel G. Utilisation and upgrading of the neutron beam lines facilities at the SAFARI-1 nuclear research reactor in South Africa. https://www-pub.iaea.org/MTCD/publications/PDF/P1575_CD_web/datasets/papers/A10%20de%20Beer.pdf, Oct 2021

62. Rush JJ, Cappelletti RL (2011) The NIST center for neutron research: over 40 years serving NIST/NBS and the nation. Special publication 1120, Aug 2011. National Institute of Standards and Technology, pp 1–70
63. Otta F, Loupiaca C, Déserta S, Hélarya A, Laviea P (2015) IMAGINE: a cold neutron imaging station at the Laboratoire Léon Brillouin. Phys Procedia 69:67–70
64. Hilger A, Kardjilov N, Strobl M, Treimer W, Banhart J (2006) The new cold neutron radiography and tomography instrument CONRAD at HMI Berlin. Phys B 385–386(Part 2):1213–1215
65. Kardjilov N, Manke I, Hilger A, Arlt T, Bradbury R, Markötter H, Woracek R, Strobl M, Treimer W, Banhart J (2021) The neutron imaging instrument CONRAD—post-operational review. J Imaging 7(11):1–16
66. Treimer W, Ebrahimi O, Karakas N, Seidel SO (2011) PONTO—an instrument for imaging with polarized neutrons. Nucl Instrum Methods Phys Res A 651:53–56
67. Calzada E, Gruenauer F, Mühlbauer M, Schillinger B, Schulz M (2009) A new design for the ANTARES-II facility for neutron imaging at FRM II. Nucl Instrum Methods Phys Res A 605:50–53
68. Rosta L, Grósz T, Hargitai T (2002) Liquid hydrogen cold neutron source in operation at the Budapest research reactor. Appl Phys A 74(Suppl.):S240–S242
69. Kis Z, Szentmiklósi L, Belgya T (2015) NIPS–NORMA station—a combined facility for neutron-based non-destructive element analysis and imaging at the Budapest neutron centre. Nucl Instrum Methods Phys Res A 779:116–123
70. HFIR. Oak Ridge National Lab, USA picture taken from https://neutrons.ornl.gov/hfir, Oct 2021
71. HFIR. Neutron imaging facility, Oak Ridge National Lab taken from https://neutrons.ornl.gov/suites/imaging, Oct 2021
72. Smith T, Bilheux H, Ray H, Bilheux J-C, Yan Y (2015) High resolution neutron radiography and tomography of hydrided Zircaloy-4 cladding materials. Phys Procedia 69:478–482
73. He L, Han S, Wang H, Wei G, Wang Y, Wu M, Liu Y, Chen D (2015) Characterization of a real-time neutron imaging test station at China advanced research reactor. Phys Procedia 69:79–86
74. Han S, Wu M, Wang H, Hao L, Wei G, He L (2013) Design of cold neutron imaging facility at China advanced research reactor. Phys Procedia 43:73–78
75. Sinha A (2000) Development of three dimensional neutron tomography system and its applications. Rev Sci Instrum 71(3):1455–1459
76. KAMINI reactor. http://www.igcar.gov.in/rfg/kaminiintro.html, Oct 2021
77. Kashyap YS, Agrawal A, Sarkar PS, Shukla M, Roy T, Sinha A (2012) Neutron phase contrast imaging beamline at CIRUS, reactor, India. Appl Radiat Isot 70:625–631
78. Shukla M, Roy T, Kashyap Y, Shukla S, Singh P, Ravi B, Patel T, Gadkari SC (2018) Development of neutron imaging beamline for NDT applications at Dhruva reactor, India. Nucl Instrum Methods Phys Res A 889:63–68
79. Fantidis JG, Nicolaou GE, Tsagas NF (2009) A transportable neutron radiography system based on a SbBe neutron source. Nucl Instrum Methods Phys Res Sect A 606(3):806–810
80. Sinha A, Bhawe BD, Panchal CG, Shyam A, Shrinivasan M, Joshi VM (1995) Exploratory studies on neutron radiography with a small neutron source using nuclear scintillation imaging technique. In: The 2nd international topical meeting on neutron radiography system design and characterization, 12–18 Nov 1995. The Shonan Village Center, Japan, pp 99–106
81. Namboodiri GN, Sambamurthy E, Sai Krupa M, Satheesh PK, Ramlet U, Gunasekaran R, Rajendran L, Thomas C, Thomas CR (2016) Challenges faced in realization of object manipulator for neutron radiography of pyro-devices. NDE-India, pp 205–213
82. Wang S, Yin W, Liu B, Li H, Sun Y, Cao C, Wu Y, Huo H-Y, Zhu S-L, Lou B-C, Wu C-L, Tang B (2021) A moveable neutron imaging facility using D-T neutron source based on a compact accelerator. Appl Radiat Isot 169:109564
83. Hokkaido University Neutron Source (HUNS) (2015) International survey on neutron imaging facilities world-wide. IAEA

84. Kiyanagi Y (2018) Neutron imaging at compact accelerator-driven neutron sources in Japan. J Imaging 4:55
85. Lehmann EH, Vontobel P, Wiezel L (2001) Properties of the radiography facility NEUTRA at SINQ and its potential for use as European reference facility. Nondestruct Test Eval 16(2–6):191–202
86. Kaestner AP, Hartmann S, Kühne G, Frei G, Grünzweig C, Josic L, Schmid F, Lehmann EH (2011) The ICON beamline—a facility for cold neutron imaging at SINQ. Nucl Instrum Methods Phys Res Sect A 659(1):387–393
87. Morgano M, Peetermans S, Lehmann EH, Panzner T, Filges U (2014) Neutron imaging options at the BOA beamline at Paul Scherrer Institut. Nucl Instrum Methods Phys Res Sect A 754:46–56
88. Energy resolved neutron imaging (ERNI) at FP5. Retrieved from: https://lansce.lanl.gov/facilities/lujan/instruments/fp-5/index.php, Oct 2021
89. Asterix. Retrieved from: https://lansce.lanl.gov/facilities/lujan/instruments/asterix/index.php, Oct 2021
90. Versatile neutron imaging instrument (under construction). Retrieved from: VENUS | BL-10 | SNS https://neutrons.ornl.gov/venus, Oct 2021
91. IMAT. Retrieved from: https://www.isis.stfc.ac.uk/Pages/IMAT.aspx, Oct 2021
92. Materials and life science experimental facility. Retrieved from: https://j-parc.jp/researcher/MatLife/en/instrumentation/ns_spec.html#bl22, Oct 2021
93. CSNS instruments. Retrieved from: http://english.ihep.cas.cn/csns/doc/4230.html, Oct 2021

Chapter 5
Basic Principles of Neutron Radiography and Tomography

Tushar Roy

5.1 Introduction

Neutron imaging was first used successfully in 1935, shortly after the discovery of neutron by James Chadwick. Since then it has developed manifold and is now used in research for a wide range of applications. With the emergence of digital detectors with better spatial and temporal resolution, neutron imaging has developed into a valuable and reliable technique for materials research.

Neutron imaging is a general term used for the process of forming an image on a detector medium using neutrons as radiation source. More commonly, conventional neutron radiography uses attenuation in the medium or absorption contrast to form the image. Neutron imaging, in general, can also be used with other interaction modalities like refraction, scattering, diffraction, etc.

5.2 Imaging with Radiation

When radiation passes through an object, it interacts with matter and suffers loss in its intensity. The beam is said to undergo attenuation in the medium. The loss in intensity is given by the well-known Beer-Lambert law:

$$I(x) = I_0 e^{-\mu x} \tag{5.1}$$

where I_0 is the incident intensity, $I(x)$ is the transmitted intensity, μ is the linear attenuation coefficient of the medium and x is the thickness of the medium. Here it is implicitly assumed that μ is constant. For non-uniform μ, Eq. 5.1 may be expressed

T. Roy (✉)
Technical Physics Division, Bhabha Atomic Research Centre, Mumbai, India
e-mail: tushar@barc.gov.in

D. K. Aswal et al. (eds.), *Neutron Imaging*,
https://doi.org/10.1007/978-981-16-6273-7_5

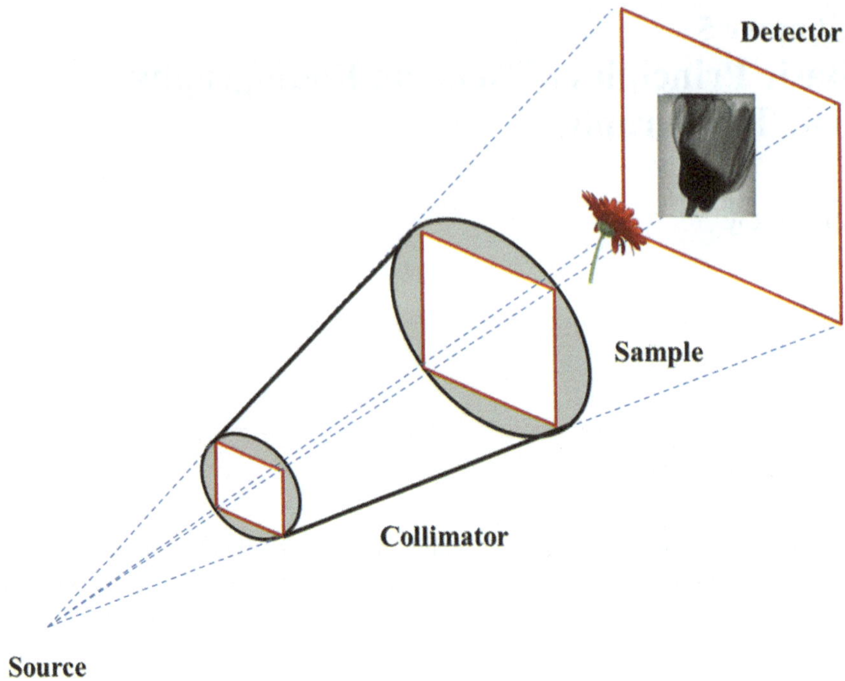

Fig. 5.1 Schematic configuration for radiography

as

$$I = I_0 e^{-\int \mu(x)\mathrm{d}x} \quad (5.2)$$

where the integral is over the path traversed by the radiation.

The Beer-Lambert law forms the basis of imaging with radiation. When the incident radiation after transmission through an object is captured on a detector/screen, it forms a two-dimensional image of the three-dimensional object. This is known as the *radiograph* of the object (Fig. 5.1).

5.3 Neutron Radiography

The interaction of neutrons [1] with nuclei of target material under investigation is governed by the cross section (σ), which is defined as the probability of neutron-nuclei interaction. The cross section for a given nuclear interaction is also dependent on other factors, such as the energy of the neutron, the type of interaction, and the stability of the target nucleus. The intensity of a neutron beam passing through a target material of thickness x is expressed as

$$I(x) = I_0 e^{-\Sigma_t x} = I_0 e^{-N\sigma_t x} \quad (5.3)$$

where I_0 is the incident intensity, I is the transmitted intensity and x is the thickness of the medium. The total macroscopic cross section (Σ_t) is defined as

$$\Sigma_t = N\sigma_t = \sigma_t \rho N_A / A \quad (5.4)$$

where N is the atom density of the target material, σ_t is the total microscopic cross section, ρ is the material density, N_A is the Avagadro's number and A is the atomic weight. Depending on the different reaction channels, the total cross-section (σ_t) defines the overall probability of neutron reaction with the target nucleus, and is expressed as

$$\sigma_t = \sigma_{\text{el}} + \sigma_{\text{inel}} + \sigma_c + \sigma_f + \cdots \quad (5.5)$$

where σ_{el} is elastic scattering cross section, σ_{inel} is inelastic scattering cross section, σ_{c} is capture cross section (may be split into individual capture reactions) and σ_{f} is fission cross section.

Comparing Eqs. 5.1 and 5.3, it may be noticed that the attenuation of neutrons in the medium is given by total macroscopic cross-section Σ_t which is a more convenient nomenclature used in case of neutrons, whereas the linear attenuation coefficient μ is conventionally used for X-ray or gamma interaction. For composite materials, Eq. 5.3 may be written as

$$I(x) = I_0 e^{-\int \Sigma_t(x)\mathrm{d}x} = I_0 e^{-\int N\sigma_t(x)\mathrm{d}x} \quad (5.6)$$

Equation 5.6 is the governing principle of neutron radiography. The transmitted neutron intensity is incident on a neutron converter screen (or neutron scintillator) which generates light photons that can be recorded on a film or digital detector like CCD.

Thermal Neutron Radiography requires a collimated beam of thermal neutrons with intensity typically in the range of 10^6–10^8 n/cm^2 s. The neutron beam interacts with the sample and is attenuated in the sample depending on their interaction. Light elements such as hydrogen (H), lithium (Li), and boron (B) and some heavy elements such as gadolinium (Gd), cadmium (Cd), and dysprosium (Dy) have a very high attenuation coefficient for thermal neutrons. On the other hand, materials like lead (Pb), iron (Fe), aluminum (Al), zirconium (Zr) have very low attenuation coefficient for thermal neutrons. Neutron radiography is therefore useful in imaging objects containing hydrogen or light elements such as plastic, rubber, polymer even when shielded by heavy elements like lead.

5.4 Line Integrals and Projection Image

In the expression for intensity (Eq. 5.2), the integral on the right is a line integral which represents the integral of the linear attenuation coefficient of the object along a line. In the context of neutron radiography, the line integral represents the total attenuation suffered by the incident neutron beam as it travels in a straight line through the object.

In Fig. 5.2, the object is represented by the function $f(x, y)$. In the co-ordinate system shown in figure,

$$\begin{aligned} s &= x\cos\theta + y\sin\theta \\ t &= -x\sin\theta + y\cos\theta \end{aligned} \tag{5.7}$$

We define the line integral as

$$P_\theta(s) = \int\limits_{(s,\theta)\text{line}} f(x, y)\mathrm{d}t \tag{5.8}$$

where the line (s, θ) is the line along which the integral is evaluated (or the path of the neutron beam through the object) and is given by the equation

$$s = x\cos\theta + y\sin\theta \tag{5.9}$$

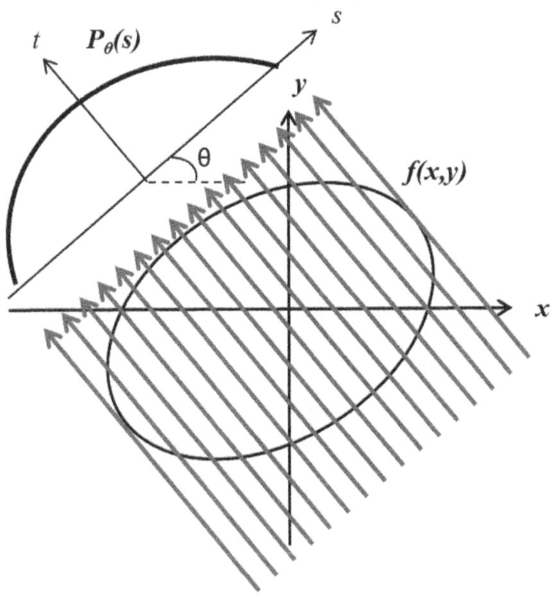

Fig. 5.2 Projection data in parallel beam geometry

Using a delta function, Eq. 5.8 can be re-written as

$$P_\theta(s) = \int_{-\infty}^{\infty}\int_{-\infty}^{\infty} f(x, y)\delta\ (x\cos\theta + y\sin\theta - s)\mathrm{d}x\mathrm{d}y \tag{5.10}$$

The set of line integrals $P_\theta(s)$ are combined to form a projection. The simplest form of projection is a collection of parallel ray integrals as given by $P_\theta(s)$ for a constant θ and is known as parallel projection. The projections can be formed for various angles θ by rotating either the object or the source-detector pair. In the case of neutrons, the source cannot be moved; hence the object (or sample) is rotated. The complete set of projections at various angles is used to deduce cross-sectional information of the object and this technique is known as *tomography.*

5.5 Principles of Neutron Computed Tomography

Tomography is derived from the Greek word '*tomos*' (which means slice or section) and '*graphia*' (which means to draw). Computed Tomography (CT) [2–4] technique is used to obtain cross-sectional information/images of objects. A conventional radiograph is a projection image of the density and thickness variation of the sample. The information along the path of the incident beam is integrated/superimposed in the radiographed and cannot be discerned as such. However, a combination of multiple projection images (radiographs) acquired at various angles can be used to extract features from the complete volume of the object. This technique is known as Computed Tomography. The cross-sectional information is extracted mathematically using various reconstruction methods. The mathematical basis for tomographic imaging was first explained by Radon [5].

Figure 5.3 shows the schematic of data acquisition for transmission tomography. A neutron beam is incident on the sample. The transmitted intensity (radiograph) is recorded by the detector. The sample is rotated using a sample manipulator and the projections (or radiographs) are recorded by the detector for different angles. The set of projections are then used to reconstruct the complete three-dimensional volume of the object on a three-dimensional grid which is discretized into unit cells known as '*voxels*' (short for volume pixels). The reconstructed volume represents the attenuation map of the object.

5.5.1 The Radon Transform

The Radon transform [5] was defined by Johann Radon in 1917. Let $\mathbb{R}^2$ denote the 2D Euclidean space (Fig. 5.4) with a point representation $\overline{x} = (x, y)$ in Cartesian co-ordinate.

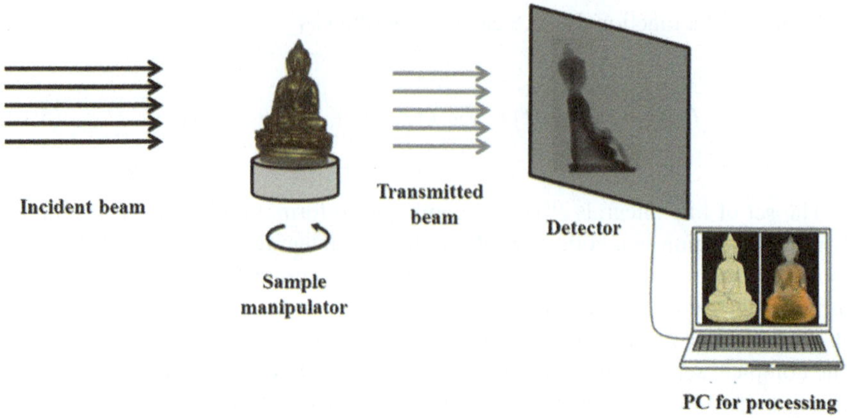

Fig. 5.3 Schematic configuration for transmission tomography

Fig. 5.4 Co-ordinate system for parallel beam projection

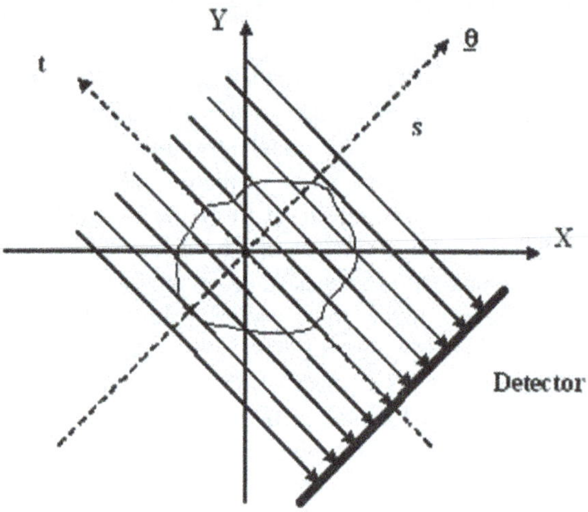

In the rotated co-ordinate system (s, t) with axes parallel to vectors $\hat{\theta}(\theta)$ and $\hat{\theta}^{\perp}(\theta)$, we have

$$\begin{pmatrix} s \\ t \end{pmatrix} = \begin{pmatrix} \cos\theta & \sin\theta \\ -\sin\theta & \cos\theta \end{pmatrix} \begin{pmatrix} x \\ y \end{pmatrix} \tag{5.11}$$

and

$$\begin{pmatrix} x \\ y \end{pmatrix} = \begin{pmatrix} \cos\theta & -\sin\theta \\ \sin\theta & \cos\theta \end{pmatrix} \begin{pmatrix} s \\ t \end{pmatrix} \tag{5.12}$$

A function $f(x, y)$ in $\mathbb{R}^2$ is denoted by $f(s, t) = f(x\cos\theta + y\sin\theta, -x\sin\theta + y\cos\theta)$ in the rotated co-ordinate system (s, t), that is the rotation of (x, y) by an angle θ in the counter-clockwise direction.

The integrals of a 2D function $f(s, t)$ along all possible lines is called the (two-dimensional) Radon Transform of $f(s, t)$

$$g(s, \theta) = \int_{-\infty}^{\infty} f(s, t)\mathrm{d}t \tag{5.13}$$

5.5.2 *The Fourier Slice Theorem*

An important property of the Radon transform is its close correspondence with the Fourier transform.

The Fourier Slice Theorem (also called Central Slice Theorem) [6] states: "The one-dimensional Fourier transform of a projection of a function $f(x, y)$, i.e. the Fourier transform of data along a line through the origin in the Radon space of $f(x, y)$, is same as the data along the same line through the two-dimensional Fourier transform $F(u, v)$ of $f(x, y)$". The principle of Fourier Slice theorem is schematically shown in Fig. 5.5.

The Fourier transform of $g(s, \theta)$ is given by

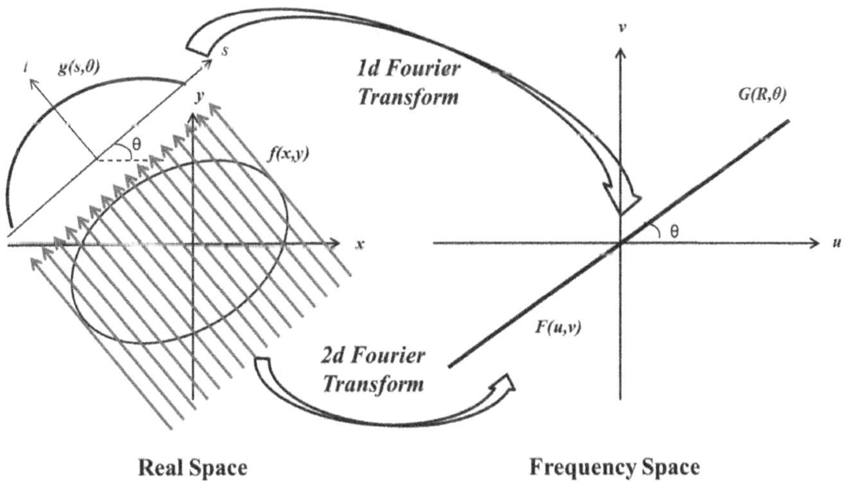

Fig. 5.5 Schematic representation of Fourier Slice theorem

$$G(R,\theta) = \int_{-\infty}^{\infty} g(s,\theta)\mathrm{e}^{-i2\pi Rs}\mathrm{d}s \tag{5.14}$$

Using Eq. 5.12, this becomes

$$G(R,\theta) = \int_{-\infty}^{\infty} \left\{ \int_{-\infty}^{\infty}\int_{-\infty}^{\infty} f(x,y)\delta(x\cos\theta + y\sin\theta - s)\mathrm{d}x\mathrm{d}y \right\} \mathrm{e}^{-i2\pi Rs}\mathrm{d}s \tag{5.15}$$

Changing the order of integration, we get

$$G(R,\theta) = \int_{-\infty}^{\infty}\int_{-\infty}^{\infty} f(x,y) \left\{ \int_{-\infty}^{\infty} \delta(x\cos\theta + y\sin\theta - s)\mathrm{e}^{-i2\pi Rs}\mathrm{d}s \right\} \mathrm{d}x\mathrm{d}y \tag{5.16}$$

The inner integral in Eq. 5.16 contributes under the condition stated in Eq. 5.11. Thus we have

$$G(R,\theta) = \int_{-\infty}^{\infty}\int_{-\infty}^{\infty} f(x,y)\mathrm{e}^{-i2\pi(x\cos\theta + y\sin\theta)R}\mathrm{d}x\mathrm{d}y \tag{5.17}$$

Substituting

$$u = R\cos\theta$$
$$v = R\sin\theta$$

Equation 5.17 becomes

$$G(R,\theta) = \int_{-\infty}^{\infty}\int_{-\infty}^{\infty} f(x,y)\mathrm{e}^{-i2\pi(xu+yv)}\mathrm{d}x\mathrm{d}y = F(u,v) \tag{5.18}$$

Thus it is shown that the 1D Fourier Transform $G(R,\theta)$ of projection data $g(s,\theta)$ is indeed equal to the two-dimensional Fourier transform $F(u,v)$ of $f(x,y)$. This is the Fourier Slice Theorem.

We can rewrite Eq. 5.18 as

$$G(R,\theta) = F(R\cos\theta,\ R\sin\theta) \tag{5.19}$$

5.5.3 Direct Fourier Method

Once $F(u, v)$ is obtained from the Fourier Transform $G(R, \theta)$ of the projection data $g(s, \theta)$ using Fourier Slice Theorem, the function $f(x, y)$ may be calculated using the Inverse Fourier Transform on $F(u, v)$. This is the Direct Fourier reconstruction.

However, there is a catch. The standard Inverse Fourier Transform requires data on a rectangular grid whereas Fourier slice Theorem gives data on a polar grid. For a practical implementation, a complicated frequency space interpolation is required [7, 8].

5.5.4 Backprojection

Let us suppose that the projection data is $g(s, \theta)$. The backprojection operator B may be defined as

$$[Bf](x, y) = \int_{-\pi/2}^{\pi/2} g(s, \theta)\mathrm{d}\theta \tag{5.20}$$

Qualitatively, the backprojection step is akin to "smearing out" the line integral data (projection data) along the same lines in the reconstructed object that produced the line integrals in the original object and adding the data for all the projected rays. However, the quality of the reconstructed image generated depends on the number of projections used. Figure 5.6 shows the effect of number of projections on the final reconstructed image. This is illustrated using a Shepp-Logan phantom [9]. Increasing the number of projections enhances the image contrast and reduces the streaks due to backprojection.

5.5.5 Filtered Backprojection Method

If the reconstruction process consists of backprojection only, the net effect is low-pass filtering which is manifested easily when we have a point object. To compensate for the loss in information related to sharp features on account of low-pass filtering, it is essential to filter the projection data with a high-pass filter before the backprojection step (Fig. 5.7). This forms the basis of filtered backprojection (FBP) scheme. FBP is the most widely used technique for 2D tomography reconstruction.

Mathematically, the high-pass filter is nothing but a ramp filter in the Fourier domain. The filtered backprojection algorithm may be derived as follows. Using inverse Fourier transform on Eq. 5.18, the object function $f(x, y)$ can be expressed as

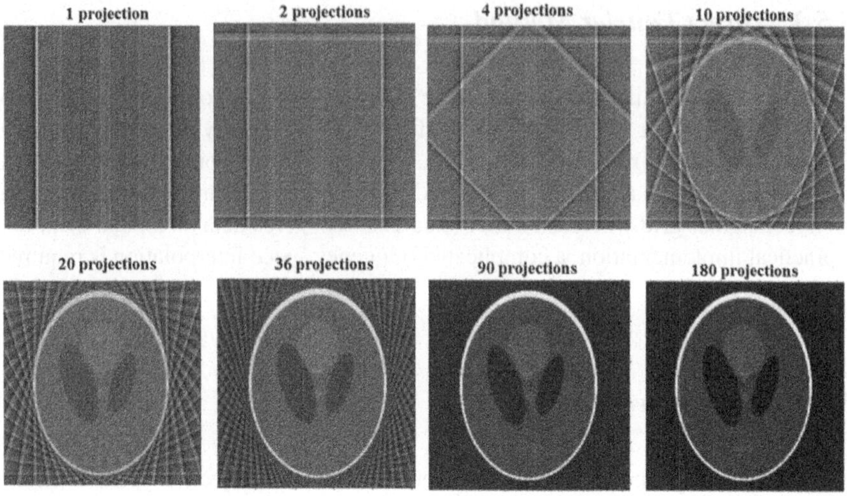

Fig. 5.6 Effect of number of projections on the backprojected image (Illustration uses a Shepp-Logan phantom)

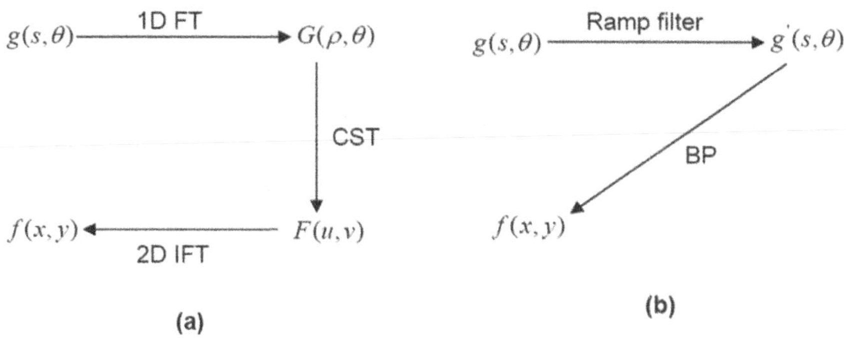

Fig. 5.7 Schematic representation of **a** Direct Fourier Transform. **b** Filtered backprojection

$$f(x, y) = \int_{-\infty}^{\infty}\int_{-\infty}^{\infty} F(u, v)e^{i2\pi(ux+yv)}\mathrm{d}u\mathrm{d}v \tag{5.21}$$

By substituting

$$\begin{aligned} u &= R\cos\theta \\ v &= R\sin\theta \end{aligned} \tag{5.22}$$

Equation 5.21 becomes

$$f(x,y)=\int_0^{2\pi}\int_0^{\infty} G_\theta(R,\theta)\mathrm{e}^{i2\pi R(x\cos\theta+y\sin\theta)}R\mathrm{d}R\mathrm{d}\theta \tag{5.23}$$

where $G_\theta(\cdot)$ is the Fourier function in polar co-ordinates. We can re-write Eq. 5.23 as

$$f(x,y)=\int_0^{\pi}\int_0^{\infty} G_\theta(R,\theta)\mathrm{e}^{i2\pi R(x\cos\theta+y\sin\theta)}R\mathrm{d}R\mathrm{d}\theta + \int_0^{\pi}\int_0^{\infty} G_\theta(R,\theta+\pi)\mathrm{e}^{i2\pi R(x\cos(\theta+\pi)+y\sin(\theta+\pi0)}R\mathrm{d}R\mathrm{d}\theta \tag{5.24}$$

Using the fact that Fourier function $G_\theta(\cdot)$ is periodic with period 2π

$$G_\theta(R,\theta+\pi)=G_\theta(-R,\theta) \tag{5.25}$$

Using the interval $0\le\theta<\pi$ for $-\infty<R<\infty$, Eq. 5.24 may be written as

$$f(x,y)=\int_0^{\pi}\left[\int_{-\infty}^{\infty} G_\theta(R,\theta)|R|\mathrm{e}^{i2\pi R(x\cos\theta+y\sin\theta)}\mathrm{d}R\right]\mathrm{d}\theta \tag{5.26}$$

Using the Fourier slice theorem, the 2D Fourier transform $F_\theta(R,\theta)$ is equal to the 1D Fourier transform of the projection at angle θ, we get

$$f(x,y)=\int_0^{\pi}\left[\int_{-\infty}^{\infty} G_\theta(R,\theta)|R|\mathrm{e}^{i2\pi R(x\cos\theta+y\sin\theta)}\mathrm{d}R\right]\mathrm{d}\theta = \int_0^{\pi}\left[\int_{-\infty}^{\infty}\left[\int_{-\infty}^{\infty} g(s,\theta)\mathrm{e}^{-i2\pi Rs}\mathrm{d}s\right]|R|\mathrm{e}^{i2\pi(x\cos\theta+y\sin\theta)}\mathrm{d}R\right]\mathrm{d}\theta \tag{5.27}$$

Equation 5.27 describes the complete filtered backprojection scheme. The projection data $g(s,\theta)$ are Fourier transformed $\left(\int_{-\infty}^{\infty} g(s,\theta)\mathrm{e}^{-i2\pi Rs}\mathrm{d}s\right)$, filtered with a ramp filter $|R|$, inversely Fourier transformed $\left(\int_{-\infty}^{\infty}[\ldots]\ldots\mathrm{e}^{i2\pi R(x\cos\theta+y\sin\theta)}\mathrm{d}R\right)$ and finally backprojected $\left(\int_0^{\pi}[\ldots]\mathrm{d}\theta\right)$.

5.6 Algebraic Reconstruction Techniques

In the algebraic reconstruction technique, the projections are expressed as a set of linear equations. The construction of the attenuation distribution is then obtained by calculating the least squares (LS) solution for the set of equations. The algebraic reconstruction techniques, such as ART, SART, and SIRT are methods for solving the LS problem using different types of iterations.

5.6.1 *Image Representation*

CT imaging can be described by the following equation:

$$g = Af \tag{5.30}$$

where $g = (g_1, g_2, \ldots, g_M)^T$ is the observable (measured) data, i.e. projections, $f = (f_1, f_2, \ldots, f_N)^T$ is unknown linear attenuation distribution in the object, and A is a $M \times N$ system matrix. Projections are 2D images of 3D objects which are not used directly for analysis. Reconstruction is a procedure used to transform projections to a meaningful 3D distribution of attenuation in the object, and then the reconstructed images are used for analysis. From the mathematical point of view, reconstruction is an inverse problem: the aim is to find unknown f using known g by solving the Eq. (5.30). Therefore, most theories of inverse problems can be applied in the field of tomographic reconstruction.

In an algebraic approach to CT problem, it is assumed that the reconstructed object consists of a matrix of unknowns, and the projection data are modeled by a set of linear equations. The solution of the reconstruction problem is then obtained by the Least Square solution for the set of equations.

5.6.2 *Observation Model*

Let us consider the vector of linear attenuation coefficients in the pixels $f = (f_1, f_2, \ldots, f_N)^T \in \mathbb{R}^{N\times 1}$. The projection data $g = (g_1, g_2, \ldots, g_M)^T \in \mathbb{R}^{MX1}$ can be expressed as a set of linear equations.

$$\begin{aligned} g_1 &= a_{11} f_1 + a_{12} f_2 + \cdots + a_{1N} f_N \\ g_2 &= a_{21} f_1 + a_{22} f_2 + \cdots + a_{2N} f_N \\ &\vdots \\ g_M &= a_{M1} f_1 + a_{M2} f_2 + \cdots + a_{MN} f_N \end{aligned} \tag{5.31}$$

where ε_i is the additive noise. In matrix notation, Eq. 5.31 can be written as

$$\begin{pmatrix} g_1 \\ g_2 \\ \cdot \\ \cdot \\ g_M \end{pmatrix} = \begin{pmatrix} a_{11} & a_{12} & \cdot\cdot & a_{1N} \\ a_{21} & a_{22} & \cdot\cdot & a_{2N} \\ \cdot & \cdot & \cdot\cdot & \cdot \\ \cdot & \cdot & \cdot\cdot & \cdot \\ a_{M1} & a_{M2} & \cdot\cdot & a_{MN} \end{pmatrix} \begin{pmatrix} f_1 \\ f_2 \\ \cdot \\ \cdot \\ f_N \end{pmatrix} \tag{5.32}$$

or

$$g = Af \tag{5.33}$$

For a 2D slice, $N = n_p^2$ is the total number of pixels and $M = n_p \cdot n_\varphi$ is the number of detectors, n_p is the number of pixels in one row of the planar image and n_φ is the number of acquisition angles. The matrix $A \in \mathbb{R}^{M\times N}$ is called the observation matrix or the probability system matrix. The matrix element a_{ij} is the path traveled by the neutron in the ith pixel when the neutron is detected by the jth projection bin.

If the matrix A is modeled correctly it includes all the physical features of the detecting system. Forming projection bins as a linear combination of attenuation coefficients is clearly a discrete version of integrating attenuation coefficients over some path length as presented in Eq. 5.31.

5.6.3 The Linear Least Square Estimation

Equation 5.33 can be solved using linear least square (LS) estimation. Our aim is to solve the estimate $\hat{f}_{LS}$ satisfying the condition

$$\hat{f}_{LS} = \arg\min \|g - Af\| \tag{5.34}$$

Knowing the projection data g and constructing the probability system matrix A, our task is to solve the LS problem denoted by Eq. 5.34. The observation model is mostly under-deterministic, that is, $N > M$. In such a case, the LS problem has no unique solution and some of the solutions must be chosen based on constraints. Most commonly, the minimum norm solution is used. Furthermore, the inverse problem of CT is usually ill-posed and some regularization may be needed in solving the LS problem. In addition, the matrix A is usually very large and the LS solution must be computed iteratively.

5.6.4 Implementation of ART

To implement the algorithm for computation, an initial guess of the solution is required. This guess, denoted by $f_1^{(0)}, f_2^{(0)}, f_3^{(0)}, \ldots, f_N^{(0)}$, may be assigned a value of zero or an average value of the projection sum to all the initial f_i's. The reconstruction process may be expressed mathematically as follows:

$$f_j^{(k)} = f_j^{(k-1)} + \frac{g_i - \sum_{j'=1}^{N} f_{j'}^{(k-1)} a_{ij'}}{\sum_{j'=1}^{N} a_{ij'}^2} a_{ij} \tag{5.35}$$

In Eq. 5.35, g_i is the measured ray-sum along the ith ray. The term $\sum_{j=1}^{N} f_j^{(k-1)} a_{ij}$ may be considered to be the computed ray-sum for the same ray based on the $(k-1)$th iteration. The correction Δg_i to the jth cell is obtained by first calculating the difference between the measured ray-sum and the computed ray-sum, normalizing this difference by $\sum_{j=1}^{N} a_{ij}^2$ and then assigning this value to all the image cells in the ith ray, each assignment being weighted by the corresponding a_{ij}.

There are different variants of ART. These algorithms differ in the manner in which corrections are applied and are presented here in brief.

5.6.5 Additive ART

The additive ART or simple ART algorithm was originally proposed by Gordon [10] for CT reconstruction. In this method, corrections are applied to all the voxels through which the ith ray passes, before calculating the correction for the next ray. The correction applied to jth cell of the object grid for ith ray is computed as [4, 11–13]

$$\hat{f}_j^{(k)} = \hat{f}_j^{(k-1)} + \lambda \frac{a_{ij} \Delta g_i}{\alpha_i} \tag{5.36}$$

where $\Delta g_i = g_i - \hat{g}_i$, $\alpha_i = \sum_{j=1}^{N} a_{ij}^2$ and λ is a relaxation parameter and $\lambda < 1$.

ART reconstruction usually tend to be noisy due to approximations introduced in the evaluation of a_{ij}'s. The reconstruction thus results in poor approximations of corresponding ray-sums. The effect of noise can be reduced by using the relaxation parameter λ.

5.6.6 Simultaneous Iterative Reconstruction Technique (SIRT)

In simultaneous iterative reconstruction technique (SIRT), the elements of the unknown function f_j for a particular cell are modified after all the correction values corresponding to individual rays have been calculated. The correction applied to jth cell of the object grid is computed as

$$\hat{f}_j^{(k)} = \hat{f}_j^{(k-1)} + \sum_{i=1}^{N_{c_j}} \lambda \frac{a_{ij}\Delta g_i}{\alpha_i} \tag{5.37}$$

The algorithm is similar to additive ART but the correction $\Delta g_i = g_i - \hat{g}_i$ is applied for all the ith rays before moving on to the next cell. This usually results in smoother reconstructions at the expense of slower convergence rates.

5.6.7 Simultaneous ART (SART)

The SART method was invented by Anderson and Kak [2, 14]. It connects advantages of ART and SIRT algorithms. It was found to be very efficient, accurate and superior in implementation [2, 11, 12, 14]. It yields good quality reconstruction with single iteration.

For each projection angle θ, the correction is applied to the jth cell sequentially for all ith rays as follows:

$$\hat{f}_j^{(k)} = \hat{f}_j^{(k-1)} + \lambda \frac{a_{i_\theta j}\left(g_{i_\theta} - \hat{g}_{i_\theta}\right)}{\sum_{j=1}^{N} a_{i_\theta j}^2} \tag{5.38}$$

One iteration is completed when all ith rays have been used exactly once. Since SART updates the values of cells in the image for each projection angle, it is well suited for GPU-based corrections. However, SART is slightly slower than ART in terms of computational time, due to the voxel-based pooling of correction terms.

5.7 Energy-Resolved Neutron Imaging

Energy resolved neutron imaging techniques comprise of experiments where specific energy is selected (using double crystal monochromator [15] or time-of-flight method [16] or mechanical velocity chopper [17]) to form an image corresponding to a single energy. In this way, neutron images can be formed single energies over the entire neutron spectrum, and we can obtain an image stack of energy-resolved images for

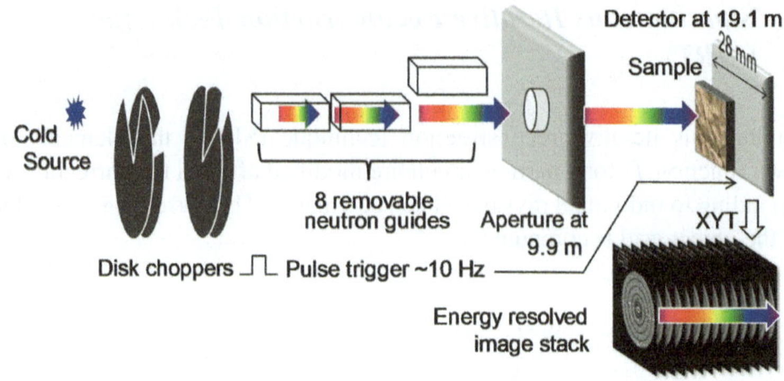

Fig. 5.8 Schematic of energy-resolved neutron imaging experimental setup at Bilby beamline, Australian centre for neutron scattering (reproduced from [18] with permission from AIP publishing)

the same object. Figure 5.8 shows the schematic of an energy-resolved experimental setup at Bilby beamline, Australian Centre for Neutron Scattering [18].

The selection of energy in a continuous neutron spectrum source (such as reactor source) can be implemented using a monochromator or velocity chopper system. However, this is achieved at the cost of reduced neutron intensity as a small fraction of the available neutrons are only utilized to achieve high wavelength resolution for a relevant bandwidth. The use of pulsed neutron sources (like spallation sources) can offer high spectral resolutions with large wavelength ranges without any additional penalty on neutron intensity.

One of the most popular energy-resolved technique is the neutron Bragg edge imaging [19, 20]. Neutron Bragg edge imaging enables spatially resolved studies of crystalline structures by exploiting the Bragg edges in the transmission spectra recorded in each pixel of an imaging detector. Bragg edges in the transmission spectrum are the signature of Bragg scattering from polycrystalline materials. For a given crystal lattice family hkl, with lattice spacing d_{hkl}, the scattering angle increases with the wavelength (λ) upto $\lambda = 2d_{\text{hkl}} \sin(\pi/2)$. Beyond this wavelength, the Bragg condition cannot be satisfied any longer, which results in a sharp drop of the attenuation of the material, called the Bragg edge.

Figure 5.9 shows the total cross-section for different polycrystalline materials and the associated Bragg edges. Thus, the analysis of the Bragg edges allows for the characterization of crystalline features of materials such as lattice strains or phase fractions. The Bragg cut-offs for many crystal materials are in this energy range that allows radiography contrast variations by applying energy-selective imaging techniques.

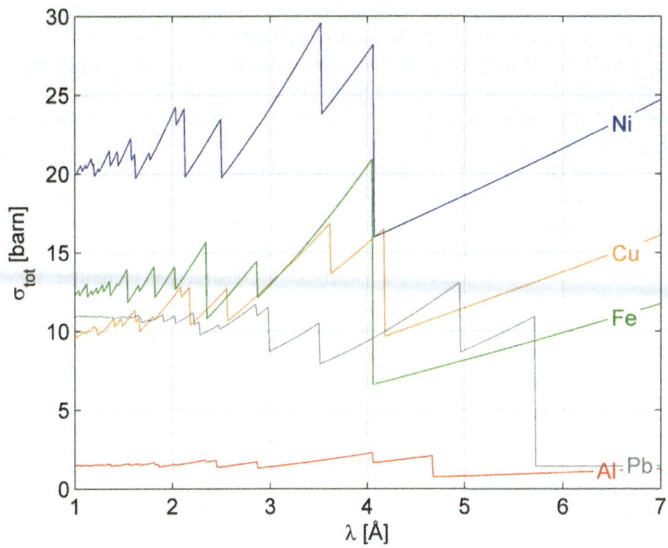

Fig. 5.9 Total cross-section and the associated Bragg edges for different polycrystalline materials

5.8 Summary

Neutron radiography and neutron tomography and its related neutron imaging techniques have established themselves as invaluable non-destructive inspection methods and quantitative measurement tools. The focus of this chapter has been to understand the mathematics of neutron image formation and methods for volumetric measurements using computed tomography and related reconstruction methods. The focus has been primarily related to the estimation of macroscopic material interaction cross-sections which is the goal of image analysis. A discussion of volumetric reconstruction techniques using analytic or iterative computed tomography algorithms has been done.

References

1. https://www.eichrom.com/wp-content/uploads/2018/02/Neutron-Attenuation-White-Paper-by-D-M-rev-2-1.pdf
2. Kak AC, Slaney M (1987) Principles of computerized tomographic imaging. IEEE Press
3. Herman GT (1980) Image reconstruction from projections: the fundamentals of computerized tomography. Academic Presss
4. Natterer F (1986) The mathematics of computerized tomography. Wiley, New York
5. Radon J (1917) On the determination of functions from their integrals along certain manifolds [in German]. Math Phys Klass 69:262–277

6. Haykin S (ed) (1985) Tomographic imaging with diffracting and non-diffracting sources. In: Array signal processing, Prentice-Hall, Englewood Cliffs, NJ
7. Stark H, Woods JW, Paul I, Hingorani R (1981) Direct Fourier reconstruction in computer tomography. IEEE Trans Acoust Speech Signal Process ASSP-29:237–244
8. Jacobsson C (1996) Fourier methods in 3D-reconstruction from cone-beam data. Ph. D. Thesis, Linkoping University
9. Shepp LA, Logan BF (1974) The Fourier reconstruction of a head section. IEEE Trans Nucl Sci NS-21:21–43
10. Rangayyan R, Prakash A, Gordon R (1985) Algorithms for limited view computed tomography: an annotated bibliography and challenge. Appl Opt 24(23):4000–4012
11. Chlewicki W (2001) 3D simultaneous algebraic reconstruction technique for cone-beam projections. Master of Science Thesis
12. Mueller K (1998) Fast and accurate three-dimensional reconstruction from cone-beam projection data using algebraic methods. Ph. D. thesis, The Ohio State University
13. Seppnen AO (2000) Correction of collimator blurring and attenuation in single photon emission a computed tomography. Master of Science thesis, University of Kuopio, Faculty of Natural and Environmental Sciences Physics Medical Physics
14. Anderson AH (1989) Algrebraic reconstruction in CT from limited view. IEEE Trans Med Imag 8(1):50–55
15. Schulz M, Boni P, Calzada E, Muhlbauer M, Schilinger B (2009) Energy-dependent neutron imaging with a double crystal monochromator at the ANTARES facility at FRM II. Nucl Inst Methods Phys Res A 605:33–35
16. Kockelmann W, Frei G, Lehmann EH, Vontobel P, Santisteban J (2007) Energy-selective neutron transmission imaging at a pulsed source. Nucl Inst Methods Phys Res A 578:421–434
17. Friedrich H, Wagner V, Wille P (1989) A high-performance neutron velocity selector. Physica B 156&157:547–549
18. Tremsin AS, Sokolova AV, Salvemini F, Luzin V, Paradowska A, Muransky O, Kirkwood HJ, Abbey B, Wensrich CM, Kisi EH (2019) Rev Sci Instrum 90:035114
19. Josic L, Lehmann E, Kaestner A (2011) Nucl Instrum Methods Phys Res 651:166–170
20. Peetermans S, Bopp M, Vontobel P, Lehmann EH (2015) Phys Proc 69:189–197

Chapter 6
Advanced Neutron Imaging

Yogesh S. Kashyap

6.1 Introduction

In traditional neutron radiography, we are primarily concerned with the intensity variations caused by inhomogeneous attenuation of the neutron beam in the object. This inhomogeneous neutron attenuation causes an intensity modulation in the transmitted beam, which is referred to as absorption contrast. It is worthwhile to point out that this absorption imaging includes both attenuation and scattering out of the beam, mostly due to incoherent scattering. This concept of attenuation-based imaging has been widely used since the discovery of X-ray by Rontgen, and later on, the same principles were adopted by the neutron imaging community. However, this approach fails to work with materials of high transmittance or when one is interested in probing the magnetic or electric field distribution inside a bulk. In order to overcome the limitations of conventional neutron imaging techniques, the concept of wave particle duality of neutrons, as defined through quantum mechanics, can be invoked. In terms of wave picture, neutron–matter interaction can then be treated on par with wave matter interaction, and hence, the concept of "phase of wave" can be implemented to improve the capabilities of existing neutron imaging techniques. The development of the neutron phase-contrast imaging technique is, therefore, an important advancement in improving the contrast and sensitivity of the existing transmission-based neutron imaging techniques [1–4].

Because phase information is often lost in traditional transmission imaging, many approaches have been developed to convert the unseen "phase" to intensity modulation. The concept of phase-contrast imaging was discovered in 1942 by Zernike, and he invented optical Zernike phase-contrast (ZPC) microscope to visualize phase undulations [5]. Zernike was awarded the Nobel Prize in Physics in 1953 for this invention. Following it, Nomarski then devised differential interference

Y. S. Kashyap (✉)
Technical Physics Division, Bhabha Atomic Research Centre, Mumbai 400085, India
e-mail: yskashyap@barc.gov.in

D. K. Aswal et al. (eds.), *Neutron Imaging*,
https://doi.org/10.1007/978-981-16-6273-7_6

contrast (DIC) microscopy in 1952, based on beam-splitting and shear interferometry concepts [6]. The disadvantage of these proposed techniques was that quantitative image analysis was not possible. In 1972, Gerchberg and Saxton [7] proposed Gerchberg-Saxton (GS) algorithm, the first iterative phase-retrieval algorithm for quantitative phase measurement, and the ideas have been successfully demonstrated in the fields of optical and X-ray microscopy. In contrast to iterative methods, Teague first proposed the idea of free-space propagation to recover phase quantitatively, in a non-iterative manner, using the transport of intensity equation (TIE) in 1982 [8, 9].

However, the widespread adoption of phase-based techniques for neutrons started only in early 2000. This long delay could be attributed to multiple reasons such as difficulties in fabrication of neutron optics, need for high mechanical stability in optical alignment and low neutron source coherence combined with low neutron flux at neutron imaging facilities. During the previous decade, new developments in fabrication technologies of neutron optics along with the availability of high brightness neutron sources have contributed towards the adoption of various phase-sensing techniques for neutrons.

On a different note, neutron spin-based imaging with polarized neutrons combines absorption (attenuation) and phase-based interactions and has become an important visualization probe for magnetic fields, domains and quantum effects such as the Meissner effect and flux trapping among others. It is achieved by obtaining the change in polarization of the neutron beam due to Larmor precession as it passes through a region of magnetic field. The neutron's deep penetration capability in most of the materials makes it a unique probe for non-destructive study of magnetic fields inside objects.

This chapter describes the basic principles of neutron phase-sensitive imaging and polarized neutron imaging, and presents an overview of the progress in this emerging field of neutron imaging in the last decade.

6.2 Basic of Phase-Contrast Imaging

The neutron–matter interaction (nuclear interaction, for simplicity) through the wave model can be best described through the complex refractive index of the matter, expressed as:

$$n = \sqrt{1 - \frac{\overline{V}}{E}} = 1 - \frac{\lambda^2 N}{2\pi}\sqrt{b_c^2 - \left(\frac{\sigma_r}{2\lambda}\right)^2} + i\frac{\sigma_r N \lambda}{4\pi} = n_r + i n_i \tag{6.1}$$

where λ is neutron wavelength, N is the mean number of scattering nuclei per unit volume and $b_c = \langle b \rangle$ is the mean coherent scattering length. The complex refraction index counts for absorption (σ_a) and incoherent scattering ($\sigma_{s,\,incoh}$) processes. (σ_r) = (σ_a) + ($\sigma_{s,\,incoh}$) is the total reaction cross section per atom. One can express the complex refractive index simply as follows:

$$n(\mathbf{r}, \lambda) = 1 - \delta(\mathbf{r}, \lambda) - i\beta(\mathbf{r}, \lambda) \tag{6.2}$$

Real part δ corresponds to the phase of the propagating wave and β represents the absorption in the medium. Furthermore, if the neutron absorption within the object is negligible, the imaginary component in the above relation may be ignored, and the complex refractive index of the object for monochromatic neutron of wavelength λ can be compactly described through the following relation:

$$n = 1 - \frac{\lambda^2 N b_c}{2\pi} \tag{6.3}$$

Consider a plane wave with amplitude A_0, and initial wave-vector $\mathbf{k}$, incident on a material with thickness $\boldsymbol{d}$ and refractive index (Fig. 6.1). Let the amplitude of the unperturbed wave after the material be $\psi = A_0 e^{i(kd-\omega t)}$; where $k = \omega/c$, and amplitude of the perturbed wave after the object is given by

$$\psi = A_0 e^{i(k'd-\omega t)} = A_0 e^{-i\frac{\omega}{c}\delta d} e^{-\frac{\omega}{c}\beta d} e^{i(kd-\omega t)} \tag{6.4}$$

where $k' = n\omega/c$. It may be noted that this equation contains a phase factor $\exp(-i\phi(d))$ with $\phi(z) = \omega\delta d/c$, which represents the phase difference between matter and vacuum (in terms of wave picture). Similarly, the amplitude attenuation of the wave is given by $\exp(-\omega\beta d/c)$. Hence, the intensity in the exit plane of the object is $I = |\psi|^2 = I_0 e^{-2k\beta d} = I_0 e^{-\mu d}$, which is nothing but Lambert–Beer's law where $I_0 = |A_0|^2$ and attenuation coefficient μ is defined as $\mu = 2k\beta = 4\pi\beta/\lambda$. This is the basic principle of conventional neutron radiography where the information based on differential attenuation or absorption is recorded. However, it can be observed that any information contained in the phase of the wave has been completely lost. The phase-contrast technique aims to detect this component of the wave by using suitable optical elements.

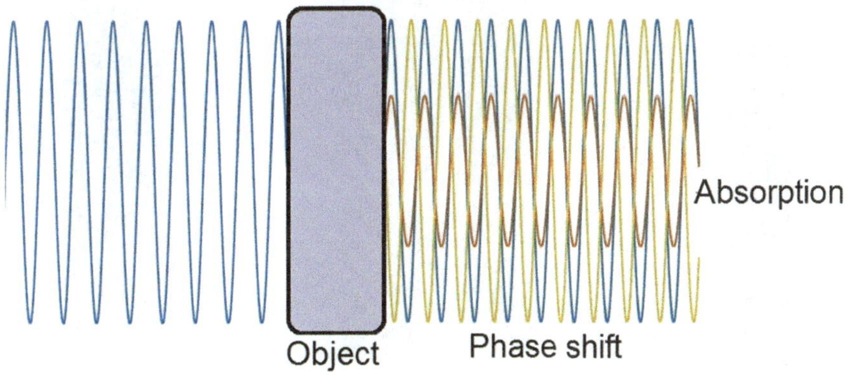

Fig. 6.1 Example of phase shift and wave attenuation as the wave passes through the object

Therefore, it can be stated that phase-contrast imaging additionally takes advantage of the real portion of the refractive index $(1 - \delta)$, whereas traditional neutron radiography takes advantage of the imaginary part (β). In contrast to X-rays, which have a refractive index of slightly less than unity, neutrons have a refractive index that can be larger than or less than unity. This is due to the fact that neutrons' coherent scattering length can be either positive or negative. The phase shift due to nuclear coherent scattering for an object of uniform thickness D for the neutrons can be expressed as $\Delta\phi = -Nb_c\lambda D$.

In order to further illustrate these conclusions and evaluate the advantage of phase-based approach over conventional absorption-based neutron imaging, consider a simulated phantom consisting of a carbon tube enclosed in a lead cylinder. Figure 6.2a shows the absorption and (b) the phase map of the simulated phantom. As observed in Fig. 6.1a, absorption map of the item is mostly attributable to lead and the carbon tube is totally hidden. These are common circumstances in which traditional radiography fails to probe the materials. The phase map is presented in Fig. 6.2b, and the phase contributions of both the lead and carbon sinkers are apparent.

Similarly, phase shift due to magnetic interaction, for a uniform length of thickness D can be expressed as

$$\Delta\phi_{\text{mag}} = \pm\frac{\mu Bm\lambda D}{2\pi\hbar^2} \tag{6.5}$$

where μ is the neutron dipole moment, B the applied magnetic field, m the neutron mass and λ is neutron wavelength.

It may be noted that phase shifts can appear in the presence of different varieties of scalar and vector electromagnetic potentials. Although not discussed here, phase shift effects can also occur due to gravitational [10], Coriolis [11], Aharonov–Cashire [12], Aharonov–Bohm [13], magnetic Josephson [14], Fizeau [15] and geometric (Berry)

Fig. 6.2 **a** Absorption and **b** phase shift distribution of a carbon tube enclosed in the lead cylinder

Table 6.1 Type of interaction and different types of phase shift due to these potentials [10–16]

Type of interaction	Type of potential	Phase shift
Nuclear	$\frac{2\pi\hbar^2}{m} b\delta^3(r)$	$-Nb\lambda D$
Magnetic	$\vec{\mu} \cdot \vec{B}$	$\frac{\mu m\lambda DB}{2\pi\hbar^2}$
Gravitation	$m\vec{g} \cdot \vec{r}$	$\frac{m_i m_g \lambda g A \sin(\alpha)}{2\pi\hbar^2}$
Coriolis	$\hbar\omega(\vec{r} \times \vec{k})$	$\frac{2m}{\hbar}\vec{\omega} \cdot \vec{A}$
Aharonov–Cashire	$-\mu \cdot (\vec{v} \times \vec{E})/c$	$\frac{2\mu}{\hbar c}\vec{E} \cdot \vec{D}$
Aharonov–Bohm	$\vec{\mu} \cdot \vec{B}$	$\frac{\mu BT}{\hbar}$
Josephson (magnetic)	$\vec{\mu} \cdot \vec{B}$	ωt
Fizeau	NA	$-Nb\lambda D\left(\frac{\omega_z}{v_z-\omega_z}\right)$
Geometric	NA	$\Omega/2$

[16] interactions, and have been verified experimentally using neutron interferometry techniques, and the same has been briefly presented in Table 6.1.

To expand on the concepts, we examine the most fundamental type of phase-contrast imaging, which involves free-space propagation of a coherent wave-field through a transparent object and then measuring the intensity modulation with a spatially resolved detector after propagating the exit wavefront over a distance z behind the object. This is the classical way of observing the diffraction effects using more coherent laser sources. In the absence of the object in the beam path, a constant intensity would be recorded on the detector. If we place the object in the beam path, make the beam propagation distance zero after the object, we have a contact image that shows only absorption-contrast. It is worth noting that a contact image of a perfectly transparent object just shows the intensity distribution in the detector plane. Nonetheless, the object shifts the phase of the wavefield, but the wave amplitude remains constant, and thus, the intensity remains constant. Refraction and diffraction create amplitude fluctuations in the propagating wavefield due to interference of wavefields, which leads to variations in the recorded intensity when the detector is moved further away from the object. Figure 6.3 shows the different regimes of the phase-contrast imaging where the detector can be placed, and the data can be recorded. This completely depends upon the object-related information that one wants to retrieve from these measurements. The advantage of the near-field regime is that there is a direct correlation between the data recorded on the detector and the phase-contrast effect. Sometimes, one need not retrieve the phase difference, and the projection image itself can be used for further analysis. However, in the far-field regime, one records the interference pattern (in reciprocal space), and hence, suitable phase retrieval algorithms need to be employed before any meaningful conclusion can be drawn from the recorded pattern. Near-field regimes are easy to implement experimentally; however, sensitivity in terms of phase difference would be higher in the far-field regime.

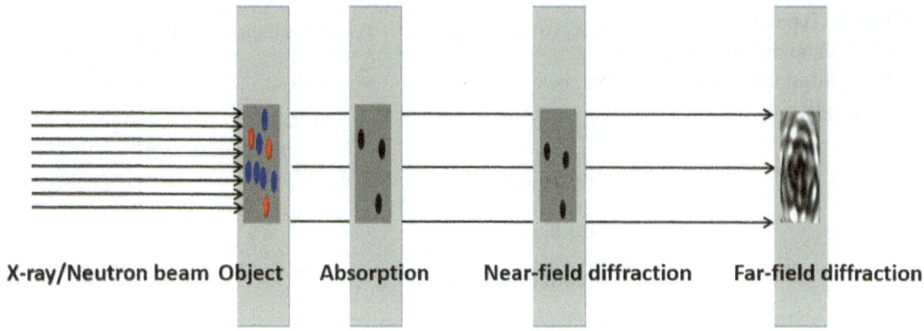

Fig. 6.3 Different regimes in the X-ray/neutron phase-contrast imaging

Figure 6.4 shows schematics of various phase-sensing techniques, which have been implemented using thermal and cold neutron sources. In what follows, we discuss in detail the basic ideas behind different phase-contrast imaging techniques and present some experimental findings using these techniques.

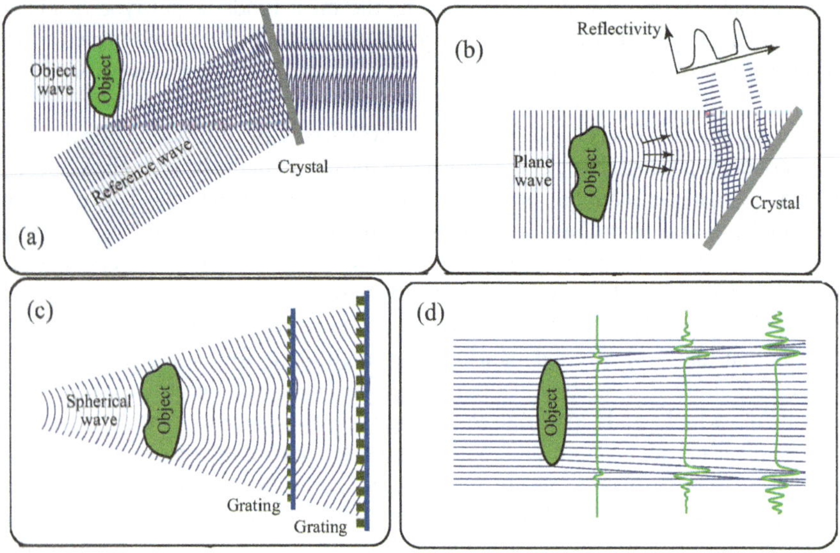

Fig. 6.4 **a** Crystal-based interferometer, **b** analyser-based setup, **c** grating-based setup, **d** propagation-based setup for neutron-based phase-contrast imaging [17]

6.3 Neutron Interferometry

6.3.1 Single-Crystal Neutron Interferometers

Interferometry is the most obvious choice when one is interested in measuring the phase shifts. Maier-Leibnitz made the first attempts to build a neutron interferometer in 1962 with coherent neutron beams. Maier-Leibnitz and Springer [18] using a pair of prisms, and up to 60 μm path separation was reported. However, the first successful X-ray interferometer, equivalent to Mach–Zehnder optical interferometers, was reported by Bonse and Hart in 1965 [19] using perfect single crystals, and later, the same technique was also adopted for neutron interferometry [20]. The interferometer is made up of three crystal slices that are carved from a single big, nearly flawless monolithic silicon crystal (Fig. 6.5). The first crystal slice serves as a beam splitter, dividing the monochromatic and well-collimated entering beam into two coherent neutron beams.

The second slice serves as a mirror, allowing the two beams in the third slice, the analyser, to recombine. To compensate for the interferometer's built-in phase patterns, a compensator is frequently used in one of the coherent beams. The sample is placed between the mirror and the analyser in one of the two beams.

Neutron interferometry can be considered as a method of measuring the phase difference induced by a sample (in the direction of the illuminating wave's propagation) across spatial locations perpendicular to the wave's propagation, modulo 2π. The intensity $I(x, y)$ recorded in the detector plane can be given as follows:

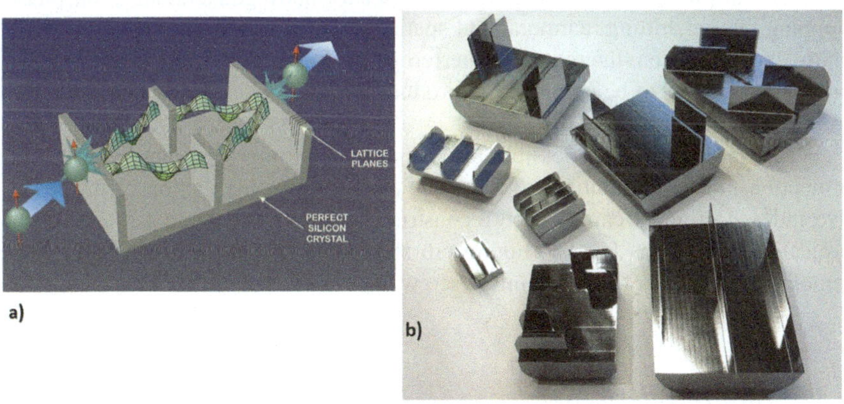

Fig. 6.5 A neutron interferometer made of a single piece of perfect silicon crystal. **a** A diagram of the interferometer: at the first plate, an incident beam is split into two beam paths, which are then recombined at the last plate. The intensity modulation of two beams leaving the interferometer is typical and depends on the relative phase of the two beams in the interferometer [21]. **b** Photographs of a variety of neutron interferometers [22]

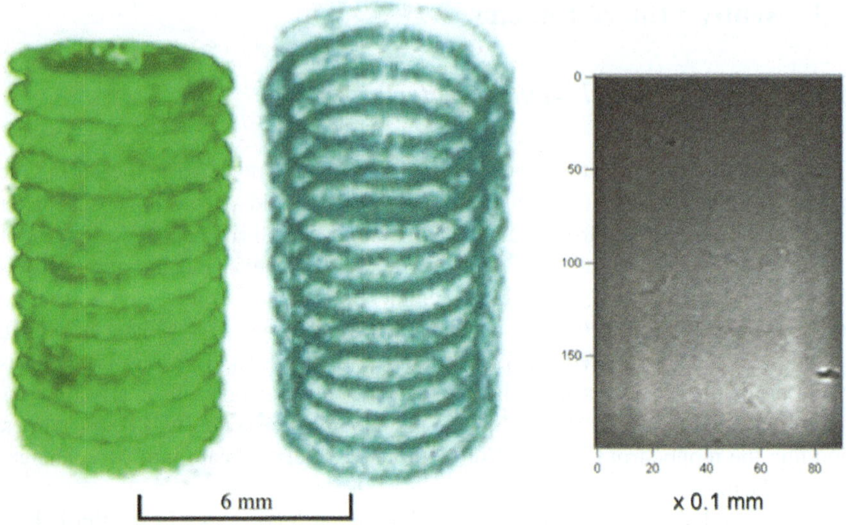

Fig. 6.6 Neutron phase tomography of aluminium screw from the phase shifts using neutron interferometry [24]

$$I(x, y) = I_0(x, y) + a(x, y)\cos[\Phi(x, y) + \phi_r(x, y)] \tag{6.6}$$

where $I_0(x, y)$ is the background illumination intensity, $a(x, y)$ is the amplitude of the beam, $\Phi(x, y)$ is the phase of the object and $\phi_r(x, y)$ is the reference phase. There are various techniques such as Fourier transform, fringe skeletonization, phase-stepping, phase-shifting, temporal and spatial heterodyning which can be used for phase extraction from the recorded interferogram. In practice, techniques of "phase unwrapping" need to be employed to produce a phase image in the range $[-\pi, \pi]$ from the interferogram.

Furthermore, it is also possible to do tomographic measurement by collecting projection images at multiple rotation angles and local distribution of the refractive index decrement can then be reconstructed using conventional filtered back-projection algorithms [4, 23]. Figure 6.6 shows the first reconstructed 3D phase tomography images of the aluminium screw [4].

6.3.2 Moire Interferometry

Although perfect crystal neutron interferometry is very sensitive to small phase shifts for thermal or cold neutrons, however, these interferometers are not only difficult to fabricate, but have narrow wavelength acceptance and also require very stringent conditions in terms of vibration isolation and incident beam collimation. In order to overcome the limitations of existing single-crystal interferometers, research has been

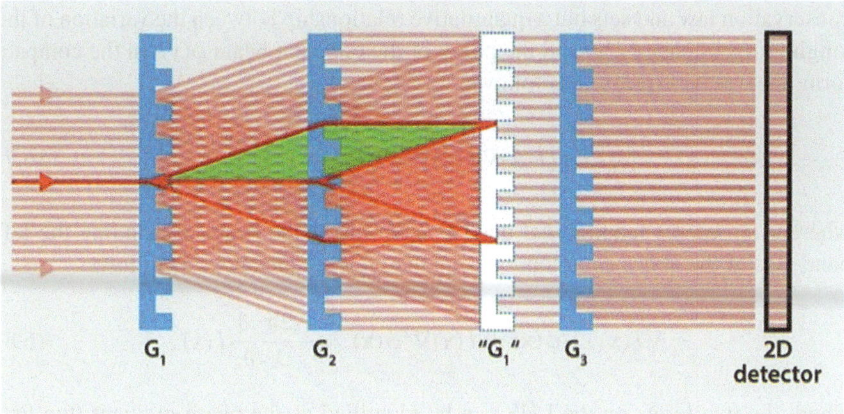

Fig. 6.7 Three PGMI schematic diagram where the third grating is offset from the echo plane to produce the Moire pattern [25]

carried out to develop grating-based interferometers. Recent advances in far-field phase grating-based Moiré interferometry have allowed neutron interferometry to be extended to medium-intensity thermal neutron sources, demonstrating the potential for neutron interferometry studies to be extensively employed. Figure 6.7 shows the schematic of the experimental technique [25]. The first grating (G_1) diffracts a neutron beam into three rays: one transmitted and two deflected. The second grating (G_2) is placed between phase grating G_1 and analyser grating G_3. This grating's principal function is to refocus diffracted waves from G_1 into a sequence of Fourier image in a specified plane downstream. The G_3 grating is positioned close to this plane in order to induce phase Moiré effects between itself and the Fourier images. The three grating setup has the advantage that large sample size (~cm^2) can be accommodated due to large possible separation between the second and third grating. Therefore, by detecting changes in the interference pattern, as compared with the empty interferometer, the microstructures within the material can be detected.

It should be noted that, unlike the near-field gratings discussed later in this chapter, phase-based Moire gratings are effective in the far-field region, producing interference fringes (~mm scale) that are orders of magnitude greater than the grating's period (~μm scale), allowing direct detection with an neutron imaging detector [26].

6.4 Near-Field Phase-Contrast Imaging

As discussed previously, there is a simple linear relation between phase shifts and intensity recorded by the detector in the near-field regime as given by the transport of intensity equation. The transport of the intensity equation (TIE), which Teague had undeniably first derived in 1982, is simply an alternative statement of the energy conservation law and sets out a quantitative relationship between the variation of the longitudinal intensity $I(x)$ and the phase of the coherent beam $\phi(x)$. In the compact form, the can be expressed as follows:

$$\nabla \cdot [I(\mathbf{x})\nabla\phi(\mathbf{x})] = -\frac{2\pi}{\lambda}\frac{\partial}{\partial z}I(\mathbf{x}) \tag{6.7}$$

where z represents longitudinal displacement along the beam. Expanding the left hand side of the above equation, one can then rewrite the above as follows:

$$\nabla I(\mathbf{x}) \cdot \nabla\phi(\mathbf{x}) + I(\mathbf{x})\nabla^2\phi(\mathbf{x}) = -\frac{2\pi}{\lambda}\frac{\partial}{\partial z}I(\mathbf{x}) \tag{6.8}$$

where the two terms on the LHS can be identified as the phase gradient (the first derivative) and phase curvature (the second derivative), respectively. Therefore, both phase gradient and phase curvature determine the longitudinal fluctuation of the intensity. Like a prism, the phase gradient causes intensity translation and constitutes a prism-like effect that transversely displaces optical energy. The transverse displacement is directly proportional to the local deflection angles (refraction).

Similarly, like a lens, phase curvatures cause intensity convergence or divergence and measures the Laplacian of the phase (Fresnel diffraction). This interpretation of the transport of intensity equation has therefore opened the door for the development of new non-interferometric phase-contrast techniques. In the following section, we describe phase-contrast techniques that measure the first derivative of phase (analyser and grating-based neutron imaging) and second derivative of phase (propagation-based phase-contrast imaging) and discuss some experimental results.

6.4.1 Analyser-Based Neutron Imaging

Analyser-based imaging is a form of imaging that utilizes the properties of a single crystal to bring out the phase gradients as per the refractive index variation in the sample [27]. This modality corresponds to Schlieren imaging in classical optics [28] and was adopted for neutron imaging using a combination of diffractive neutron optics [29, 30]. The angular deviation due to refraction of neutron beam as it transverses the object, gives rise to local phase gradient, lending the name differential phase to this contrast modality. Due to the weak nature of neutron–matter interaction, angular deviations due to local phase gradients are within a few arcsec or microradians. This matches with the angular acceptance or the Darwin width of perfect single crystals (as an example, for Si(111) the Darwin width is ~8 arcsec for thermal neutrons). This implies that only the neutrons travelling within this narrow angular width will be transmitted or reflected by the crystal analyser, and that too with variable degree of reflectivity. Usually, a monochromatic neutron beam, selected by using a neutron monochromator, is incident upon a sample, and the transmitted beam is then reflected by an analyser single crystal. The intensity recorded by the detector after transmission through the analyser in each position can be expressed as follows

$$I = I_0 R(\theta_a + \alpha(x, y)) \tag{6.9}$$

where $\alpha(x, y)$ is the refraction angle, I_0 is the apparent absorption intensity, and $R(\theta)$ is the reflectivity of the analyser crystal at the angle θ. The analyser is aligned at the Bragg angle as that of the monochromator, and by rotating the crystal analyser around the Bragg angle, data at multiple points on the rocking curve is collected. This data can be subsequently processed to separate both absorption and refraction contribution from the object. Furthermore, one can repeat the measurement steps at different sample orientation and three-dimensional tomography can be carried out. A tomographic reconstruction based on the standard filtered back projection method (FBP) and utilising the linear filter function (Ram-Lak filter) would, however, not result in a proper reconstruction of the real component of the refractive index. Using the Fourier derivative theorem, a new filter function defined as follows can be derived [2].

$$F(k(\nu)) = \begin{cases} -i\ \mathrm{sign}(\nu), & |\nu| \le 1/2\Delta y' \\ 0, & |\nu| \ge 1/2\Delta y' \end{cases} \tag{6.10}$$

where ν is represents spatial frequency component and sign(ν) is the sign function.

Spatially unresolved, disorderly or partially arranged sample microstructures may lead to the small angle neutron scattering around the refracted ray. The unresolved microstructure may broaden the refracted neutron beam, and the same can be recorded in the transmitted direct beam. The method, therefore, can be used to obtain signatures of small-angle scattering contrast because the relevant angular range of refraction matches well with small-angle scattering from structures in the micrometre and sub-micrometre range (100 nm–10 μm). The following line integral may be used to quantify and approximate the small-angle scattering-induced width of the refracted beams or their angular distribution [31]

$$\Delta\theta^2 = \int \frac{\sigma(x, y)N(x, y)}{R^2(x, y)}\mathrm{d}l \tag{6.11}$$

where σ is the scattering cross-section, N the particle density and R is a parameter with the dimension of a length specifying an average size or correlation length in the scattering object. In contrast to the phase shift, with this definition of small-angle scatter signal, one can reconstruct slices using the conventional linear filter. Using double single-crystal setup, the first attempts to probe the spatial distribution of small-angle scattering signal using a spatially resolved neutron imaging detector have been reported [31]. Figure 6.8 shows three-dimensional phase tomographic reconstruction of the aluminium cylinder with asymmetric hole along with different cut-away sections using analyser-based phase-contrast imaging. Figure 6.9 shows

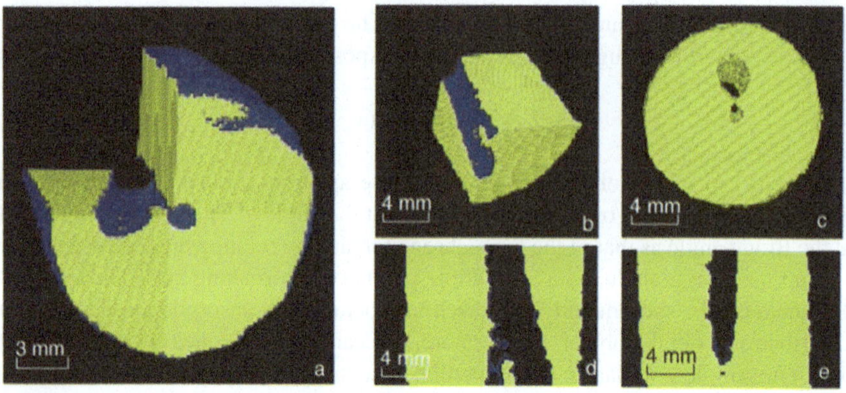

Fig. 6.8 Three-dimensional refraction contrast reconstruction of the sample volume (aluminium cylinder with asymmetric hole drilled); cut-open (**a**, **b**); view from top (**c**); sagittal cut (**d**); frontal cut (**e**). Reproduced from [30] with permission from © Elsevier 2004

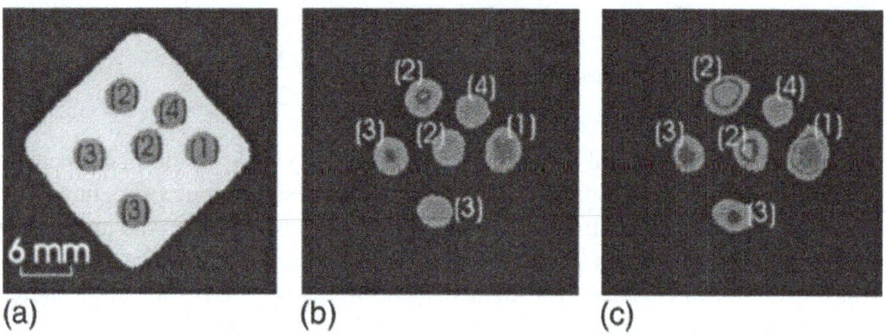

Fig. 6.9 Reconstructed sample cross section of Al matrix filled with different concentrations of β-carotene in D_2O: (1) 23.0 (5) wt%, (2) 11.6 (5) wt%, (3) 5.8 (5) wt%, (4) 3.8 (5) wt%; **a** refraction contrast; **b** absorption contrast; **c** USANS contrast. Reproduced from [31] with permission from Copyright © 2004 American Institute of Physics

ultrasmall angle X-ray contrast tomography slice images of different concentrations of β-carotene in D_2O inside various holes in a square Al matrix.

6.4.2 Near-Field Grating-Based Neutron Imaging

When a plane wave illuminates a periodic transmission mask, coherent wave propagation in the near-field causes a periodic self-image of the transmission mask. If the plane wave and the diffraction mask were both infinitely long, these recurrences would continue indefinitely. This phenomenon, known as the Talbot effect, was first observed in 1836 by Talbot and is a natural consequence of Fresnel diffraction.

The Talbot effect is used in neutron grating interferometer imaging, which requires a phase grating to create a near-field Talbot diffraction pattern and an analyser grating to analyse the interference pattern [2]. Simply put, it is a multi-collimator that converts local angular deviations into variation in locally transmitted intensity that can be easily detected with a neutron imaging detector [32].

Typically, an absorbing mask with transmitting slits, also known as a source grating, is placed near the neutron source (pin-hole) to create an array of line sources, with each line source meeting the differential phase-contrast image process coherence requirements. The need for the periodic absorption mask arises, as usually, the neutron sources are highly incoherent, and therefore, there is a need to generate a coherent source for generating near-field diffraction pattern. The period and distances are chosen so that at the chosen Talbot distance, the interference patterns due to different individual beams from the source grating are superimposed constructively. The relationship between the different relevant parameters can be expressed as: $p_s = p_a D_1 / D_2$, where p_s is the source grating period and D_1 is the distance between source grating and phase grating, D_2 is the distance between phase and analyser grating, and p_a is period of analyser grating. To separate the phase information from the recorded Moire pattern, a phase stepping approach or Fourier transform approach can be used. Just like the analyser-based imaging, both the phase gradient and the attenuation image can be reconstructed using a set of the recorded interferogram.

Similar to the analyser-based imaging, another application of grating-based imaging setup is to detect dark-field image contrast, through analysis of the decrement visibility of interference pattern. The visibility is defined as $V = (I_{max} - I_{min})/(I_{max} + I_{min})$, where I_{max} and I_{min} are the maximum and the minimum intensity of a modulation period across the beam. The loss of visibility can be either due to small-angle scattering from sub-microscopic structures or when magnetic features affect spin-up and spin-down components of the neutron [32]. Likewise as in the analyser-based imaging, one can carry out three-dimension tomographic reconstruction of the real part of the refractive index using Hilbert filter followed by back projection [2].

The dark-field contrast, which is caused by local convolution of the small-angle scattering function with the amplitude of the interference pattern, has a logarithmic dependence on the sample thickness, allowing for tomographic reconstruction similar to conventional tomography algorithm [33]. This differs from dark-field contrast in an analyser-based setup, which is quantified as a broadening of the angular intensity distribution. Although the functionality of grating interferometers is quite similar to that of analyser-based imaging, the key advantage is that it can accept a significantly higher neutron wavelength spread and input divergence.

Figure 6.10 shows a tomographic reconstruction of a piece of aluminium with several drilled holes, displaying the refractive index distribution, dark-field contrast and three-dimensional rendering of a sediment found in the dark-field tomogram [33].

Figure 6.11a shows a (110)-oriented iron silicon (FeSi) single-crystal disc using dark-field neutron imaging technique. The structures in the dark-field image arise due to strongly degraded neutron wave-front on account of multiple refraction at the domain walls, leading to enhanced scattering of neutron wave. Figure 6.12b shows

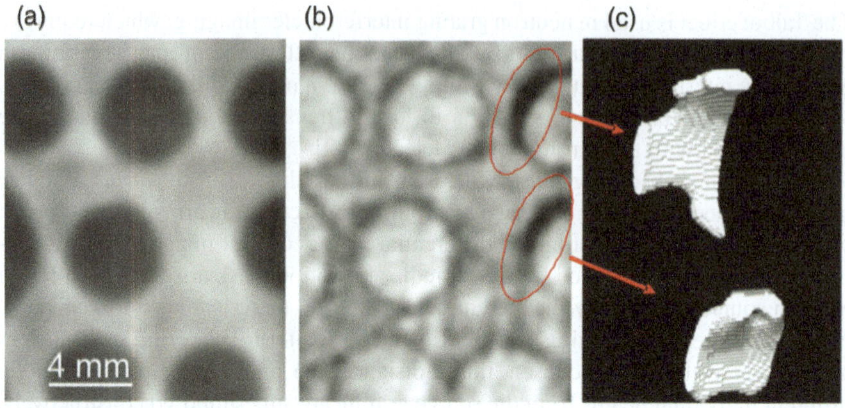

Fig. 6.10 A tomographic reconstruction of a piece of aluminium with 4 mm drilled holes **a** the refractive index distribution (from differential phase contrast data), **b** dark-field contrast (displaying image contrast due to sediments) and **c** a three-dimensional rendering of a sediment found in the dark-field tomogram. Reproduced from [33] with permission from Copyright © 2008 American Institute of Physics

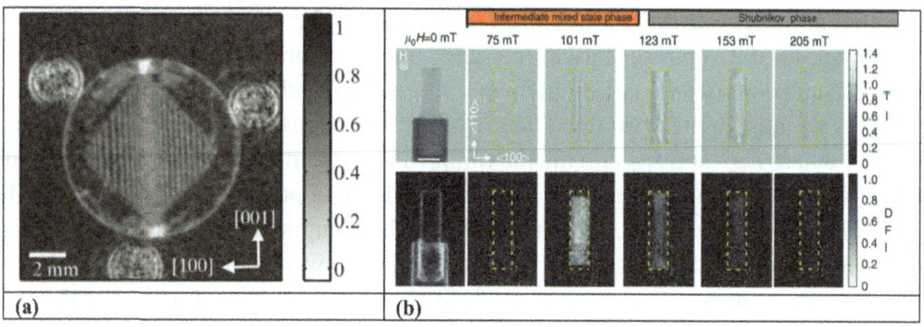

Fig. 6.11 **a** Dark-field image of a 1 mm monocrystalline FeSi plate. Reproduced with permission from [35] Copyright © 2008 American Institute of Physics. **b** Neutron transmission and dark-field images of ultra-pure niobium as a function of magnetic field showing the flux line lattice within the vortex domains and the vortex lattice domain formation [34]

3D neutron dark-field imaging of morphology of vortex lattice domain structure in the Type-II superconductor (ultra-pure niobium) as a function of magnetic field [34]. Thus, the neutron phase-contrast imaging provides a powerful non-destructive method in direct visualization and better understanding the magnetic domains within the bulk of a magnetic material as first postulated by Weiss.

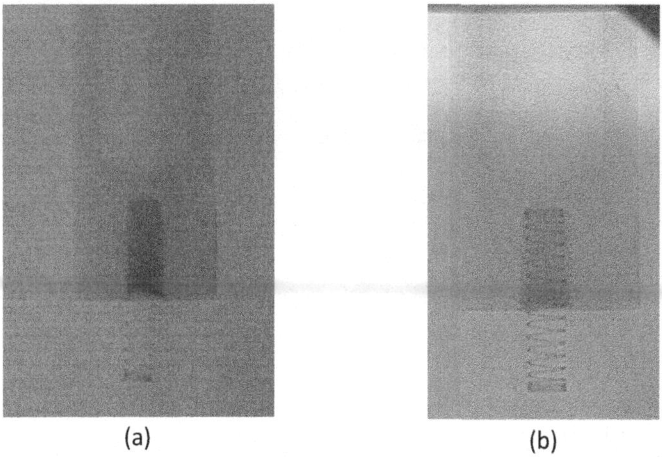

Fig. 6.12 Neutron radiograph of iron spring within an aluminium matrix **a** absorption image, **b** phase-contrast image

6.4.3 Free-Space Propagation-Based Neutron Imaging

If the analyser crystal in the analyser-based phase-contrast imaging setup is removed from the beam path, the neutron beam originating from the sample at various angles will propagate through free space until they reach the detector. A traditional absorption image can be obtained by placing the detector directly behind the sample. However, if the detector is moved away from a well collimated coherent neutron beam, Fresnel or near-field diffraction occurs [36]. To elaborate, the rays that do not pass by the edges of the object remains undeflected. Those that go through it are slightly deflected, and become out of phase when compared to the undeflected ones. The various sets of wavefronts superimpose and interfere at some distance behind the sample because they originate from a single coherent source. The "image" is formed due to the superposition of the distorted wavefront with the undistorted incident wavefront. This superposition gives rise to interference fringes at the edges or feature boundaries as there is a discontinuity in phase at these edges. These fringes improve edge visibility (contrast). In practice, the blurring of the image caused by the divergence of the neutron beam limits the maximum sample-to-detector distance [37]. Figure 6.12 depicts an iron spring encased in an aluminium matrix. It may be noted that thermal neutrons have an extremely small absorption cross-section in aluminium. Even for springs constructed of iron, the absorption image (Fig. 6.12a) shows very weak contrast. The phase contrast image (Fig. 6.12b) on the other hand, even for the aluminium matrix, displays substantially improved contrast, and the structure within the matrix is obvious due to the edge enhancement [38].

The edge-enhancement effect at the edges is illustrated in Figs. 6.13 and 6.14 using a conical piece of lead and iron syringe. Lead is high-Z element but has low neutron attenuation cross-section and high coherent scattering cross-section. The

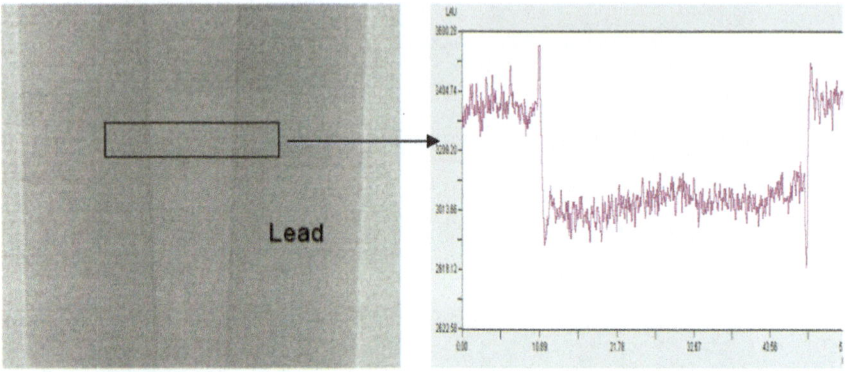

Fig. 6.13 *Left* Neutron phase radiograph of lead sample with conical hole and *Right* edge profile across highlighted area

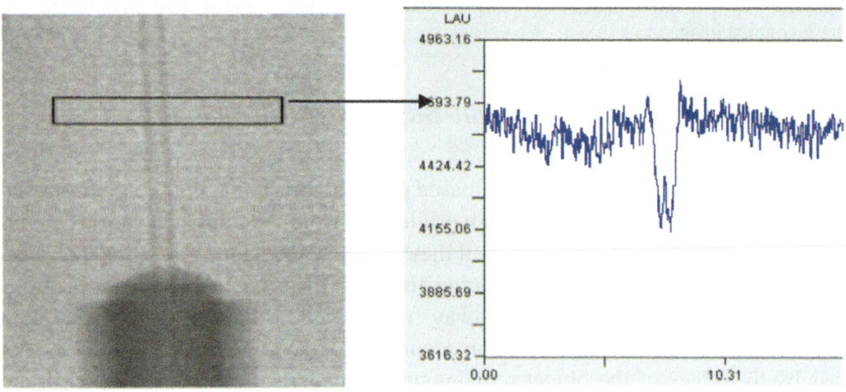

Fig. 6.14 *Left* Neutron phase radiograph of iron syringe and *Right* edge profile across the highlighted area

profile map across the drilled hole in the lead is shown in Fig. 6.13. In the plotted profile, the enhancement of intensity across the edges, owing to the phase effects is obvious. A similar enhancement effect across the edge of the iron syringe is also clearly visible in Fig. 6.14. These results conclusively demonstrate that this approach of phase-contrast imaging clearly produces superior images as compared to conventional neutron radiographs. One of the critical demands of this technique is that one need to have very coherent beam of neutrons. This demand increases data collection time as effective coherence property of the neutron source is achieved using a small aperture (~1 mm). Similarly, high-resolution detectors are needed to record the increased visibility at the edges or across the discontinuities within the sample.

The neutron beam must have a high degree of transverse spatial coherence, as characterized by the cross-correlation between two points in the wave (at all times), in order to obtain the phase-induced intensity variation in the image. The coherence area, whose width is given by the transverse coherence length, characterizes the extent of spatial coherence ($l_c = \frac{\lambda R}{a}$). It is dependent on the neutron wavelength λ, neutron source dimension a and the distance between source and object R. To improve phase effects in phase-contrast neutron imaging, the following must be achieved: (a) maximization of the R/a of the system and (b) maximization of the neutron beam effective wavelength. These goals must be achieved while meeting the accompanying design restrictions, such as a sufficient neutron flux on the image plane and an appropriate SNR for neutron radiography.

Like the previous phase-contrasting approaches, computed tomography can be performed again in order to re-build quantitatively the 3D distribution of the of the object refractive index. As discussed previously, phase tomography reconstruction algorithms can be divided into two classes. In the first case, one retrieves the phase using earlier described approaches and then uses the conventional filtered back projection (FBP) algorithm to reconstruct the real part of the refractive index. This is a two-step methodology and the process of phase retrieval and reconstruction are decoupled with each other. If the projections or the recorded images contain information about the second derivative of the some function $g(x, y, z)$, or more specifically, the Laplacian of the phase shift generated by the object, then the same (line-integral) can be expressed as:

$$L(y', z, \theta) = \int_{-\infty}^{\infty} \left(\frac{\partial^2 g(x', y', z)}{\partial^2 y'} + \frac{\partial^2 g(x', y', z)}{\partial^2 z} \right) \mathrm{d}x' \tag{6.12}$$

Therefore, the tomographical reconstruction utilizing the traditional FBP will not lead to an appropriate reconstruction of the original object function. Using the Fourier derivative theorem, a new filter function defined as follows can be derived [39]

$$F(k(\nu)) = \begin{cases} \frac{|\nu|}{\nu^2+\omega^2}, & |\nu| \le 1/2\Delta y' \\ 0, & |\nu| \le 1/2\Delta y' \end{cases} \tag{6.13}$$

where ν and ω are frequency components in two orthogonal direction in the Fourier space. This one-step approach, although less accurate than the earlier discussed two-step method, is easy to implement computationally. This algorithm was first derived by Bronikov in the context of X-ray phase-contrast tomography. We illustrate the same using an example of a carbon tube inside the lead cylinder, as discussed previously.

Bronikov version of back-projection technique was used for tomographic reconstruction of the lead (Pb)-containing carbon sinker sample in phase-contrast mode. We have generated radiography data over 180° in the step of 1°, and it was used as input to the reconstruction algorithm. Figure 6.15b shows the reconstructed image at the midplane of Fig. 6.15a, which is nothing but the phase radiograph of the object. The edge-enhancement effects in the phase-contrast mode help to increase contrast and makes it possible to image these materials using neutrons. Figure 6.15b brings

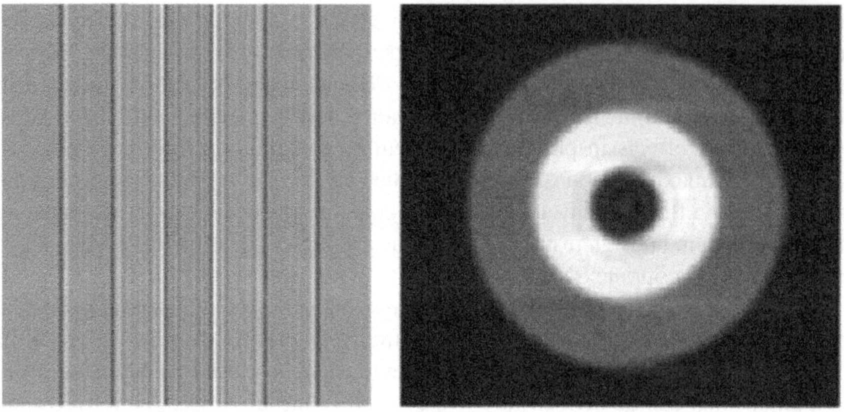

(a) (b)

Fig. 6.15 **a** Phase contrast image of carbon sinker enclosed in lead matrix, **b** tomographically reconstructed slice image

out the spatial distribution of the carbon sinker enclosed within lead cylinder. It may be noted that this technique is able to provide the distribution of coherent scattering length density without any need of neutron optics.

6.5 Polarized Neutron Imaging

It may be recalled that neutrons can interact with a magnetic field by the virtue of spin, and hence, are an excellent probe for studying magnetic field distributions. Therefore, the distribution of magnetic fields, and even electric fields within solid samples can be examined and visualized in three-dimensions, which is not possible with any other available experimental technique. Over the last decade, a number of experimental techniques utilising polarized neutron beam have been developed, allowing for the spatially resolved investigation of magnetic field distribution [40–47]. The most common methods for polarizing neutrons are as follows:

(i) Total external reflection from polarizing magnetic multi-layer-based supermirrors such as Fe/Si-based polarizing supermirror. This technique is most suitable to obtain the polarized neutron in cold energy region.
(ii) It is preferred to use Bragg reflection by single crystals such as Cu_2MnAl (Heusler crystals), to obtain polarized neutrons in thermal neutrons.
(iii) Polarized He-3 filters, which rely on the spin-dependent absorption of neutrons by ^3He; anti-parallel spins have a significant absorption cross-section, whereas all neutrons with parallel spins pass through the filter cell. The benefit of the filter is that it works well over a wide range of neutron energies.

It is worth recalling that the neutron's spin, which is oriented anti-parallel to its magnetic moment, will experience Larmor precession around the field, as it transverses region of magnetic field of intensity *B*. Polarized neutron imaging is based on measuring the precession angles of a spins of polarized monochromatic neutron beam that is transmitted through a magnetic field in combination of neutron imaging detector. The use of an area detector allows for the measurement of a spatially resolved magnetic field distribution. When working with a beam of neutrons, which is an ensemble of many neutrons, it is more convenient to deal with the polarization vector rather than individual spins. Precession of the polarization vector can thus be used instead of neutron spin (**S**). The polarization vector (*P*) may be thought of as the normalized average spin orientation of the whole beam and defined as expectation values of the Pauli spin matrices (σ) as $P = \langle\sigma\rangle = \frac{2}{\hbar}\langle S\rangle$.

That precession of the polarization vector as the neutron transverse through a region of stationary magnetic field (*B*) can be described by Bloch-equation as follows:

$$\frac{\mathrm{d}\vec{P}}{\mathrm{d}t} = \gamma_n[\vec{P}(t) \times \vec{B}(t)] \tag{6.14}$$

where γ_n is neutron's gyromagnetic ratio. As the polarization vector precesses around the magnetic field *B*(*t*), its instantaneous angular velocity of the precession is given by:

$$\omega(t) = \gamma_n\left|\vec{B}(t)\right| \sin\theta \tag{6.15}$$

where the angle θ is the angle between the vectors *P*(*t*) and *B*(*t*) and is independent of orientation of the magnetic field vector. It is important to note that if there are magnet field components perpendicular to the original polarization direction, neutron spins will fluctuate between the two spin-states, and consequently, the polarization vector will oscillate as well. The following formula may be used to calculate the rotation angle of the polarization vector around the magnetic field *B*(*t*):

$$\mathrm{d}\phi = \omega(t)\mathrm{d}t \tag{6.16}$$

Substituting the neutron path variable d*s* (d*t* = d*s*/*v*) for the time variable d*t*, one can obtain following relation:

$$\mathrm{d}\phi(s) = \omega\frac{\mathrm{d}s}{v} = \left(\frac{\lambda m}{h}\right)\omega\mathrm{d}s = \gamma_n\frac{\lambda m}{h}\left|\vec{B}(s)\right| \sin\theta\mathrm{d}s \tag{6.17}$$

where *v* and λ are the neutron velocity and wavelength, respectively, *m* is the mass of neutron, *h* is the Planck constant and d*s* is a short section of neutron path. Therefore, total angle of spin precession around the vector *B* can be easily obtained by integrating Eq. (6.17) across the neutron path:

$$\phi = \int \gamma_n \frac{\lambda m}{h} \left| \vec{B}(s) \right| \sin\theta \mathrm{d}s \tag{6.18}$$

The spin precession is dependent on the magnitude and direction of the magnetic field, as well as the neutron wavelength, as can be seen from the preceding equation. As a result, a monochromatic neutron beam with a longer wavelength (cold neutrons) is preferred to precisely investigate the magnetic field. It is worth noting that the precession angle can be linked to the integrated field along the neutrons path through the magnetic field, provided monochromatic neutron beam is used for carrying out the experiments. However, it is also possible to utilize polychromatic neutron beam, if the energy of neutron can be measured, and therefore, its wavelength distribution is known or can be precisely measured. Multiple observations with all conceivable combinations of neutron beam polarization vectors and in combination with polarization analyser are required for a complete three-dimensional reconstruction of the magnetic field vector *B*. A typical experimental arrangement for carrying out polarized neutron imaging experiments is shown in Fig. 6.16.

As seen from Fig. 6.16, the signal recorded by the two-dimensional neutron detector placed after the neutron analyser is nothing but a two-dimensional inhomogeneous intensity distribution (projection image), and the same can be expressed as a product of the contrast due to polarization vector rotation, $I_{\mathrm{p}}(x, y)$, and the traditional attenuation image, $I_{\mathrm{abs}}(x, y)$, as follows:

$$I(x, y) = I_{\mathrm{a}}(x, y) \times I_{\mathrm{p}}(x, y) = I_{\mathrm{a}}(x, y)\frac{1}{2}(1 + \cos\alpha(x, y)) \tag{6.19}$$

where

$$I_{\mathrm{a}}(x, y) = I_0(x, y)\mathrm{e}^{-\int \mu(x,y,z)\mathrm{d}z} \tag{6.20}$$

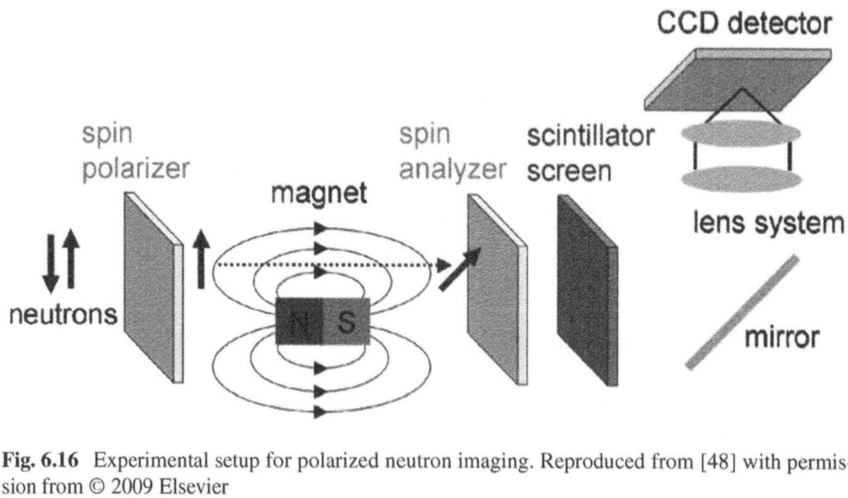

Fig. 6.16 Experimental setup for polarized neutron imaging. Reproduced from [48] with permission from © 2009 Elsevier

where $\mu(x, y, z)$ is the sample's linear attenuation coefficient as a function of spatial coordinates (x, y, z), $I_0(x, y)$ is the incident beam's intensity, and α is the angle the analyser maximum direction of transmission and neutron polarization vector. It should be emphasized that, as can be shown from Eq. (6.19), the measured intensity is only dictated by the sample attenuation if there are no magnet field components perpendicular to the initial polarization direction ($\phi = 0$ because no rotational precession is present for $\theta = 0$). Alternatively, if the magnetic field of the sample can be turned off (e.g. magnetic field created by a current), the impact of sample attenuation may be evaluated first and subsequently utilized for normalization and to isolate the magnetic field component from Eq. (6.19). Another option is to rotate the polarizer 180° and repeat the measurement for two spin components separately. The degree of beam depolarization can be reconstructed precisely [8] using spin up and down measurements for both with the sample and without sample, and the attenuation attributable to the sample can be easily isolated.

The other directions of magnetic field components can be studied by realigning the spin polarizer and analyser along the other axes. By varying the orientations of the spin polarizer and analyser, a series of nine measurements may be acquired, allowing the complete three-dimensional magnetic field distribution to be reconstructed. Figure 6.17a shows an example of distribution of trapped magnetic field in a Niobium superconducting radio-frequency cavity, and influence of AC external magnetic field on flux trapping behaviour by varying their amplitude, frequency and offset was investigated. The spatially resolved magnetic field distribution within the cavity shows that in contrast to type-I superconductor, the trapped flux seems to be homogeneously distributed. Similarly Fig. 6.17b shows three-dimensional distribution of magnetic field within the type-I superconductor lea at 4.3 K. This shows that polarized neutron imaging can be used to visualize and quantify an arbitrary magnetic vector; fields within bulk materials can be visualized and quantified in 3D

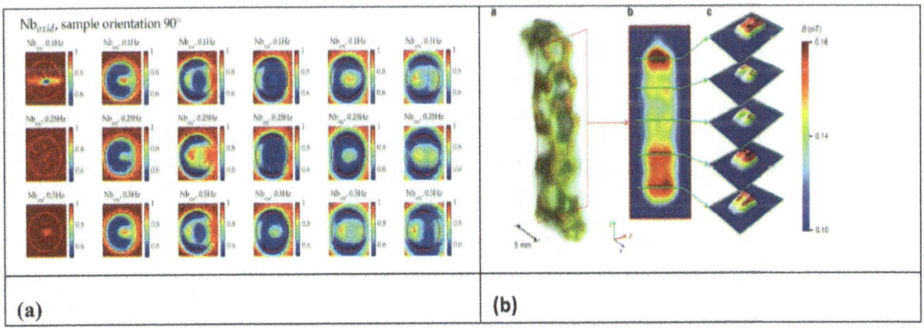

Fig. 6.17 **a** Nb_{oxid} sample (orientation 90°), B_{offset} = 0 mT … 5 mT (horizontal images) and increasing AC-frequency f = 0.1, 0.25 and 0.5 Hz (vertical images), blue ~0.5 and red ~0.9 normalized intensity [50]. **b** Tensorial neutron tomography measurement of the three-dimensional distribution of the y-component of the magnetic vector field inside the bulk of the superconductor at 4.3 K. **a** 3D visualization of the magnetic flux density, **b** colour-coded magnetic flux density (y-component) in the vertical cross section (yz-plane) and **c** magnetic flux distribution in selected horizontal planes as marked by the arrows [49]

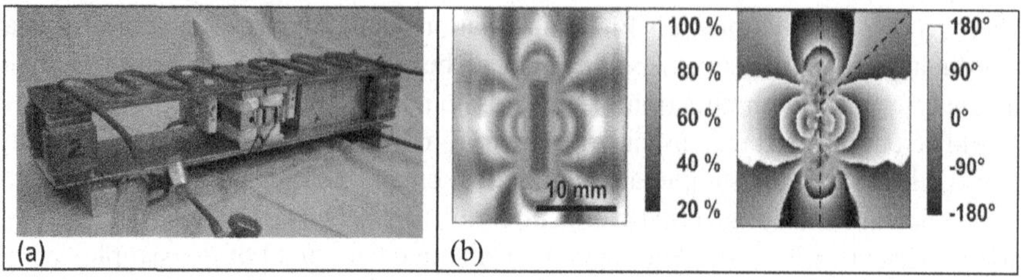

Fig. 6.18 **a** Photograph of the Ramsey apparatus. Reproduced from [51] with permission from Copyright © 2009 Elsevier. **b** A radiography image of 9 mm-long cylindrical ferromagnetic rod obtained using Ramsey technique showing magnetic field lines and corresponding phase shifts. Reproduced from [52] with permission from Copyright © 2009 American Physical Society

using a set of nine spin-polarized neutron imaging measurements and using novel tensorial multiplicative algebraic reconstruction techniques [49].

Another approach enabling polarized neutron imaging is to use Ramsey's technique of separated oscillating fields in conjunction with a neutron imaging detector, which is comparable to other spin-echo techniques like pulsed nuclear magnetic resonance or neutron spin-echo (Fig. 6.18). A neutron Ramsey apparatus consists of a region homogeneous steady magnetic field through use of permanent magnets or coils, and the spins of the (monoenergetic) polarized neutron beam is non-adiabatically flipped twice by 90° by a combination of two phase-locked fields oscillating perpendicularly the direction of steady magnetic field. A spin analyser is positioned between the neutron imaging detector and the second spin flipper, and by successively scanning the oscillating field around the Larmor frequency, quantitative distribution of the magnetic fields may be obtained through careful analysis of obtained Ramsey pattern.

6.6 Neutron Spin-Echo Imaging

As discussed in previous section, one can use different types of neutron optical elements to introduce modulation in the neutron beam. These structured modulations of the neutron beam intensity in presence and absence of the object helps to retrieve phase gradients that can be related to the real part of the neutron scattering length. The basic idea of neutron spin-echo imaging technique is to manipulate the polarization state of beam, and generate controlled spatial modulation of the neutron beam intensity by creating Larmor precession phase differences. Such modulations have been already used in small-angle scattering experiments.

Figure 6.19 shows the fundamental concept of the experimental setup. The setup is realized by two triangular magnetic field sectors with equal apex angles, in which the spins of a polarized neutron beam precesses around the external magnetic field. Magnetic fields within these triangular sectors are of opposing sign and strength, and these values are modified based on the distances between these devices and the

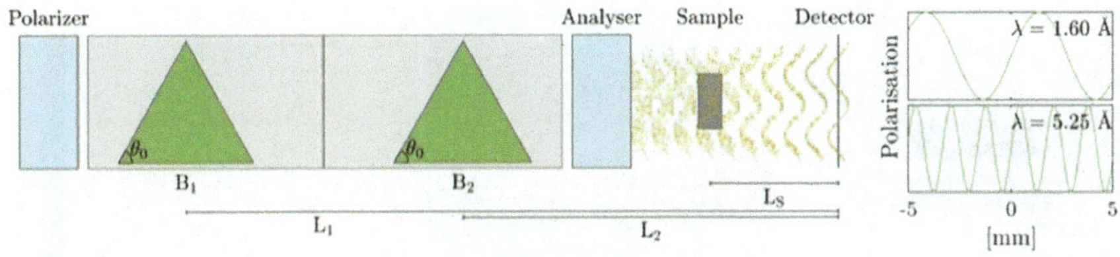

Fig. 6.19 Schematic illustration of experimental setup for carrying out neutron spin-echo imaging [53]

detector. The precession of neutron spin in the first magnetic field is fully compensated by the second magnetic field for those neutrons that arrive at the detector plane along the optical axis of the triangle magnetic field areas, fulfilling the requirement of a spin-echo of the precession in each device [47].

The precession of the neutron spin, along a path parallel to the optical axis, is proportionally to the relative distance from the optical axis. The precession of neutron spin in the triangular region of magnetic field B, as shown in Fig. 6.19, is dependent only on the height of the neutron path, say y, and be expressed as follows:

$$\phi = \frac{2\lambda m \gamma_n B}{h \tan \theta_0} y \tag{6.21}$$

where θ_0 is the inclination of the precession field surfaces to the beam, λ the wavelength and γ_n and m the gyromagnetic ratio and the mass of the neutron and the h is the Planck constant. For the neutron travelling along another path and arriving at the same height from the optical axis will likewise have a spin-echo, if $L_1 B_1 = L_2 B_2$, where B_1 and B_2 are the magnetic fields in the precession devices and L_1 and L_2 are the distances to the detector. The resultant spin precession is only dependent on the location of the neutron (y-coordinate) at the detector which can be simply written as

$$\phi = \frac{2\lambda m \gamma_n}{h \tan \theta_0} (B_1 - B_2) y \tag{6.22}$$

This precession results at the lateral modulation in the neutron beam intensity are recorded at detector with a period:

$$p = \frac{\pi h \lambda \tan \theta_0}{2m\gamma_n (B_1 - B_2)} \tag{6.23}$$

The period of the modulation is determined by the wavelength employed, the magnetic field settings and the field inclination. Figure 6.20 depicts an example of neutron spin-echo-based imaging. The object consist of two cuvettes kept on top of each other, while the top cuvette contained a magnetic metal powder of few micron grain size, and the lower cuvette was filled monodispersed polystyrene nanoparticles of 136 nm diameter suspended in D_2O solution. The pixel-by-pixel analysis of data

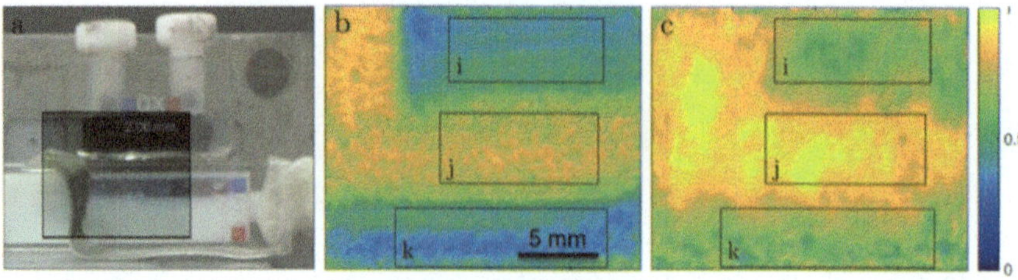

Fig. 6.20 **a** Sample setup photo with the exposed area highlighted. **b** Attenuation contrast image of exposed region. **c** Dark-field SEMSANS image displaying the visibility of the spin-echo modulation at a certain spin-echo length. Three areas of interest are highlighted in **b**, **c**: powder sample (*i*), empty beam area (*j*) and PS dispersion (*k*) [53]

showed good agreement with the complimentary SESANS measurements and theory curves describing the structural features of 1 μm and 136 nm for the metallic powder and the PS dispersion, respectively [53].

The data was analysed pixel by pixel and found to be in good agreement with complementary SESANS measurements and generated theoretical curves. The data could be modelled and explain properly with dimensions of 1 μm and 136 nm metallic powder, thereby validating the proposed approach.

6.7 Summary

We have outlined various advanced phase neutron imaging techniques, which can be easily implemented in a conventional neutron imaging facility. The phase-contrast imaging extends the use of neutrons for new class of materials like metal composites, metal foams, where the contrast in the attenuation-based imaging is expected to be quite weak. We believe that the phase-based neutron imaging will be useful in increasing utilization of neutron imaging for industrial problems. In parallel, neutron spin-dependent interaction with the matter can be used to map the electric and magnetic field within the bulk of the materials. Due to high penetration depth of neutrons in most of the materials, the polarized neutron imaging is uniquely placed. One can even combine the phase-sensitive interaction with polarized neutrons to obtain anisotropic distributions in magnetic fields.

References

1. Allman BE et al (2000) Phase radiography with neutrons. Nature 408(6809):158–159
2. Pfeiffer F et al (2006) Neutron phase imaging and tomography. Phys Rev Lett 96(21):215505
3. Treimer W et al (2003) Refraction as imaging signal for computerized (neutron) tomography. Appl Phys Lett 83(2):398–400

4. Dubus F et al (2005) First phase-contrast tomography with thermal neutrons. IEEE Trans Nucl Sci 52(1):364–370
5. Zernike F (1942) Phase contrast, a new method for the microscopic observation of transparent objects. Physica 9(7):686–698
6. Nomarski GM (1955) Differential microinterferometer with polarized waves. J Phys Radium Paris 16:9S
7. Gerchberg RW (1971) Phase determination for image and diffraction plane pictures in the electron microscope. Optik (Stuttgart) 34:275
8. Teague MR (1982) Irradiance moments: their propagation and use for unique retrieval of phase. J Opt Soc Am 72(9):1199–1209
9. Teague MR (1983) Deterministic phase retrieval: a Green's function solution. J Opt Soc Am 73(11):1434–1441
10. Colella R, Overhauser AW, Werner SA (1975) Observation of gravitationally induced quantum interference. Phys Rev Lett 34(23):1472–1474
11. Werner SA, Staudenmann JL, Colella R (1979) Effect of earth's rotation on the quantum mechanical phase of the neutron. Phys Rev Lett 42(17):1103–1106
12. Cimmino A et al (1989) Observation of the topological Aharonov-Casher phase shift by neutron interferometry. Phys Rev Lett 63(4):380–383
13. Allman BE et al (1992) Scalar Aharonov-Bohm experiment with neutrons. Phys Rev Lett 68(16):2409–2412
14. Badurek G, Rauch H, Tuppinger D (1986) Neutron interferometric double-resonance experiment. Phys Rev A 34(4):2600–2608
15. Klein AG et al (1981) Neutron propagation in moving matter: the Fizeau experiment with massive particles. Phys Rev Lett 46(24):1551–1554
16. Wagh AG et al (1997) Experimental separation of geometric and dynamical phases using neutron interferometry. Phys Rev Lett 78(5):755–759
17. Momose A, Takano H, Wu Y, Hashimoto K, Samoto T, Hoshino M, Seki Y, Shinohara T (2020) Recent progress in X-ray and neutron phase imaging with gratings. Quant Beam Sci 4(1):9. https://doi.org/10.3390/qubs4010009
18. Maier-Leibnitz H, Springer T (1962) Ein Interferometer für langsame Neutronen. Z Phys 167(4):386–402
19. Bonse U, Hart M (1965) An X-ray interferometer. Appl Phys Lett 6(8):155–156
20. Rauch H, Treimer W, Bonse U (1974) Test of a single crystal neutron interferometer. Phys Lett A 47(5):369–371
21. http://www.epsnews.eu/2015/04/a-neutron-optical-approach-to-explore-the-foundation-of-quantum-mechanics/. Accessed 15 Aug 2021
22. https://www.nist.gov/image/neutron-interferometer-crystalsjpg. Accessed 15 Aug 2021
23. Dubus F et al (2002) Tomography using monochromatic thermal neutrons with attenuation and phase contrast. In: International symposium on optical science and technology, vol 4503. SPIE
24. https://www.ndt.net/article/wcndt2004/pdf/radiography/141_zawisky.pdf. Accessed 15 Aug 2021
25. https://physics.aps.org/articles/v11/26. Accessed 27 Aug 2021
26. Sarenac D et al (2018) Three phase-grating Moiré neutron interferometer for large interferometer area applications. Phys Rev Lett 120(11):113201
27. Davis TJ et al (1995) Phase-contrast imaging of weakly absorbing materials using hard X-rays. Nature 373(6515):595–598
28. Hecht E (2002) Optics. Addison-Wesley, Reading, MA
29. Strobl M et al (2004) Neutron tomography in double crystal diffractometers. Phys B 350(1):155–158
30. Strobl M, Treimer W, Hilger A (2004) First realisation of a three-dimensional refraction contrast computerised neutron tomography. Nucl Instrum Methods Phys Res Sect B 222(3):653–658
31. Strobl M, Treimer W, Hilger A (2004) Small angle scattering signals for (neutron) computerized tomography. Appl Phys Lett 85(3):488–490

32. Pfeiffer F et al (2006) Phase retrieval and differential phase-contrast imaging with low-brilliance X-ray sources. Nat Phys 2(4):258–261
33. Strobl M et al (2008) Neutron dark-field tomography. Phys Rev Lett 101(12):123902
34. Reimann T, Mühlbauer S, Schulz M et al (2015) Visualizing the morphology of vortex lattice domains in a bulk type-II superconductor. Nat Commun 6:8813
35. Grünzweig C et al (2008) Bulk magnetic domain structures visualized by neutron dark-field imaging. Appl Phys Lett 93(11):112504
36. Wilkins SW et al (1996) Phase-contrast imaging using polychromatic hard X-rays. Nature 384(6607):335–338
37. Kardjilov N et al (2004) Phase-contrast radiography with a polychromatic neutron beam. Nucl Instrum Methods Phys Res Sect A 527(3):519–530
38. Kashyap YS et al (2012) Neutron phase contrast imaging beamline at CIRUS, reactor, India. Appl Radiat Isot 70(4):625–631
39. Bronnikov AV (2002) Theory of quantitative phase-contrast computed tomography. J Opt Soc Am A 19(3):472–480
40. Manke I et al (2010) Three-dimensional imaging of magnetic domains. Nat Commun 1(1):125
41. Schlenker M et al (1980) Imaging of ferromagnetic domains by neutron interferometry. J Magn Magn Mater 15–18:1507–1509
42. Podurets KM et al (1994) Neutron radiography with depolarized contrast. J Tech Phys 39:971
43. Leeb H et al (1998) Neutron magnetic tomography: a feasibility study. Aust J Phys 51(2):401–413
44. Strobl M et al (2007) Magnetic field induced differential neutron phase contrast imaging. Appl Phys Lett 91(25):254104
45. Dawson M et al (2009) Imaging with polarized neutrons. New J Phys 11(4):043013
46. Kardjilov N et al (2008) Three-dimensional imaging of magnetic fields with polarized neutrons. Nat Phys 4(5):399–403
47. Strobl M et al (2011) Polarized neutron imaging: a spin-echo approach. Phys B 406(12):2415–2418
48. Manke I, Kardjilov N, Hilger A, Strobl M, Dawson M, Banhart J (2009) Polarized neutron imaging at the CONRAD instrument at Helmholtz Centre Berlin. Nucl Instrum Methods Phys Res Sect A 605:26–29
49. Hilger A et al (2018) Tensorial neutron tomography of three-dimensional magnetic vector fields in bulk materials. Nat Commun 9(1):4023. https://doi.org/10.1038/s41467-018-06593-4
50. Treimer W, Junginger T, Kugeler O (2021) Study of possible frequency dependence of small AC fields on magnetic flux trapping in niobium by polarized neutron imaging. Appl Sci 11:6308. https://doi.org/10.3390/app11146308
51. Piegsa FM, van den Brandt B, Hautle P, Konter JA (2009) A compact neutron Ramsey resonance apparatus for polarised neutron radiography. Nucl Instrum Methods Phys Res Sect A 605:5–8
52. Piegsa FM, van den Brandt B, Hautle P, Kohlbrecher J, Konter JA (2009) Quantitative radiography of magnetic fields using neutron spin phase imaging. Phys Rev Lett 102:145501
53. Strobl M et al (2015) Quantitative neutron dark-field imaging through spin-echo interferometry. Sci Rep 5(1):16576. https://doi.org/10.1038/srep16576

Chapter 7
Safety, Regulations, Metrology and Standards in Neutron Imaging

P. S. Sarkar

7.1 Basic Requirement and Implementation Methodology

Everybody is exposed to some kind of radiation either natural (terrestrial, cosmic, etc.) or man-made sources (accelerator, isotopic sources, nuclear reactors, etc.). The radiation exposure due to occupation needs to be controlled. The exposures to individuals need assessment, and guidelines must be set towards enabling safety practices. ICRP clearly mentions the development of special dosimetric quantities related to radiation exposures taking into account their variation of effectiveness as well as variation in the biological effectiveness from one organ/tissue to the other. IAEA, through international cooperation, has adopted safety standards for the protection of health and environment keeping in mind the necessities. The safety guides must be followed, in addition to those framed by national regulators, to ensure minimum radiation hazards as far as possible on human health. The 9-key point mentioned in the Occupational Radiation Protection Call-for-Action by IAEA in 2014, in collaboration with other international organizations, clearly states the avenues to focus towards ensuring best practices for workers and safe use of ionizing radiation. Any laboratory or Institute using radiation source must have a qualified expert such as health physicist or radiation safety officer who individually or in coordination with a safety committee designs the best practices for human safety with minimum risk towards achieving the desired goals. Three TDS principles in implementing minimization of radiation hazards are (a) minimize time (T) in radiation areas, (b) maximize distance (D) from the radiation source or area and (c) employ proper shielding (S) of the source.

Radiation Quantities and Units: A unit is necessary for the measurement of any physical quantity. Radiation units are frequently reviewed and updated from time to time by the international commission on radiation units and measurement (ICRU).

P. S. Sarkar (✉)
Technical Physics Division, Bhabha Atomic Research Center, Mumbai 400085, India
e-mail: pss@barc.gov.in

D. K. Aswal et al. (eds.), *Neutron Imaging*,
https://doi.org/10.1007/978-981-16-6273-7_7

ICRP Publication 74 provides an extensive and authoritative set of data linking the operational quantities defined by ICRU with the dosimetric and protection quantities defined by ICRP. Some of the important radiation protection quantities are discussed below in brief.

The quantities of interest are Activity, Air Kerma, Exposure, Absorbed dose, Equivalent Dose and Effective Dose.

Activity: It is defined as the number of disintegration per second from a radioactive material.

Unit of activity in SI is Becquerel (Bq), where 1 Bq = 1 disintegration per second. Its old unit was Curie (Ci), where 1 Ci = 3.7×10^{10} Bq = 37 GBq.

KERMA (K): The radiation field of indirectly ionizing radiations at any point in matter is given by the quantity KERMA (K), an acronym for "kinetic energy released per unit mass". It is defined as the sum of the initial kinetic energies of all charged particles liberated by uncharged radiation in a material of mass 1 kg and its unit is J/Kg. Its unit in SI is named as Gray (Gy), where 1 Gy = 1 J/kg. When Reference material is air, the quantity is called as Air Kerma.

EXPOSURE(*X*): It is defined as the absolute value of the total charge of the ions of one sign produced in air when all the electrons liberated by photons per unit mass of air are completely stopped in air and is denoted as *X* and is defined by Eq. (7.1)

$$X = \mathrm{d}Q/\mathrm{d}m, \tag{7.1}$$

where dQ is the amount of charge developed in dm quantity of mass. It is expressed in SI unit as *C*/kg. The old unit was Roentgen (R) and expressed as 1 R = $2.58*10^{-4}$ C/kg (air) or 1 C/kg = 3876 R. The exposure rate (X/t) is defined as exposure produced per unit time.

Dose (*D*): The effect of radiation (either physical, chemical, biological or their combination) on a medium mainly depends on the energy absorbed by the medium. The amount of energy absorbed per unit mass of matter at the point of interest is called as absorbed dose or simply dose and is denoted in SI unit as 1 Gy = 1 J/kg. The old unit was rad, and it is related as 1 Gy = 100 rads.

Equivalent Dose (H_T): Effects of radiation depend on the type of ionizing radiation and on the medium on which it is calculated. It is as per the energy loss per unit length of the medium. Hence, the biological damage caused by the same dose from different radiations may be different as the energy losses are different for different radiations. 1 Gy of alpha dose is found to be more effective than 1 Gy of gamma dose. Hence, in radiation protection, to account for this variation in the effectiveness of different types of radiations, a factor called radiation weighting factor (W_R) is accounted for the absorbed dose due to each type of radiation towards the production of stochastic effects. The weighted absorbed dose in any tissue is called the equivalent dose H_T and expressed as $H_T = \Sigma D_R\ W_R$, where D_R is the average dose over the tissue or organ due to radiation and the summation is accounted over all radiations (R). Since W_R is a dimensionless quantity, the unit of equivalent dose is J/kg. In SI units, it is denoted as Sievert (Sv) and 1 Sv = 1 J/kg. Earlier its unit was "rem" and related to Sv as 1 Sv = 100 rem. Table 7.1 provides the values of W_R for different types of radiation.

Effective Dose: The effective dose to the whole body is calculated by taking into account the doses received by individual organs or tissues considering their radiosensitivity, seriousness of damage and curing susceptibility varies from one organ or tissue to the other. This is accounted by a weighting factor, called tissue weighting factor (W_T), defined as per the extent of the stochastic effect in the organ/tissue. The effective dose (E) is defined as,

$E = \Sigma W_T \cdot H_T$, where W_T is the weighting factor for organ or tissue T and H_T is the equivalent dose in organ or tissue T and the summation is accounted for all the tissues (T) affected. Organ/tissue-specific W_T is provided in Table 7.2.

Ambient Dose Equivalent ($H^*(10)$): This is defined as the dose equivalent at a point, lying at a depth 10 mm in the ICRU sphere, in a radiation field that would be produced by the field expanded and aligned in the direction of the measured point. The ICRU

Table 7.1 Radiation weighting factor (W_R) for various types of radiations (ICRP 103, 2007)

Type	Energy range	W_R
Photons	All	1
	All	1
Neutron	<1 MeV	$2.5 + 18.2 * e^{-([\ln(E)]^2)/6}$
	1–50 MeV	$5 + 17 * e^{-([\ln(2E)]^2)/6}$
	>50 MeV	$2.55+3.25*e^{-([\ln(0.04E)]^2)/6}$ (the maximum is at 1 MeV) E is the energy of the neutron
Protons, other than recoil protons	All	2
Alpha particles, fission fragments and heavy nuclei	All	20

Table 7.2 Organ/tissue weighting factor

Organ/Tissue	W_T	Organ/Tissue	W_T
Bone marrow	0.12	Lung	0.12
Bladder	0.04	Liver	0.04
Bone surface	0.01	Oesophagus	0.04
Brain	0.01	Salivary glands	0.01
Breast	0.12	Skin	0.01
Colon	0.12	Stomach	0.12
Gonads	0.08	Thyroid	0.04
Remainder	0.12	Total	1.00

sphere, a phantom, is made up of 30 cm tissue-equivalent plastic (density ~ 1 g/cm^3), made from oxygen, carbon, hydrogen and nitrogen with 76.2%, 11.1%, 10.2% and 2.6% mass composition, respectively. Ambient dose equivalent was conceptualized towards developing a unified metrology standard for all types of radiation, relatable and directly measurable through, in some cases using simplified phantom models of the human body, considering different radiation qualities towards ensuring adequate radiation protection. The reference depth of 10 mm was envisioned by ICRU as maximum dose equivalent happens in this depth for penetrating radiation [1].

Directional Dose Equivalent ($H'(d, \Omega)$): This is defined as the ambient dose equivalent in a specified direction Ω.

Personal Dose Equivalent ($H_p(d)$): This is defined as the dose equivalent in a soft tissue, for both weakly and strongly penetrating radiation, at a depth "*d*" below a specified point on the body. As per the latest ICRU guidelines, *d* is taken as 0.07 mm for skin and 3 mm for eye. $H_p(10)$ is the operational radiation quantity which is measurable and gives the estimate of effective dose having annual dose limits.

Operational Limits

The International Commission on Radiological Protection (ICRP) sets up guidelines on the safety standards in the use of ionizing radiations taking into account all the current scientific information available on radiation effects [2].

System of Dose Limitation: ICRP has recommended a system of dose limitation based on the following principles:

1. **Justification**: No practice shall be adopted unless its introduction produces a positive net benefit.
2. **Optimization of protection**: All exposure shall be kept **A**s **L**ow **A**s **R**easonable Achievable (**ALARA**), economic and social factors being taken into account.
3. **Application of dose limits**: Dose to individuals shall not exceed recommended Limits.

Dose limits are designed to achieve the following:

1. Prevent deterministic effects.
2. Minimize the stochastic effects to levels which are acceptable.
3. Ensure that practice involving radiation exposures is justified.

As per IAEA Basic Safety Standard (BSS), exposures can be categorized into two types, normal exposure arising out of certain medical or industrial practices having predictable magnitude and potential exposure arising out of unexpected circumstances such as accident and equipment malfunction. The former type can be controlled by using strict restrictions towards only the required dose delivery such as in medical imaging or radiotherapy. The later type can be controlled through design and implementation of proper shielding, administrative control and operating procedure. Radiation exposures can be divided into three categories: occupational, public and medical. Table 7.3 details the recommended dose limits as per ICRP.

Table 7.3 Dose limits recommended by ICRP

Application	Dose limit	Public
	Occupational workers	
Whole body (effective dose)	20 mSv per year, averaged over the defined period of 5 years. With no more than 50 mSv in a single year. The limit set by AERB in India is 30 mSv	1 mSv in a year, averaged over 5 year period
Parts of the body (equivalent dose)		
Lens of the eye	20 mSv per year	15 mSv per year
Skin, hands and feet	500 mSv* per year	50 mSv per year
Equivalent dose to surface of abdomen of pregnant women	1 mSv after declaration of pregnancy and up to term	

*Averaged over areas of not more than 1 cm^2, regardless of the area exposed

7.2 Human Safety

Human safety is at the foremost important part, and all possible measures must be employed to ensure minimum radiation risk to the working personnel involved in any radiation facility. Radiation safety, electrical safety and other mechanisms such as interlock, administrative control and warning signs are the important mechanisms implemented to ensure human safety.

7.2.1 Radiation Safety

The biological effects of ionizing radiation [3] upon humans can be either deterministic or stochastic. When the radiation dose is above some threshold (>0.1 Gy), the effects shortly arising after exposure to the radiation dose is called deterministic. Acute radiation sickness (such as nausea, vomiting, tissue burns) and chronic radiation sickness (such as cataract, cancer and mutation) are two types of deterministic effects. In case when there is no dependence on threshold and severity but on the probability of occurrence, the effects are termed as stochastic. Somatic and genetic are two types of stochastic effects.

Radiation Shielding for Neutron Beamline

Neutrons interact with material through capture and scattering mechanism. The parameter defining the interaction probability with any material is called its cross-section. Since neutrons, irrespective of its energy, interact with the nucleus, they have the potential of changing the nuclear property of the materials upon interaction. Also, any neutron source will also have some gamma emission associated with it. Both neutrons and gamma possess radiation hazard for all personnel working with it or around the facility generating neutrons. This is why shielding of any neutron source is dealt with strictly as per the laid down procedures given in the guides and codes of regulatory authority. The emphasis is to design the experimental station involving neutron beam in such a manner and use suitable materials so that neutrons along with gamma (produced by neutron-induced interactions) should not exit the shielded enclosure. Also, care should be taken to provide shielding to the components used in the experimental station as to increase their operational life and minimizing activation due to prolonged exposure to neutrons. Neutrons interact with materials through elastic scattering, inelastic scattering, radiative capture and multiplication. So the choice of materials for shielding and their construction depends upon the neutron source spectrum.

Generally, low Z elements interact better with neutrons for elastic scattering due to their high cross-section. Materials such as hydrogenous compound or graphite, through multiple collisions reduce the energy of fast neutrons, and hence, their use is widespread for neutron shielding material in facilities generating fast neutrons. Since the mean free path of neutrons in graphite is approximately 6 times than hydrogen, hydrogenous materials are the best choice towards compact shielding design. Depending upon the neutron spectrum, the gamma produced upon their interaction with shielding materials needs to be taken into account even though their numbers are less. For example, thermal neutron capture in hydrogen and nitrogen gives 2.2 and 10.8 MeV gamma rays. Since the interaction cross-section of most materials with fast neutrons (~1 MeV or more) are couple of orders less than the thermal neutrons, they need to be slowed down through multiple collisions to the thermal neutron energy range before being absorbed. Moreover, fast neutrons interacting with carbon, nitrogen and oxygen through inelastic scattering produces 4.43, 5.15 and 6.12 MeV gamma lines respectively. To summarize, neutron shielding must have two parts: moderator and absorber.

To perform any shielding design, understanding of the depth of penetration of neutron of all the energies (either in binned or un-binned manner) present in the spectra is of prime importance. Neutron transport through materials can be described using the Boltzmann transport equation, and the methods to solve it are Monte Carlo method and deterministic method. In Monte Carlo method, the transport equation is not solved directly but history of every particle interaction is tracked through energy and change in direction till its existence in the material becomes insignificant. MCNP, GEANT4, FLUKA, PENELOPE are some commercially available Monte Carlo simulation packages. Deterministic methods are based on various approximations, and among them multi-group approximation is widely used [4]. Interested readers may delve deeper into specific books related to these subtopics.

In reactor-based imaging facilities, the experimental hutch shielding consists of layers of high-density borated polythene and lead in an interlocking manner. The boron part takes of the absorption of thermal neutrons producing 0.42 MeV gammas, and this can be shielded by suitable thickness of lead. These slabs are held in position using heavy metallic structures, preferably grouted to the floor. The portion of the shielding facing the beam port needs extra shielding as it faces directly the neutrons which are attenuated a little by the presence of components in its path used for experimentation such as sample, imaging unit and accessories. For example, in the neutron imaging beamline [5, 6] at Dhruva reactor, India, 400 mm thick borated polyethylene (BPE) blocks and 200 mm thick Pb has been used for the walls and 600 mm BPE in combination with 200 mm Pb has been used for the shielding hutch door facing the beam port. The total weight of Pb and BPE for the shielding motorized door is approximately 12 tonnes. This has been detailed in Chap. 6.

External shielding can also be provided by BPE and concrete slabs in an interlocking mechanism to avoid leakage of radiation by any means. This has been implemented in the TREAT (Transient REActor Test) radiography facility [7] used for imaging of irradiated fuels, which are highly radioactive and that is the reason for using concrete blocks. Figure 7.1 shows the picture of the TREAT neutron radiography facility having shielding walls made up of BPE and concrete. In case of neutron imaging in high radiation environment, the imaging system shielding also needs special attention and BPE in combination with Pb blocks is used for reducing the neutron and gamma background which creates hot spots in the image, thus reducing image quality [8]. Care should also be taken to remove the detector electronics heat through forced air or similar mechanism for reducing detector noise as well as maintaining its life. Figure 7.2 shows the new improved neutron imaging system designed for radiography stations at the Idaho National Laboratory where two-mirror mechanism, and enhanced shielding has resulted in lesser background and hence realization of better image quality. Implementation of vent and exhaust has resulted in lowering thermal effects in the camera system.

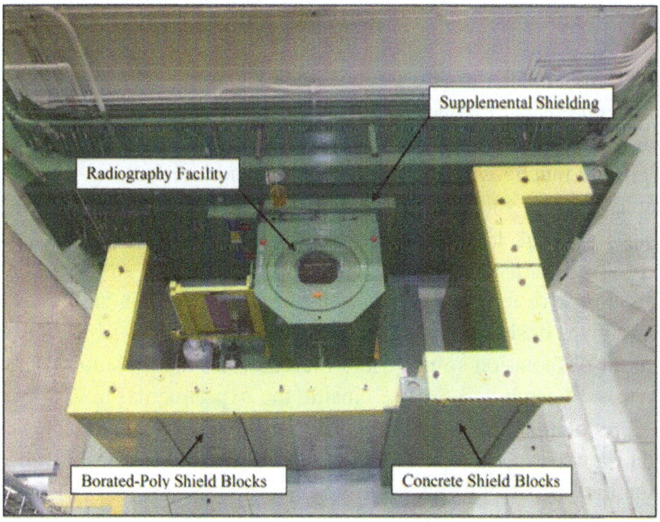

Fig. 7.1 Neutron radiography facility at the TREAT [7]. Reproduced with permission from Rightslink CCC

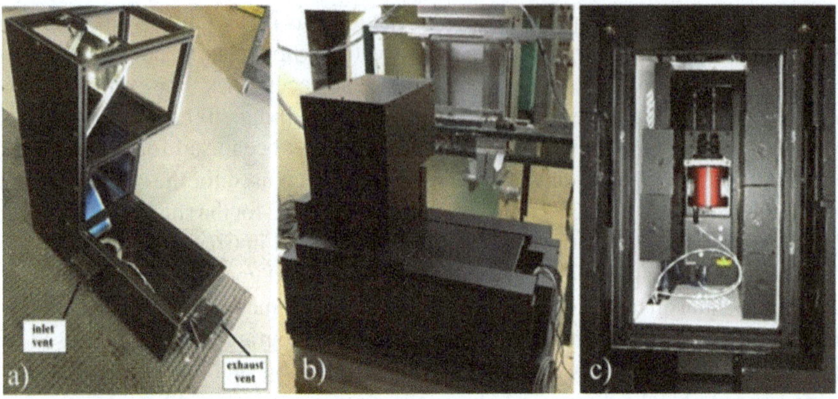

Fig. 7.2 Photographs of the improved imaging system realized at ANTERES where **a** shows the utilization of two mirrors to move the camera further away from the beam towards reducing background, incorporation of vents to reduce thermal effects in the camera, **b** robust shielding provided to the imaging system and, **c** incorporation of motorized mechanism towards online focusing remotely [8] (Open Access)

Safety Mechanisms Incorporated in Facilities Towards Implementing Best Practices

Interlock: Interlock is a means of Engineered Control and is implemented to control radiation by shutting off radiation or not allowing radiation-related risk in any event of emergency or malfunction of any related equipment to avoid accidental exposure to the personnel involved in the facility. In the case of a reactor neutron imaging beamline, various radiation area monitors (RAM) are placed, some of them are located inside the experimental hutch (EH), and some are outside. The RAM inside the EH is interlocked with the gate shutters and EH door to decide whether EH is safe for personnel entry as per the facility safety guidelines [9]. Neutron has the property of activating various materials used in the beamline. These activations will produce some background much higher than permissible, and interlocking with the EH door will prevent its opening and entry of personnel till the background radiation subsides to the permissible limit. For accelerator-based facilities, interlocks are provided to inhibit the production of radiation by various means such as stopping the ion beam reach the target, inadvertent entry door opening, vacuum system fault, cooling system fault, increase in the dose level more than the permissible value inside the accelerator hall on account of radiation source yield increase, fire alarm, smoke detection, etc. Options such as "Search and Secure" inside the experimental hutch or accelerator facility are a recommended practice before closing the entry/exit door and start of the radiation for experimental purpose. This procedure must be followed every time after opening the facility entry/exit door.

Administrative Control: Strict rules regarding restrictions and ensuring follow of safety practices in controlled areas are collectively called as administrative controls. Facility-specific protocols are designed towards allowing personnel access, working methodology, supervision and checking of written documents for all instances during the operation of the facility. Following safe work practices will avoid radiation risk and if in case of any untoward incidence, critical review of the laid down procedures will help in analysing the situation towards implementing further rules in avoiding such instances.

Warning Signs (Audio/Visual): Any facility dealing with radiation must have the warning symbols both in the form of picture or sign and also in written in languages familiar to all the personnel working in the facility. Signs such as "Radiation Area", "High Voltage", "Caution", "Hot Surface", "RF Source", "Mind Your Steps" and "Heavy Shielding Door" are common. Visual display by illuminating signs can also be used. Audible warning system incorporating flashing light should be integrated into the facility operating system to warn all the persons present at the start as well as during the operation of the system. Radiation sources must be labelled properly mentioning the type of source, name, contact dose and the activity with date signed by the respective authority.

Radiation Area Monitors/Survey Instruments: Radiation area monitors are an integral part in any radiation facility in order to provide local radiation level towards ensuring safe practices to be followed. They are generally fixed at locations intended to be monitored for dose levels and can relay the signal to the observer location. Modern RAMs provide display and audible alarm with respect to the set threshold. They are available for different types of radiation environment and dose levels. They can be easily interlocked with the system and ensure no tolerance to radiation risk. Survey metres on the other hand are portable instruments used to evaluate radiation levels at intended locations such as isotope source assembly and activated materials after neutron irradiation. These types of instruments are based on ionization/proportional/Geiger–Muller/scintillation detectors and are mostly used for dose or activity measurements on X-ray or gamma ray. For neutron radiation measurement, the neutrons are either scattered elastically to generate protons which are then detected through scintillation mechanism or moderated (using polyethylene), absorbed in specific materials having high absorption cross-section (^{6}Li, ^{10}B, ^{157}Gd) and finally detect the scintillations. Thermal neutron survey metres (e.g. Ludlum make) have ^{3}He or BF_3 detector housed in high-density poly-ethylene moderator. Since the availability of ^{3}He has been scarce, various other methodologies have been adopted towards developing neutron radiation survey metres. Gas counter having mixture of methane and nitrogen embedded in bulk polythene are commercially available from Fuji Electric. These instruments can measure neutron dose in the range of 0.1 nSv to 100 mSv. They are to be periodically calibrated and maintained. Survey instruments providing the dose and spectrum of the radiation field are also available and are helpful in designing safety-related protocols.

Personnel Dosimetry: Radiation dosimeters are passive or active detectors intended to record total amount of external radiation received by an individual. These are mandatory in recording the dose received by an individual involved in a radiation generating facility in due course of time and ascertain the safe work practices in the facility as well as his /her lifetime dose. Various types of personal dosimeters are available such as thermoluminescent detectors (TLD), optically stimulated luminescent detectors (OSL), film badges, electronic personnel dosimeter (EPD) and pocket ion chamber (PIC). Dose reading recorded by TLD, OSL and film badges are processed after specific monitoring period such as quarterly or monthly. EPD- and EPC-type dosimeters are capable of providing real-time dose data.

Towards calibration of routine dosimetry systems for radiation processing ISO/ASTM 51261:2013 specifies the requirements for calibrating routine dosimetry systems for use in radiation processing, including establishing measurement traceability and estimating uncertainty in the measured dose using the calibrated dosimetry system.

7.2.2 Electrical Safety

All equipment using electricity for its operation has the potential to cause harm in terms of electrical shock, fire or arc flash, if proper care is not taken to resist it. The severity may depend upon various factors such as magnitude of current, time exposure and humidity conditions. Some important points regarding electrical safety are as follows:

- Ensure proper safety ground to shunt away any dangerous currents arising out of malfunction or accident. All the wiring must be carried out with the highest professionalism, and there must not be any open.
- Since accelerator facilities involve very high voltages and currents, requisite earthing and periodic maintenance of earth pits must be ensured.
- All electrical points must be labelled with warning signs related to its nature.
- In case of high voltage equipment, safety mechanisms must be in place to restrict personnel from accessing them when they are in ON condition. This is done by making enclosure with interlock. Care must be taken to ensure complete de-energizing them prior to their use after they have been put OFF.
- Only qualified persons should handle the electrical equipment, in case they are of special requirements.
- If imaging systems working on electrical power are sensitive to humidity conditions, necessary precautions must be taken into account to ensure their performance.
- Laid down facility-specific procedures must be followed while putting ON / OFF the systems.
- Short circuit, overloading, loose electrical connections, earth fault, etc. are major contributors towards the cause of electrical fire and must be avoided.
- IEC standards such as 60059, 60060, 60073, 60204, 60231, 60243 and 60270 are to be applied wherever required.

7.2.3 *Material Safety*

Neutron facilities, small or big, have heavy shielding materials in and around them to minimize radiation risk. Local shielding inside a facility is done by stacking one above the other as well as sideways in order to provide adequate thickness of specified materials. Right from the design to the installation various protocols must be followed towards ensuring safety. These are

- Design must incorporate the neutron source type, energy, yield and available area.
- ISO 14152 must be adhered to while choosing materials for neutron and gamma shielding.
- Shielding material must retain its physical characteristics in the facility environment and must have resistance to impact, wear, etc.
- If possible, materials must be painted with permissible paints to minimize corrosion or oxidation if any.
- Must be fire resistant and should not provide any thermal load to the facility
- Load distribution must be ensured as per facility floor (or on which the shielding is laid) load characteristics
- Interlocking of heavy materials ensuring their stability in case of any seismic activity
- Materials should not be placed so as to obstruct movement of personnel. In case, if it is unavoidable, proper sign must be displayed at appropriate locations.
- Overhead cranes if used for erection of the shielding must be operated by trained and qualified operator.
- Since neutrons activate materials such as iron, stainless steel aluminium which is generally used in beamline hutch, activity measurements are a must ensuring safety of the working personnel.
- Beam dump must ensure no radiation leak outside the hutch above the permissible radiation level. In case where the entry/exit door also serves the purpose of beam dump, being heavy, due considerations must be followed towards its smooth operation.
- Toxic and corrosive materials must be avoided or kept as per the stipulated methodologies; otherwise, they can be a major concern for fire. As, for example, chromium, cobalt, iron, manganese, nickel and zinc react with acids. Similarly, aluminium and zirconium react with caustic alkalis. In such cases of fire, water must be avoided as there is a chance of severe steam explosion.

7.2.4 *Radiation Hazard and Precautions Towards Handling DD/DT Neutron Generator (NG)*

Radiation safety measures with proper shielding and standoff distance for remote operation are the principles to be followed to ensure best practices. Neutron and gamma shielding is generally taken care of by high-density concrete walls, whose thickness depends upon the neutron energy and yield. Additional arrangements of borated high-density polyethylene blocks may be used if required. In any case, radiation dose in the control room and accessible areas around the building or laboratory should not increase more than the permissible limit set by AERB. Typical thickness of shielding to reduce neutron flux and dose rate by a factor of ten is about 38cm of water or concrete. Failsafe safety mechanisms utilizing interlocks must be employed such that any breach to the radiation area should shut down the system without fail. As a matter of practice, health physicists are supposed to check radiation dose prior to entry to the NG area after shutting OFF the system and ensure follow of safety guidelines. Safety protocols followed for different types of DD/DT NG are summarized below.

(i) In case of DD NG, there is no radioactivity issue as no radiation emanates from it when the generator power supply is OFF. But regulations laid by national authorities must be followed when storing or disposing off the deuterium target.

(ii) In the case of DT NG, since tritium is radioactive, its handling needs special attention if it is an open type system or there is a mechanical breakage in the case of a sealed type system. The escape of tritium is very rare but cannot be ruled out. Hence, the following precautions must be followed.

(ii-a) Open type systems:

- The vacuum system of the accelerating column should be connected to a stack maintained at lower pressure.
- The change of tritium target and suitable packing of the used target have to be strictly carried out in a glove box. The glove box needs to be connected to the target assembly during this process. The used and eroded targets should be disposed-off as per country laid regulations.

(ii-b) Sealed systems:
In case of sealed systems, the barriers towards shielding are provided by the housing module and the target metal (generally zirconium, titanium, vanadium in hydride form) in which the tritium is adsorbed. In the case of mechanical breakage of the sealed system, the following protocols must be maintained.

- do not panic.
- ensure that the high voltages are put OFF.
- carefully decouple the NG from the connected cables after proper check of the radiation level.
- seal the housing NG assembly in a plastic bag and follow the safety regulations for either disposing it or repair by the manufacturer.

Since NGs can activate samples, they are to be handled with due safety regulations. Severity of activation depends upon various factors such as material, flux level and time of irradiation. Even the NG components (SS, Cu) exposed to the neutrons can get activated, and necessary checks are required before accessing them after irradiation/experimentations.

7.3 Neutron Metrology (NM)

Metrology is defined as the science of measures or measurements. According to Lord Kelvin, "To measure is to know" and as per D.L. Mendeleev, "Science begins when they start to measure". In any sort of development, starting from basic needs of life to the far extragalactic objects in space, measurements and the methodologies to perform them in an un-ambiguous manner, needs careful procedures to be laid down with validity and scientifically proven knowledge. For performing any measurement, procedural reasons are carried out which is then standardized in a form acceptable to all concerned so that the fruits are shared with all with uniformity. This has manifold advantages such as removes language barriers, enhances knowledge base, development of confirmatory techniques towards increasing trade commerce and so on.

The Comite International des Poids et Mesures (CIPM) is the world's highest authority in metrology. Bureau International des Poids et Mesures (BIPM) was set up by Metre Convention signed on 20 May 1875. CIPM is involved in all metrological-related work with all member state and discusses with BIPM under the delegated authority of CGPM.

BIPM has been tasked in establishing fundamental standards and scales, comparing national and international standards, ensuring coordination between different measurement techniques and coordinate measurements of the fundament of physical constants. The initial activities of BIPM were limited to measurement and metrological studies related to length and mass. Later on they were extended on standardizing electricity (1927), Photometry and Radiometry (1937), Ionizing radiation (1960), time scale (1988). Since 1965, an international journal, "Metrologia" under the guidance of CIPM, deals with publication related to development of standards, upgrading-related activates, decision and recommendation as per metre convention. In 1969, the consultative committee for standards of Ionizing radiation (CCEMRI) established four sections and Section III was for Neutron Measurements. CCEMRI was later and, in 1991, was renamed as consultative committee of ionizing radiation (CCRI). Neutron metrology (NM), though initially associated with nuclear industry recently added, accelerator-based neutron sources used for application in basic research, applied research and medical therapy. Its scope has broadened with more challenges. Accredited national laboratories from different countries work in unison with BIPM towards progress and implementation of developed standards in their respective nations. CIPM Mutual Recognition Arrangement (CIPM MRA) is a framework through which National Metrology Institutes demonstrate their measurement standards to match the international equivalence and the calibration and measurement certificates they issue.

International Commission on Radiation Units and measurements (ICRU) was conceived in 1925 in London with an objective towards proposing a unit of measurement of radiation as applied to medicine. ICRU report 85a (revised) defines fundamental quantities and units for ionizing radiation.

Regarding the development of internationally accepted recommendations, the objectives of ICRU are towards:

1. Quantities and units of radiation and radioactivity.
2. Procedure suitable for the measurement and application of these quantities in various fields and activities related to radiation.
3. Assurance of uniformity in reporting ICRU works closely with other similar organizations such as US National Council of Radiation Protection (USNCRP), IAEA, WHO, UNSCEAR, ISO, BIPM and CIPM.

A network of Secondary Standard Dosimetry Laboratories (SSDL) has been set up by IAEA in collaboration with WHO in 1976 [10]. These SSDLs are designated by individual IAEA Member States. They provide linkage regarding national and international dosimetry and measurement standards having reference to BIPM. They were also made responsible towards the dissemination of S.I. quantities and units through the proper calibration of field instruments. The IAEA maintains the Dosimetry Laboratory in Seibersdorf, Austria, which is the central laboratory of the IAEA/WHO Network of Secondary Standards Dosimetry Laboratories. The quantities used for radiation dose measurements are established in Primary Standards Dosimetry Laboratories (PSDL).

Primary standards for neutron metrology at SSDL and as follows:

1. Source emission rate (Mn bath)
2. Thermal neutron fluence (Au foils)
3. Fast neutron fluence (0.5–10 MeV): Neutron Telescope.

Secondary standard at SSDL should be established as per the guidelines set by PSDL. Radiation standards section of RSSD, BARC is recognized as the designated institute for ionizing radiation metrology in India. IAEA recognizes BARC as a secondary standard dosimetry laboratory. As and when required, BARC has developed primary, secondary and tertiary standards regarding neutron source yield and neutron fluence rate measurement.

ICRU report 66(2001) involves the determination of operational dose equivalent quantities for neutrons. ISO 8529 deals with reference neutron radiation fields and having energy range from 25 MeV to 20 MeV, used for calibration of neutron measuring devices and determining their response as a function of energy for radiation protection purposes. ISO 8529 is based on the technical report series 285 of IAEA (1998).

Components involving neutron imaging are neutron source, related collimation, shielding materials, detectors and finally analysis methods. The design of neutron imaging system or facility is primarily based on neutron spectrum as the shielding material including the detection, and imaging methodology will be different from fast, thermal, epithermal or cold neutron source. So, to start with one needs to have complete idea of neutron spectrum, and it needs measurement and quantification as per national and international standards. This importance can also be stressed as the health and safety of both instruments and personnel are linked to it. Depending on this, neutron fluence and dose rates can be ascertained using high purity metal foils, activating single or in groups and analysing the activation related γ-signature provides neutron fluence details. ASTM [11] E261-16 details standard practise for determining neutron fluence, fluence rate and spectra by Radioactivation technique, whereas ASTM E262-17, specifies the same for thermal neutrons. Various foils like indium, gold, dysprosium, cobalt are used depending upon the neutron fluence of 10^3 to 10^{20} neutron/cm^2.

7.3.1 *Basic Physical Quantities: Definitions and Methodologies*

Fluence: This is the ratio of the particle (dN) incident or crossing an area da (in cm^2) and the area da; $\Phi = \mathrm{d}N/\mathrm{d}a$ (unit: cm^{-2}).

Fluence Rate: This is defined as fluence per sec and has unit of cm^{-2} s^{-1}. Fluence rate can also be differential in energy or solid angle or both.

The distribution of fluence with respect to energy is given by $\Phi_E(E) = \mathrm{d}\Phi/\mathrm{d}E$, (cm^{-2}/MeV).

Though the general term for number of neutrons crossing a particular area per unit time is flux, ICRU prefers the use of fluence or fluence rate.

Neutron Spectrum: This denotes the neutron energy histogram or in other terms the different energies of neutron present in the beam along with their effective relative weights, the total weight is 1. Since the absorption cross-section with hydrogenous materials is higher for cold neutrons than thermal neutrons, cold neutron sources using high flux reactor or spallation sources are the main workhorses for high resolution and high contrast imaging studies such as polarized or spin-echo neutron imaging.

Collimation Ratio: This parameter is also known as *L*/*D*, where *L* is the length of the divergent beam tube used to allow the neutrons out of the generating location to the imaging site and *D* is its input diameter. *L*/*D* decides the quality of the neutron radiograph as far as spatial resolution is concerned. The higher the *L*/*D* the better is the image quality but the drawback is that as *L* increases, the neutron fluence falls off by its square. The effective collimation ratio is provided by taking into account the scattering contribution arising out of the collimator wall [12]. ASTM standard E803-20 provides guidelines for measuring the effective collimation ratio of a neutron beam. One can correlate this parameter to the focal spot in case of an X-ray imaging system. Higher *L*/*D* can be achieved in nuclear reactor as well as spallation sources. Divergent collimator is easy to construct than the earlier concept of soller or multitube collimator though the later one provides better image sharpness.

Cadmium ratio: This parameter provides the approximate ratio between the thermal neutron component with respect to the epithermal part in the neutron beam, ignoring the fast neutron component as their numbers are much lower in comparison to the other two in a reactor neutron radiography facility. The cadmium ratio (Cd-ratio) is experimentally found out by making activation analysis on standard samples in bare and cadmium foil-covered conditions. The radiography exposure time decreases with the increase in the Cd-ratio but at the expense of reduced sharpness. The beamline parameters, excluding the detector and related parameters, responsible for determining the image quality are Cd-ratio, *L*/*D* and the associated gamma component.

Neutron Beam Purity: This is in general used in direct thermal neutron radiography practices for quantifying the thermal neutron purity in the beam. As an image quality indicator, the beam purity indicator determines the quality of image in terms of the neutron beam and imaging system employed. It is measured as per ASTM E2003-10 standard.

In order to standardize the thermal neutron fluence rate measurement, one needs a standard methodology for calibrating instruments in the first stage. Recommended radionuclide neutron sources as per ISO 8529:2001 are ^{252}Cf (D_2O moderated), ^{241}Am-B and ^{241}Am-Be. Out of these, the preferred one is ^{252}Cf sources as its size is small, spectrum well-documented, very low γ-rays associated with it and can be prepared for any quantity. The main problem with ^{252}Cf is that they have a relatively lesser half-life (2.65 years), and it needs frequent calibration/replacement. In such cases, Am-Be sources are helpful because their half-life is large (432 years). As these sources are relatively bigger in size, in comparison with ^{252}Cf, effect of their size should be taken into account towards calculation. To produce thermal neutron from these isotopic sources (energy range from Kev to 10 MeV), they are placed in an aqueous solution of $MnSO_4$ (~20 g/l) kept in a spherical vessel. The vessel diameter can be from 50 to 150 cm. Neutrons are thermalized in water (aqueous media), subsequently absorbed by inactive manganese and will become active. This induced activity is measured which in turn is correlated with the neutron source strength.

For carrying out fast neutron fluence rate measurement, the primary standard is neutron telescope (NT). NT or also known as proton recoil telescope (PRT) is based on elastic scattering of neutron in a thin (1–2 mm) hydrogenated target and detection of recoil protons by using CsI scintillator detector. Liquid scintillators (LS) are also used for neutron spectra measurements but they are bulky. LS have the property to distinguish between n and γ. Neutron and gamma generate different rise time pulses which are analysed by pulse shape discrimination (PSD) technique. Neutrons interact via (n,p) and γ through Compton scattering, and using time of flight method neutron energy is determined.

In case where the source emission or its anisotropy is not known or possess a different spectrum than those reported is ISO 8529:1 (2001), then the secondary standard should be established by means of transfer instrument which are moderated ambient dose equivalent metre (rem counter), Bf3-based counter encapsulated by borated HDPE moderator Bonner spheres having either ^{3}He or ^{6}LiI (Eu) scintillator.

BARC in India, has a Mn bath setup (Vessel-Stainless steel tank of 1 m diameter) for thermal neutron fluence rate measurement. The NT setup at BARC has been established for 14 MeV neutrons. The secondary standard for neutron yield and fluence rate in BARC is the precision long counter and consists of BF3 counters place inside two consecutive cylinders separated by a neutron absorber. The primary standard or thermal neutron fluence rate measurement at BARC consists of six members of Am-Be neutron sources embedded in graphite pile. The thermal neutron flux in the central cavity has been standardized using gold and manganese foil activation method.

7.3.2 Neutron Fluence/Flux Measurement

The mechanism to determine neutron yield varies upon the energy of the neutron as their interaction cross-section with materials is energy-dependent. Here, by "materials" we mean on which the neutron interaction is carried out. Neutron activation analysis is the technique using which the neutron flux is estimated. The process is divided into two parts. First a foil having stable nuclei is irradiated with neutrons, having suitable neutron interaction cross-section at that energy or range and are made radioactive. In the next step, the signature gamma emissions correlated with the decay of the radioactive species are detected using a high resolution (in terms of energy) detector (say high purity germanium detector). From this data, one can infer with some simple calculations the thermal neutron flux at the measured location. From the flux, one can calculate the source yield incorporating the solid angle subtended by the foil with the source. The choice of irradiating foils is as per the energy or range of the incident neutrons. In general, though the neutron spectra are continuous upto 10 MeV, the beamlines, incorporated with Bismuth and Sapphire plugs get rid of gamma and fast neutrons, respectively, providing thermal neutrons for imaging applications such as radiography and tomography. In case of accelerator-based sources, the generated neutrons have varying energies depending upon the process employed.

(a) **Low Energy Neutrons**: The underlying principle of (n, γ) reaction for this region is based on the assumption that the absorption cross-section follows a $1/v$ law, where v is the speed of neutron, upto an energy of 0.4 eV with resonance peaks superimposed on the $1/v$ curve for neutron energies above 0.4 eV (Hogdahl convention). In addition, it is also assumed that neutrons having energy below and above 0.4 eV will be completely stopped or transmitted by a 0.25 mm thick cadmium foil. The formula derived by Westcott, Walker and Alexander [13] is used for finding out the induced activity. In the case, where the epithermal neutron flux component is not proportional to $1/E$, necessary changes in the formula have been proposed by Ryves [14].

The standard technique for measuring thermal neutron fluence rate of any thermal neutron facility is through foil activation. High purity gold, manganese and indium foils are mainly used for such purposes. The generalized equation governing correlation between the activity and the fluence rate is [15]

$$A_i = N_i \varepsilon_1 \varepsilon_2 \varepsilon_3 (1 - \exp(-\lambda t)) \exp(-\lambda T)\, f_s f_d \int \sigma i(E) * \varphi(E) dE$$

where

A_i is the count rate of ith detector in the counting setup, N_i is the number of the atoms, ε_1 is the isotopic abundance, ε_2 is the abundance of the radiation counted, ε_3 is the efficiency, f_s is the self-shielding factor for the foil, f_d is the flux depression factor of the foil for that medium, $\sigma_i\,(E)$ is the activation cross-section at the energy E, $\varphi(E)$ is the neutron fluence rate of energy E, t is the irradiation time, T is the decay time.

Gold (^{197}Au) Foil Activation

The reaction is given in Eq. 7.2. The main reasons for choosing gold are as follows: (a) Gold being monoisotopic in nature, highly pure form can be obtained, (b) Its absorption cross-section is high ~98.6 ± 0.3 barns, (iii) activation time is rather small owing to its activated product half-life of 2.696 days and (iv) the deviation from 1/v below cadmium cut-off is small [16].

Pure gold foils of 99.95% or better purity, thickness ~0.025–0.1 mm and diameter ~4–8 mm are generally used for irradiation and activation purposes.

$$^{197}\mathrm{Au} + n \rightarrow {}^{198}\,\mathrm{Au} \rightarrow {}^{197}\mathrm{Hg} + \mathrm{e}^- + \gamma \tag{7.2}$$

The β-decay half-life of ^{198}Au is 2.6947(3) days with well known ^{198}Hg γ intensities and energies of 411.8 keV for the main transition, 675.9 and 1087.7 keV at lower intensities [17]. The cross-section is 98.8 barns for 25 MeV neutrons. The gamma lines are measured with a high purity germanium (HPGe) detector, and the analysis is carried out taking into account the integrated counts, decay characteristics, mass of the foil and volume of the foil irradiated. It is important to emphasize that the choice of the thickness of the foil depends upon the thermal neutron yield with an optimization that the activated gamma lines should be able to emit and reach the HPGe detector for counting.

The equation governing the yield is given by Eq. 7.3.

$$Y = \left[(4pi\, L^2)\lambda A C_p \exp(\lambda t_2)\right] / \left[N_a m\, a_i\, P_y \sigma \right.$$
$$\left. (1 - \exp(-\lambda t_1)(1 - \exp(-\lambda t_3))\right], \tag{7.3}$$

where λ is decay constant (s^{-1}), L (cm) is distance between neutron source (target) and foil set, A (g) is atomic mass number of sample, C_p is the photopeak area (counts at 411.8, 675.9 and 1087.7 keV) for ^{197}Au, N_a is Avogadro number, m is atomic weight (g) of sample, a_i is the isotopic abundance, ε is photopeak efficiency of HPGe, P_y is gamma abundance, σ is reaction cross-section, $t1$, $t2$ and $t3$ are irradiation, cooling and counting time, respectively.

Using this Y and dividing by $4piL^2$, one can find out the flux at a distance L from the source. The measured value provides yield of neutrons below 0.4 eV. When activation is carried out in cadmium covering (~0.5 mm thick), it gives the measurement of all neutrons other than thermal. Difference of these two provides actual thermal neutron yield of the source being measured. In general, not one but multiple foils are irradiated to better the result [18–20].

Manganese (^{55}Mn) Foil Activation

Mn used in either pure metal or alloy form (Mn-Ni) has 13.3 barns cross-section with thermal neutrons, and the reaction is ^{55}Mn(n,γ)^{56}Mn. The resultant nucleus has a half-life of 2.58 h. The reaction yields beta and gamma having energies of 2.85 MeV and 0.847 MeV, respectively, in coincidence. The absolute measurement can be carried out using 4π beta-gamma coincidence method.

New developments: A new technique for neutron flux estimation at highly collimated monochromatic cold neutron beam has been established using a boron-10 enriched B_4C block and measuring the neutron-induced 478 keV gamma by a high-resolution HPGe detector [21]. The simplicity of this technique is that it does not involve neutron cross-sections, energy or branching ratio.

(b) **Fast Neutrons**: On account of the relative smaller cross-sections and hence long free paths associated with fast neutrons, the materials are chosen for activation analysis are different from the ones used for thermal neutrons. Different foils are used depending upon the reaction threshold.

For DD neutrons, the flux determination is carried out by neutron activation analysis on indium foils (^{113}In: 4.3% and ^{115}In: 95.7%). The reaction channels for ^{115}In are

$$^{115}\mathrm{In}\left(n,n'\right)^{115\mathrm{m}}\mathrm{In}, \tag{7.4}$$

$$T_{1/2}\left(^{115m}\,\mathrm{In}\right) = 4.486\ \mathrm{h}, \quad \gamma = 336.2\ \mathrm{keV}$$

and

$$^{115}\,\mathrm{In}(n,\gamma)^{166\,m}\,\mathrm{In} \tag{7.5}$$

Reaction (7.4) has high cross-section for fast neutrons and is used as the DD neutron flux estimation. As per ENDF database, the cross-section is 0.346 barns for 2.7 MeV neutrons and the DD neutron energy is 2.54 MeV. Various similar ^{115}In foils can be placed with respect to the source position to estimate the neutron fluence rate profile.

^{39}Al(n, p)^{27} Mg reaction having threshold of 3.25 MeV and cross-section 113 $\pm$ 1.5 mb has also been suggested [22]. In this radioactivation, photopeaks of 844 and 1014 keV are considered [23]. ^{63}Cu(n, 2n)^{62}Cu and ^{65}Cu(n, 2n)^{64}Cu reactions giving rise to 9.7 min and 12.8 h activity for ^{62}Cu and ^{64}Cu, respectively, are extensively used as 14 MeV neutron flux monitor [24]. The former has been extensively used for short irradiations, about half an hour and the later for longer irradiations. As an alternative ^{19}F(n, 2n)^{18}F reaction having threshold of 12 MeV has also been utilized for 14 MeV fluence measurements [25, 26]. Also, the $^{28}Si(n, p)^{28}Al$ reaction [27] has been shown to be useful for monitoring the emission of 14 MeV neutrons from *d-t* fusion plasmas.

7.3.3 Digital Image Archiving, Processing and Transfer Protocol

The ever-growing healthcare sector with a niche towards medical imaging and its widespread utilization from hospitals to tele-radiology prompted the developers to standardize the imaging modalities and bring it under an internationally adopted convention DICOM, an abbreviation of Digital Imaging and COmmunication in Medicine. Defined protocols towards the generation, archiving, processing and transfer of radiographic images in DICOM format [28] are practised by one and all in the healthcare sector. In the same line of work, all digital images generated through non-destructive techniques are taken care by the protocol named as DICONDE, an abbreviation for Digital Imaging and COmmunication in Non-Destructive Evaluation [29]. This was first introduced by ASTM in the year 2004 though it was established way back in 1993. Sharing NDE images from various techniques such as CR, DR, visual inspection, ultrasonic, eddy current, using a standard format, their study and examination by experts speeds up the whole evaluation process. These files contain metadata regarding the image in addition to the image. Metadata includes parameter such as process of examination involved/imaging modality, system used, study conditions (kV, mA, exposure, etc.), material description study performed by and many more including option towards report generation. The data is organized in a hierarchical way—study—series—instances. All the NDT technique, except neutron radiography and tomography, are widely used and hence had the support of the developers towards and DICONDE.

7.4 Internationally Adopted Standards for Neutron Imaging

Some of the important ASTM standards [30] related to neutron source and neutron imaging facilities are given in Table 7.3.

Table 7.3 Important ASTM standards related to neutron source and neutron imaging facilities

S. No.	ASTM-	Description
1	E261-16	Standard practice for determining neutron fluence, fluence rate and spectra by radioactivation techniques
2	E262-17	Standard test method for determining thermal neutron reaction rates and thermal neutron fluence rates by radioactivation techniques
3	E263-18	Standard test method for measuring fast-neutron reaction rates by radioactivation of iron
4	E264-19	Standard test method for measuring fast-neutron reaction rates by radioactivation of nickel
5	E265-15 (2020)	Standard test method for measuring reaction rates and fast-neutron fluences by radioactivation of sulphur-32
6	E266-17	Standard test method for measuring fast-neutron reaction rates by radioactivation of aluminium
7	E481-16	Standard test method for measuring neutron fluence rates by radioactivation of cobalt and silver
8	E496-14	Standard test method for measuring neutron fluence and average energy from 3H (d, n) 4He neutron generators by radioactivation techniques
9	E523-16	Standard test method for measuring fast-neutron reaction rates by radioactivation of copper
10	E526-17	Standard test method for measuring fast-neutron reaction rates by radioactivation of titanium
11	E545-19	Standard test method for determining image quality in direct thermal neutron radiographic examination
12	E748-19	Standard guide for thermal neutron radiography of materials
13	E803-20	Standard test method for determining the l/d ratio of neutron radiography beams
14	E1297-18	Standard test method for measuring fast-neutron reaction rates by radioactivation of niobium
15	E2003-20	standard practice for fabrication of the neutron radiographic beam purity indicators
16	E2023-19	Standard practice for fabrication of neutron radiographic sensitivity indicators
17	E2861-16 (2020)	Standard test method for measurement of beam divergence and alignment in neutron radiologic beams
18	ASTM E2662-15	Standard practice for radiographic examinations of flat panel composites and sandwich core materials used in aerospace applications

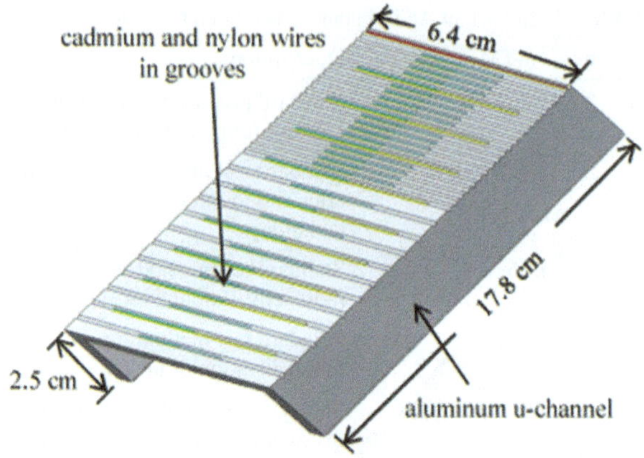

Fig. 7.3 Schematic view of a NU device showing the wire layout [12]. Reproduced with permission from Rightslink CCC

7.4.1 Effective Collimation Ratio Using No Umbra (NU) Device as per ASTM E803-20 (ISO 12721)

This standard [30] specifies the methodology to determine experimentally the *L*/*D* of the beam where *L* is the length of the neutron beam from its source to the imaging plane and *D* is the source inlet diameter. The basic approach is to find out the distance (say *d*) between the object having very high neutron absorption cross-section and the image plane which produces the sharpest image of the object. The condition is to use very thin wires having diameter much less than *D*. The NU device is made up of "*U*"-shaped aluminium plate having regularly cut grooves filled with very fine nylon and cadmium wires (Fig. 7.3). Imaging of the device is carried out by placing it 45° with respect to the beam plane. Measurements from the image are made for the width of the umbra, penumbra, and at minimum penumbral condition, the ratio between the diameter of the wire and the distance "*d*" is the measure of the *L*/*D* of the beam. ISO 12721 [31] specifies formulae for calculating the *L*/*D* for cases such as $L/D < 100$ and $L/D > 200$ in addition to find out the accuracies of the measurements.

7.4.2 Beam Purity Indicator (BPI) as per ASTM E2003-20

BPI [30] provides quantitative information related to neutron beam and direct neutron radiographic examination carried out using film-based detectors. This is an image quality indicator and is fabricated using of boron nitride discs, lead discs and cadmium/gadolinium wires tightly fitted in holes in a highly crystalline linear polymer (as shown in Fig. 7.4). ASTM E2003-10 specifies the minimum purity conditions of the materials to be met for BPI measurements. Radiographic examinations are to be done as per the procedure mentioned in ASTM E545. Densitometric measurements on the image will provide quantitative data related to the materials

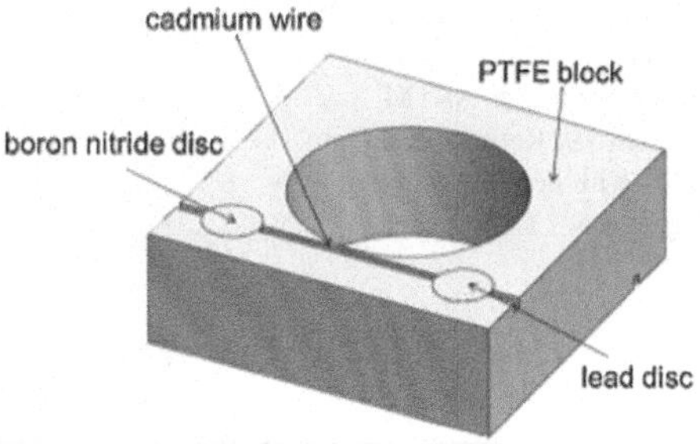

Fig. 7.4 Typical schematic of a BPI. Reproduced from [12] with permission from Rightslink CCC

used in the BPI. Since cadmium/gadolinium has appreciably high thermal neutron absorption cross-section, its image should be sharp. If that is not the case the effective collimation is not sufficient for performing this experiment. Since lead is almost transparent to thermal neutrons, image of lead with some contrast with the base material will certainly suggest the presence of gamma in the beam. If that is not the case then the gamma contribution is less and the design of the beam is said to be good. E2003-20 provides necessary formulas to determine mainly the thermal neutron content, gamma component along with the scattered neutron component towards characterizing the beam. Various parameters such as optical density of film, exposure, object to film distance must be as stipulated in the standard and multiple images are to be acquired along the film, at the centre and on the sides of the film.

7.4.3 Sensitivity Indicator (SI) for Neutron Radiography as per ASTM E2023-19

The contrast that an image is able to reproduce of the internal features in a sample is of utmost importance as it provides quantifiable value regarding the quality of the whole imaging chain. The idea is to characterize the ability of the imaging chain towards reproducing minor transparency differences, as per the probe radiation, as features in the sample in terms of image. Sensitivity contrast in an image is defined by the minimum discernible variation detected. Same feature in a sample will provide different contrast for different materials and also with varying thickness. As an example, if an object has some features say holes with varying diameter, very small to large either through its depth or upto some length, the contrast for each feature will be different and can be deduced using simple mathematics and image data values. ASTM E2023-19 [30] standard is available regarding the performance of such an exercise in finding out the minimum detectable sensitivity the imaging chain can faithfully produce in terms of image. According to the standard, a test piece (shown in Fig. 7.5) made up of highly crystalline linear polymer, lead and aluminium having step wedges of different thickness with gaps and holes of known dimensions are imaged in a neutron radiography facility using direct technique. Lead is an optional and can be replaced by acrylic. Various parameters such as optical density of film,

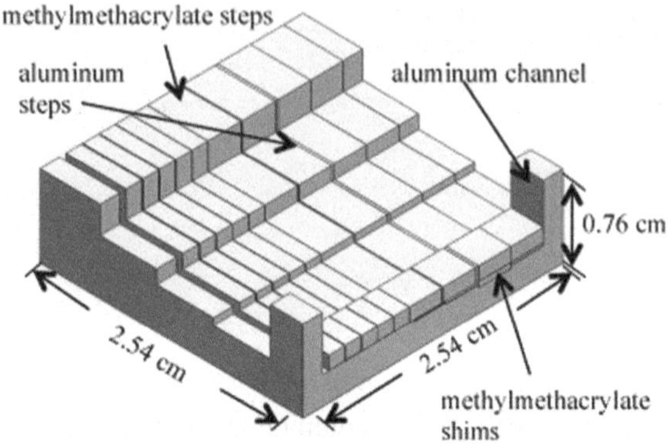

Fig. 7.5 Typical schematics of a typical sensitivity indicator as per ASTM E2023-19. Reproduced from [12] with permission from rightslink CCC

exposure, object to film distance must be as stipulated in the standard and multiple images are to be acquired along the film, at the centre and on the sides of the film.

7.4.4 Beam Divergence and Alignment Measurement of Radiography Beams as per ASTM E2861-16 (2000)

Divergent collimators serve as beam guides for neutron to travel from the source to the image plane in any neutron radiography facility. For medium to high flux sources such as a reactor and spallation, the *L*/*D* is large and the divergence is less. But in case of low flux sources such as thermal neutron assembly using DD/DT neutron generator and isotopic sources large *L*/*D* is not realizable as the thermal neutron fluence rate reduces drastically with increasing *L* or decreasing *D*. In such cases, imperfect beam alignment and divergence may introduce errors in the image analysis. In this direction, ASTM E2861-16 [30] has been developed towards determining beam centreline and its divergence. This was developed as an extension to ASTM E803 related to L/D evaluation. Since the divergence and alignment of the beam, which has circular cross-section at any point, are estimated from the image of the test piece, its shape is also circular. The test piece, as shown in Fig. 7.6, is a circular aluminium plate having very thin cadmium wires (<1 mm) in circular and radial fashion in machined grooves on one side of the plate. The radial ones forms an "*X*" aligned with the centre of the plate. The design is done to make the test piece symmetric for ease of the fabrication process as well as image calculation. Also it has 5 numbers of symmetrically placed cylindrical Al rods including one at the centre of the plate, with cadmium wire inserts at their centre, fixed perpendicular on the other side of the plate. The details regarding the preparation of the test piece are provided in Section 10 of the standard. The test piece must be placed centrally in the beam and perpendicular to the beam direction with the cylindrical Al rods facing the imaging device. The cadmium wires provide enough attenuation and provide good contrast. Measuring the Cd wire positions in the image with respect to the centre of the circular plate, imaging device misalignment

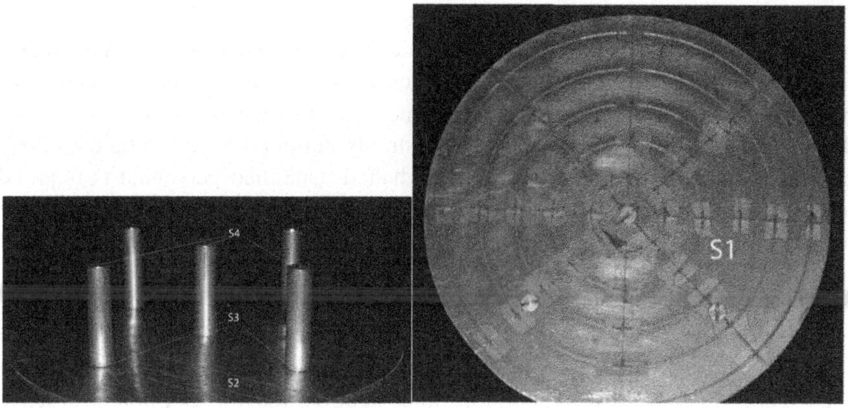

Fig. 7.6 Images of the two surfaces of the circular Al plate (left) with Al cylinder rods having Cd inserts and (right) symmetric arrangement of circular as well as radial cd wires in machined grooves on the other side of the plate. Reproduced from [30] with permission from Rightslink CCC

with the beam can be estimated and corrected if required. The positional measurement of the radial cadmium wires at the intersection with the circularly symmetric ones will provide information regarding the divergence in the beam. Necessary formulas are provided in the document E2861-16 (2000).

7.5 Personnel Qualification

Neutron imaging can be carried out using neutrons from research reactors, isotopic sources and accelerator-based such as neutron generators and spallation sources. Neutron interacts with the elements via nuclear interaction mechanism. Being energetic and having high interaction cross-section with low Z materials, their use is guided by rules and regulations to avoid any untoward incidence causing risk to human life, environment and property. On an average, 60% of the human body is composed of water. Neutrons are highly attenuated by hydrogenous materials such as water and are more prone to deposit their energy in it. As the radiation weighting factor for neutron ranges 5–20 times in comparison to X-ray/gamma rays, the effect of neutrons on humans is much more severe. Experimental evidences [32] have suggested that neutrons affect human DNA sequence leading to cell death.

Here lies the requirement of qualified personnel in avoiding such hazards. Since neutron production is inherently associated with gamma ray production either inherent with source or neutron-induced interactions with surrounding materials, shielding of such sources is also important towards human safety. In case of using reactor beamline for neutron-related investigations, there are various safety mechanisms integrated such as inner gate, outer gate, beamline hutch entry/exit door, interlocked towards implementing the safety mechanism. Also in the case of accelerator-based neutron production facilities, safety mechanisms such as interlocks for beam, door, cooling, vacuum and emergency are strictly put in place and in failure of any of one of them the neutron production is halted. Qualified personnel is required to carry out such multiple checks ensuring the highest safety standards. But while handling isotopic sources (Am-Be, Pu-Be), extra care must be taken as they emit neutrons all the time, depending upon the source emission rate, and also isotropically. Though they are always encapsulated by neutron-absorbing materials such as Borax or borated polythene block, except while using care must be taken to ensure no radiation exposure to any personnel present nearby. Hence, knowledge of handling such sources is of utmost importance. Since neutron imaging is a non-destructive procedure, the personnel working in this domain including dosimetry is trained as per the guidelines of standards available for non-destructive testing (NDT).

Available standards for training, certification and working in the NDT domain are as follows.

ISO 9712:2012—Non-destructive testing, qualification and certification of NDT personnel.

ANSI/ASNT CP-189 1995—American national standard for qualification and certification of personnel.

EN 45013: (1989)—General criteria for certification bodies operating certification of personnel.

ISO/IEC 17024: 2012—conformity assessment—General requirement for bodies operating certification of persons.

Recommended practice M: SNT-TC-IA 1996—published by American Society for Non-destructive testing, USA.

IAEA TECHDOC 628 (1991): Training guidelines in NDT techniques, Published by the IAEA, Vienna, Austria.

ISO 9712:2012 has been prepared by technical committee ISO/TC 135, subcommittee SC7. This certification covers radiography testing including other NDT topics. Basics to in-depth knowledge along with practical hands or experience are provided by accredited bodies through professionals in this domain to entail the trainees to handle radioactive sources, radiation facilities and related instruments including their calibration perform management activities related to radiation protection and, database management. In addition, specific laboratories can provide in-depth knowledge of their work and make them competent. Upon successful completion of the training expertise of the candidate must be established to enable them handle and supervise radiation-related jobs.

All radiation facilities from university to private to government laboratories based on internationally approved guidelines make their own framework to choose/training/accredit personnel to perform radiation-related jobs as radiation safety officer (RSO)/ health physicist (HP)/radiography testing personnel (levels-1/2/3) [33].

According to the American Academy of Health Physics and Radiation Safety Operation Section of the Head Physics Society [34], first and foremost, the university management must establish RSO qualification, certification, licensing, previewing procedures along with sufficient resource and support to enable their sustainability as and when their programme demands.

For national laboratories, in accordance with the norms laid upon by IAEA and the national body responsible for such personnel are trained and provide authorization to render service. For private bodies in a country, the responsible body takes care of the training, certification and authorization of personnel in the respective domain.

Atomic Energy Regulatory Board (AERB), a Government of India body in the sole competent authority in India towards authorization of license, inspection and operation of equipment/facilities and personnel involving radioactive material. Certification of Radiography inspection personnel and site-in-charge is done by AERB through Radiological Physics and Advisory Division of Bhabha Atomic Research Centre (BARC), Mumbai, India. It may be noted that RT levels-1, 2, 3 certification is independent of AERB and does not cover licensing.

In India, IS 13805: 2004 specifies the general standard for qualification and certification of non-destructive testing personnel for various NDT methods such as acoustic emissions testing (AT), Eddy current testing (EDT), liquid penetrate testing, leak testing (LT), magnetic particle testing (MPT), neutron radiography testing (NRT), radiography testing (RT), infrared testing (IRI), ultrasonic testing (VT), vibration analysis testing and visual testing (VT).

In India, Radiography testing level-1 (RT-1) started way back before 1994. Some of the topics included in the training courses were general concepts of RT, equipment and radiation sources, photographic and non-photographic recording, radiography technique selection, standard and test methods, radiation safety including regulatory aspect, energy response plan and preparedness. On successful completion of RT-1 course, the candidate is allowed to work as a Certified Radiographer, subject to the approval of AERB.

Radiography testing level-2 (RT-2), training in being handled by Isotope and Radiation Application Division of BARC. BARC RT LEVEL-2 extensively covers Radiation safety as specified in AERB/RF/Training-Syllabi/2012 approved by AERB. Successful completion of this training course will qualify the candidate, subject to approval/authorization of AERB, as designated Radiological Safety Officer (RSO) and to procure radiation sources/equipment for radiography practise in the country. The guidelines are set by ISO 9712: 2012/BIS: IS-13805. Eligibility criteria for BARC-RT-LEVEL-2 training and related information is available in www.barc.gov.in/careers/information-brochure-rtl2.pdf all course structures are guided by Recommended Practice no: SNT-TC-IA which provides guidelines for employees to establish in-home certificate programme for the qualification and certification of NDT personnel.

In India, the training programme centres for RT (1, 2 and 3) are as follows [35].

RT 1 and 2: IXAR, industrial X-ray and allies' radiographers PVT Ltd. (www.IXAR.in) Jogeshwari (w) Mumbai, Maharashtra, 400012.

RT 1 and 2: Quality NDT services, Pune, Maharashtra, 411026.

RT 1: IRC Engineering services, New Delhi, 110019 (www.ircengg.com).

RT 1 and 2: Indian society for Non-destructive Testing, Tiruchirappalli, chapter, Trichy-14 (www.isnttrichy.org).

RT 3: ISNT, Mumbai chapter, Andheri (E), Mumbai—400059.

The America society of non-destructive testing (ASNT) offers personnel training publication pertaining to radiography testing (RT). ASNT conducts NDT level-2 and Level-3, IRRSP (Industrial Radiography Radiation Safety personnel), ACCP (ASNT central certification programme) level-2 and level-3 training courses. Code of Ethics for NDT Personnel has been clearly mentioned in ISO 9712–2012 and equivalent standards mentioned in IAEA TECDOC 626 and IAEA TECDOC 628/Rev. 3 (2013) and are to be followed with integrity and professional competence as a rule [36].

References

1. https://inis.iaea.org/collection/NCLCollectionStore/_Public/36/097/36097704.pdf?r=1&r=1. Accessed date: 13/05/2021
2. ICRP (2007) The 2007 recommendations of the International Commission on Radiological Protection. ICRP Publication 103. Ann ICRP 37(2–4).
3. Choudhary S (2018) Deterministic and stochastic effects of radiation, cancer therapy. Oncol Int J Short Commun 12(2)
4. Radiation shielding of fusion systems (2010) Thesis by Andrew Davis. University of Birmingham Research Archive, Apr 2010
5. Shukla M et al (2016) Neutron imaging beamline for advanced imaging applications at Dhruva reactor, India. Non-destructive evaluation 2016, page 226–231
6. Shukla M, Roy T, Kashyap Y, Shukla S, Singh P, Ravi B, Patel T, Gadkari SC (2018) Development of neutron imaging beamline for NDT applications at Dhruva reactor, India. Nucl Instrum Methods Phys Res Sect A: Acceler Spectrometers Detect Associat Equipm 889:63–68
7. Jensen SR, Craft AE, Papaioannou GC, Empie WW, Ward BR, Batt LA (2019) Restart of the transient reactor test (TREAT) facility neutron radiography program. Nucl Technol 205(10):1325–1335. https://doi.org/10.1080/00295450.2019.1605780
8. Craft A, Schillinger B, Chuirazzi W, Papaioannou G, Smolinski A, Boutlon N (2020) First neutron computed tomography with digital neutron imaging systems in a high-radiation environment at the 250 kW neutron radiography reactor at Idaho National Laboratory. In: Materials research proceedings, vol 15, pp 42–47. https://doi.org/10.21741/9781644900574-7 (Open Access)
9. https://www.osha.gov/ionizing-radiation/control-prevention. Accessed date: 19/05/2021
10. https://www-pub.iaea.org/MTCD/publications/PDF/Newsletters/SSDL-NL-37.pdf. Accessed date:19/05/2021
11. www.astm.org. Accessed date: 19/05/2021
12. Morgan S, King J (2013) Beam characterization at the neutron radiography facility. In: Nuclear engineering and design, vol 265, pp 639–653
13. Westcott CH, Walker WH, Alexander TU (1958) Proceedings of 2nd conference on peaceful uses atomic energy, Geneva, vol 16, p 70
14. Ryves TB (1969) A new thermal neutron flux convention. Metrologia 5:119
15. Sathian V, Tripathi SM, Ghodke S, Singh Y, Shaiju L (2008) The performance evaluation studies of neutron detectors used in Indian nuclear reactors. BARC Report BARC/2008/I/010

16. Hargrove CK, Geiger KW (1964) A new thermal neutron flux density standard. Canad J Phys 42
17. Evaluated Nuclear Structure Data File (ENSDF) (2014) National nuclear Data Center. Brookhaven National Laboratory. www.nndc.bnl.gov/ensdf/
18. Shukla M, Roy T, Kashyap Y, Shukla S, Singh P, Ravi B, Patel T, Gadkari SC (2018) Development of neutron imaging beamline for NDT applications at Dhruva reactor, India. Nuclear Inst Methods Phys Res A 889:63–68
19. Craft AE, Hilton BA, Papaioannou GC (2016) Characterization of a neutron beam following reconfiguration of the neutron radiography reactor (NRAD) core and addition of new fuel elements. Nucl Eng Technol 48(1)
20. Becker H et al (2015) Neutron production and thermal moderation at the PSIUCN source. Nucl Instrum Methods Phys Res Sect A: Accel Spectrom Detect Assoc Equip 777:20–27. ISSN 0168-9002
21. Yue AT, Anderson ES, Dewey MS, Gilliam DM, Greene GL, Laptev AB, Nico JS, Snow WM (2018) Precision determination of absolute neutron flux. Metrologia 55(4)
22. Csikai J (1987) Handbook of fast neutron generators. CRC Press Inc., Boca Raton, vols I, II
23. Melent'ev VI, Ovechkin VV (1978) The ^{27}Al(n, p)^{27}Mg cross section for 14.9-MeV neutrons. Energy 44:183. https://doi.org/10.1007/BF01124710
24. Jarjis RA (1981) Some activation measurements with ^{3}H(d, n$^{)}$4He neutrons. Nucl Inst Methods 184(2–3):439–444
25. Shiokawa T, Yagi M, Kaji H, Sasaki T (1968) The ^{19}F(n,2n)^{18}F reactions as a 14 MeV neutron flux monitor. J Inorg Nucl Chem 30:1–4
26. Ikeda Y, Verzilov YM, Filatenkov (KRI) AA, Smith (ANL) DL, Maekawa F, Oychma Y (1995) Recent activities on (n, a) cross section measurements in the 14 MeV region at JAERI FNS, IAEA-RCM (Sendai, 1995), Attachment3
27. Sadler G, Jarvis ON, van Belle P, Pillon M (1990) Use of the 28Si(n, p)28Al reaction for the measurement of 14 MeV neutrons from fusion plasmas. Rev Sci Instrum 61:3175. https://doi.org/10.1063/1.1141678
28. https://www.dicomstandard.org/
29. ASTM E 2339: Standard practice for DICONDE (https://www.astm.org/Standards/E2339.htm)
30. www.astm.org
31. https://www.iso.org/standard/2271.html
32. Pang D, Berman BL, Chasovskikh S, Rodgers JE, Dritschilo A (1998) Investigation of neutron-induced damage in DNA by atomic force microscopy: experimental evidence of clustered DNA lesions. Radiat Res 150(6):612–618
33. RSO (2004) Training course on safety aspect in the research application of Ionizing radiation. In: Atomic energy (radiation protection) rules, by RPAD, BARC. https://www.aerb.gov.in/english/rso-training
34. www.aahp-abhp.org/sites/default/files/2018-06/SQP-001.pdf. Accessed on 24/05/2021
35. Sahoo S, Viswakarma RS, Palani Selvam T, Babu DAR (2014) Training and certification status of BARC-1 course. In: NDE 2014, Dec 4–6, Pune
36. Guidelines on training, examination and certification in digital industrial radiology testing (RT-D). IAEA training course series, No. 60. https://www-pub.iaea.org/MTCD/Publications/PDF/TCS-60_web.pdf

Chapter 8
Neutron Imaging for Aerospace Applications

M. Nallaperumal, Girish N. Namboodiri, and Tushar Roy

8.1 Introduction

In the aerospace sector, neutron imaging has become vitally important. Various critical components used in aircraft and spacecraft such as turbine blades, components used in stage separation mechanisms, cockpit ejection device and explosive stimuli transfer mechanisms must be carefully tested to meet the optimal design standards as any flaw or defect in the samples can lead to mission failure. Neutron radiography (NR) has become a mandatory inspection technique for aerospace applications in recent years. It has become an indispensable technique in the field of quality control and inspection of explosive-based mechanical devices used in aerospace industry.

Pyrotechnic devices or energetic devices (explosives) are used to power the ejection mechanism in fighter plane. It uses a series of small explosive charges that explode in tandem to actuate the pilot ejection mechanism. Similarly, in space launch vehicles used to inject satellites into orbit, pyrotechnic devices (or pyro devices, in short) containing explosive charges are used for various mechanisms such as solid rocket motor ignition, stage separation, bolt cutters and payload faring separation system.

Another area of application of neutron imaging in aerospace sector is identifying flaws in turbine blades. One of the largest industrial users of neutron imaging is the turbine blade manufacturing industry, especially in the USA, where most of the turbine blades used in jet engines are manufactured.

M. Nallaperumal · G. N. Namboodiri
Vikram Sarabhai Space Centre, ISRO, Thiruvananthapuram, India
e-mail: m_nallaperumal@vssc.gov.in

G. N. Namboodiri
e-mail: girish_nm@vssc.gov.in

T. Roy (✉)
Technical Physics Division, Bhabha Atomic Research Centre, Mumbai 400085, India
e-mail: tushar@barc.gov.in

D. K. Aswal et al. (eds.), *Neutron Imaging*,
https://doi.org/10.1007/978-981-16-6273-7_8

A vital aspect of these applications is an easy access to neutron imaging. Though reactor-based imaging facilities still remain the first choice, many a times the samples need to be tested on-site at the manufacturing plant. Hence, some of these applications use non-reactor-based neutron imaging facilities as well based on D-T neutron source or electron Linac-based photoneutron sources.

8.2 Turbine Blades

Turbine blades are cast in ceramic moulds out of strong and lightweight metal, such as Ti–Al alloys, that has a melting point lower than the operating temperature of the engine. The turbine blades need to be properly cooled as they are subjected to lot of heat and stress. Flaw in them might result in engine failure and potentially lead to plane crash and loss of life. Hence, turbine blades have to undergo rigorous quality checks due to their high cost of failure.

To cool the turbine blades, airflow is provided through channels running through the internal structure of the blades. This prevents the blades from melting or failing during operation. Due to their design and manufacturing process, it is possible that some of the remnants of ceramic moulds find their way into the air-cooling channels and get trapped. These remnants impede and/or redirect the air flow through the channels, thereby creating hot spots and inducing stress in the channels leading to mechanical failure. Such defects need to be identified during the manufacturing process. A rigorous inspection and regulation mechanism is required to ensure that turbine blades suffering from these flaws are not installed in jet engines.

Neutron imaging is an effective and reliable NDT method for detecting low Z elements which in this case are ceramic remnants in turbine blades. It is superior to X-ray imaging due to its high contrast in neutron attenuations between ceramic remnants and the blade.

Another advantage of neutron imaging is in the use of gadolinium as a contrast agent. Gadolinium has a high thermal neutron cross section and can be efficiently utilized in improving contrast in neutron imaging. If a turbine blade is immersed or washed in liquid gadolinium solution, the gadolinium settles inside cracks and pores in it. In particular, ceramic remnants are quite porous and soak up the gadolinium solution. This process of immersing a component in gadolinium is known as gadolinium tagging. During neutron imaging, gadolinium atoms absorb neutron due to its high capture cross section and the region containing gadolinium atoms shows up as more opaque in the radiograph, thereby increasing the image contrast. Very small remnants of ceramic, which miss detection in normal neutron imaging, show up clearly with gadolinium tagging. Figure 8.1 shows the neutron radiograph of turbine blades before and after gadolinium tagging. It can be seen that the neutron radiograph of the turbine blade shows the presence of tiny ceramic blockages in the air channels when tagged with gadolinium.

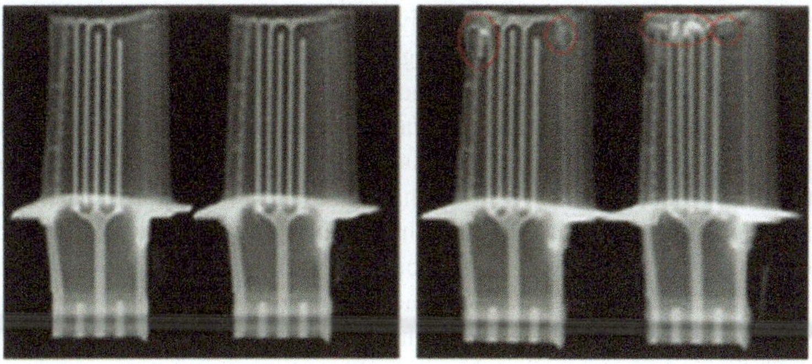

Fig. 8.1 Neutron radiograph of turbine blades inspected with neutrons (left) and after gadolinium tagging (right) [1]

8.3 Pyrotechnic Devices or Energetic Devices

Pyro devices generally use solid, hydrogen-rich chemical explosive encased in denser metal shells or enclosures (Fig. 8.2). Pyro devices are mission critical mechanical

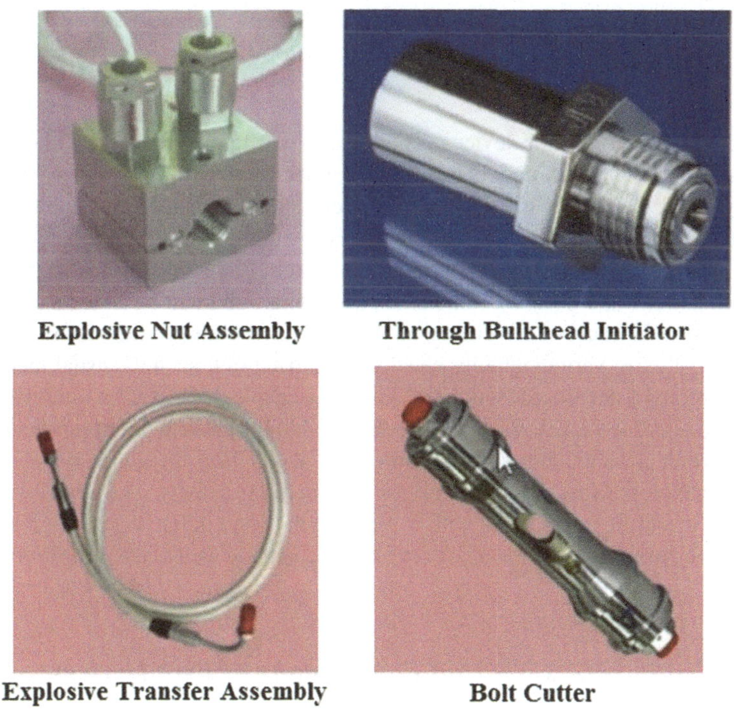

Fig. 8.2 Typical pyro devices used in space launch vehicles

devices actuated with the help of explosives. These devices perform a mechanical action such as cable cutting and bolt cutting which will aid in various flight operations such as stage separation, satellite deployment and heat shield separation with the help of explosive loaded cartridges. When initiated by a suitable current in the circuit, the explosive train gets activated which results in the release of large volume of gases in a short time or produces a sudden shock that makes a piston and anvil mechanism to perform the desired mechanical action. The reliability of these devices is very high, and mostly, they are positioned with sufficient redundancy so that if one pyro device fails, the other circuit will complete the activity. Pyro devices play a major role in liquid engines as well as cryo engines, wherein the flow of fuel and oxidizer has to be regulated as and when required. Normally closed or opened pyro valves are used in such scenarios to ensure the flow gets controlled as per programme.

Flaws or defects in pyro devices, such as voids, cracks, gaps and inclusions, can denote breakdowns in the chemical composition that can lead to misfires. X-ray imaging does not work well on these devices because X-rays will not be attenuated by low Z explosive material inside the metallic enclosure and will be unable to create sufficient contrast towards flaw detection. However, the high attenuation of neutrons by hydrogen in the explosives makes neutron radiography a suitable candidate for imaging such devices.

8.3.1 Quality Analysis

Pyro devices are one-shot devices. This means that the pyro device to be used in launch vehicle or satellite cannot be tested on ground before use. Hence, pyro devices are produced in batches in sufficient numbers and few of them are identified for ground tests to ensure their performance. Few numbers from each batch will be stored so that by any chance the pyro device used in flight has reached its permitted lifetime, and the one from the same batch kept on ground will be tested and provide life extension. Because of this critical nature, quality of these products is to be ensured with utmost care. Various quality checks are carried out, such as visual inspection and recording of the process sequence, loading of explosives, etc., by quality control (QC) personnel to ensure that the pyro device is fabricated without any flaws. All these pyro devices undergo X-ray radiography to confirm the general assembly of the internal parts as well as to check for any material defects. The devices in their cartridge level (the item loaded with explosive chain and initiators) undergo neutron radiography to ensure the presence of explosive charge as well as its filling uniformity.

8.3.2 X-ray Versus Neutron Radiography of Pyro Devices

X-ray radiography is carried out in pyro elements to confirm general assembly, presence of voids in glass seal, cracks in ceramic cup, bend in terminal pins, damages

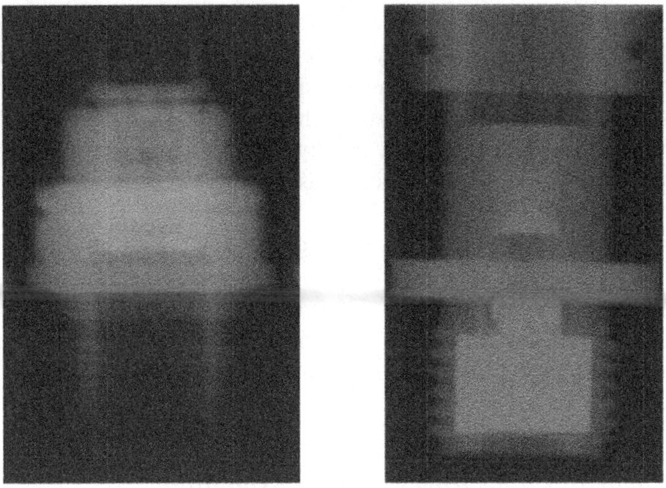

Fig. 8.3 (Left) X-ray and (Right) neutron radiograph of two pyro devices

to shear pins, etc. The pyro charges that are filled in pyro devices are mostly RDX, PETN, HMX, HNX, B/KNO_3, lead azide, zirconium potassium nitrate, etc. Some explosives such as lead azide and zirconium/potassium nitrate are detectable using X-ray radiography which confirms its presence during structural inspection itself. However, to detect the presence of explosives such as RDX, PETN, HMX, HNX and B/KNO_3, it is essential to carry out neutron radiography. Since most of these explosives are hydrogenous in nature or contains boron in it which are highly neutron attenuating in nature, NR will be able to detect their presence quite easily.

Figure 8.3 shows X-ray and neutron radiographs of two pyro devices where the explosive charge is located at the bottom. The explosive part at the bottom (white region) is clearly observed in neutron radiograph whereas the metallic parts are highlighted in the X-ray radiograph.

8.4 Neutron Imaging for Indian Space Programme

Neutron imaging has been extensively used by Indian Space Research Organisation (ISRO) for imaging and quality assurance of explosive laden pyrotechnic devices used in Indian space programme. In the subsequent sections, we shall discuss both reactor and non-reactor facilities used by ISRO as examples of neutron radiography for space applications.

NR of pyro devices for launch vehicle application started during the initial days at Apsara reactor at BARC. The first NR facility in Indian Space Research Organisation (ISRO) was developed at Satish Dhawan Space Centre, SHAR (Sriharikota Range), Sriharikota [2], using 15 MeV electron Linac. Subsequently, neutron imaging activity was shifted to reactor-based neutron imaging at KAMINI (KAlpakkam MINI) reactor facility of IGCAR Kalpakkam, India. In recent years, due to increased production of pyros to meet the tight launch schedules as well as difficulty in transportation of explosive loaded pyro devices to IGCAR each time, there was a strong requirement to have an in-house neutron radiography facility. This led to setting up a neutron radiography facility using a deuterium–tritium-based neutron generator with the technical support of BARC, Mumbai.

In short, three different modes of neutron sources—namely reactor neutron source, electron Linac-based photoneutron source and D-T neutron source—have been used for neutron imaging of pyro devices. The subsequent discussion will highlight the features of each of these facilities which will serve as representative examples of different neutron imaging sources which can be used for space application.

8.4.1 Electron Linac-Based Neutron Imaging

The neutron imaging facility at SHAR consists of a 15 MeV electron Linac with tungsten target which produces an unflattened X-ray output of 6000 rads. There are four sub-systems, namely (i) target assembly, (ii) moderator, (iii) collimator and (iv) shielding assembly. The target assembly consists of a number of beryllium (Be), uranium (U) and polyethylene discs. Photoneutrons are produced in Be target via (γ-n) reaction. The moderator is a polyethylene cylinder surrounding the target assembly to thermalize the neutrons produced in the target assembly. The thermal neutron beam is extracted from the central isotropic high flux region through a cadmium-lined aluminium divergent collimator. The schematic of the system is shown in Fig. 8.4 [2], and main characteristics of NR facility are given in Table 8.1.

Neutron radiography is carried out using dysprosium foils of optimum thickness (150 μm) with irradiation time of 60 min and transfer time of 12 h. Figure 8.5 shows some sample NR images taken in the facility.

8.4.2 Reactor-Based Neutron Imaging

Reactor-based neutron radiography for space application was carried out at KAMINI reactor facility of IGCAR Kalpakkam. KAMINI reactor is a ^{233}U fuelled, demineralized light water moderated and cooled, beryllium oxide reflected, low-power (30 kW) nuclear research reactor with a thermal neutron flux of 10^{12} n/cm^2 s at core centre. It has facilities for carrying out neutron radiography, neutron activation analysis and neutron shielding experiments. The *L*/*D* of the collimator is about 160, and the aperture size is 220 mm × 70 mm. Thermal neutron flux at the outer end of the beam tube is ~10^6—10^7 n/cm^2 s. The availability of high neutron flux coupled with good collimated beam provides high-quality radiographs with short exposure time [3].

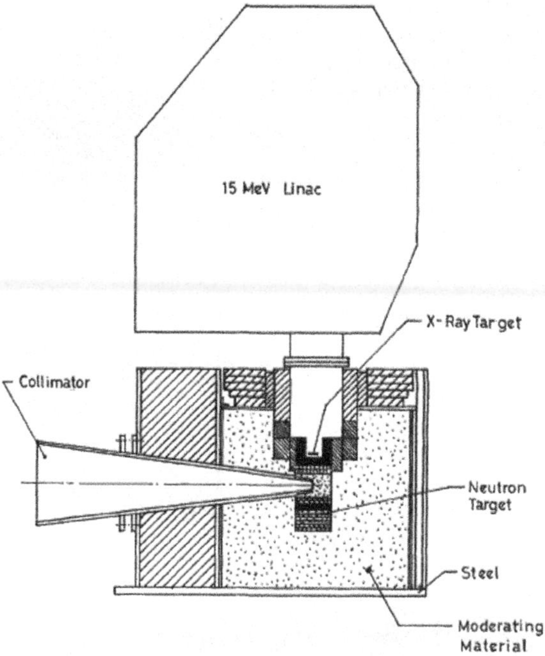

Fig. 8.4 Neutron radiography facility using 15 MeV Linac system at SDSC, SHAR. Reproduced from [2] with permission from Elsevier

Table 8.1 Major characteristics of the NR facility at SHAR [2]

Parameters	Value
Thermal neutron flux	1×10^6 n/(cm^2 s)
L/D ratio	18
Cadmium ratio	5
Neutron/gamma ratio	1×10^4 n/(cm^2 s mR)
Image plane size (field-of-view)	225 mm × 225 mm

Figure 8.6 shows the radiography beam port at KAMINI reactor and the assembled pyro device kept for NR.

Neutron radiography of pyro devices is carried out by transporting the pyro devices from VSSC to IGCAR Kalpakkam after completion of X-ray radiography. The pyro devices are loaded in special fixtures with provisions to position the dysprosium foils behind the object. Figure 8.7 shows the mechanism for assembly of multiple pyro devices on a fixture for imaging. The fixture is designed in such a way as to hold the pyro devices strongly without chances of falling during handling of the same. A stepper motor-controlled carriage drive and indexing mechanism (Fig. 8.8) is provided for neutron radiography. It has provisions for moving the carriage up/down and for lifting the indexing mechanism. Limit switches provided in the carriage drive mechanism indicate the position of the carriage inside the guide tube. The cassette drive mechanism has provisions for arranging ten cassettes and indexing is done

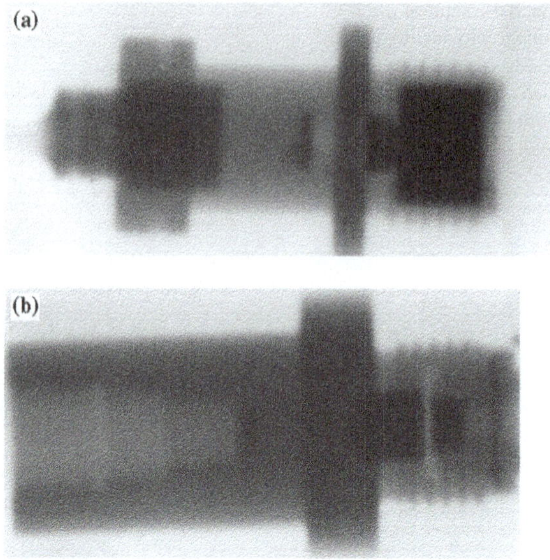

Fig. 8.5 Typical NR image of pyro device tested at SDSC, SHAR. Reproduced from [2] with permission from Elsevier

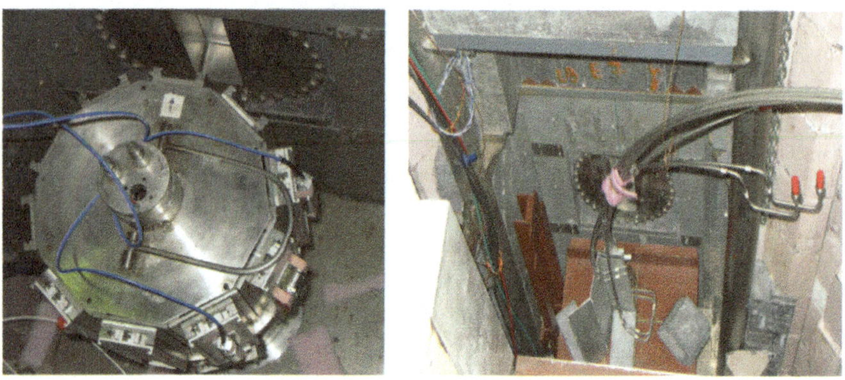

(a) (b)

Fig. 8.6 **a** Radiography beam port at KAMINI reactor and **b** assembled pyro device for NR

remotely from the control room. Indexing is provided to bring the pyro devices to be tested in front of the neutron beam each time.

Neutron radiography is carried out on film by transfer technique (Fig. 8.9). The dysprosium sheet is positioned behind the sample in such a way that the foils could be taken out easily without any damage and could be transferred to the dark room in suitable containers where it will be placed in between radiographic films tightly and allowed to expose the films for more than 6 h. The films are later processed in the dark room and interpreted.

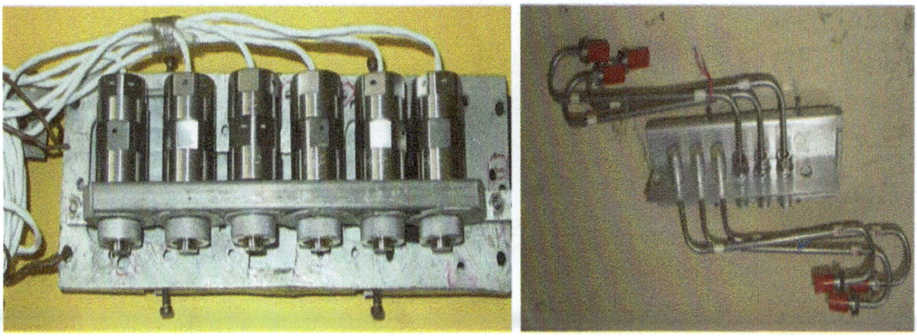

Fig. 8.7 Assembly of pyro devices for NR

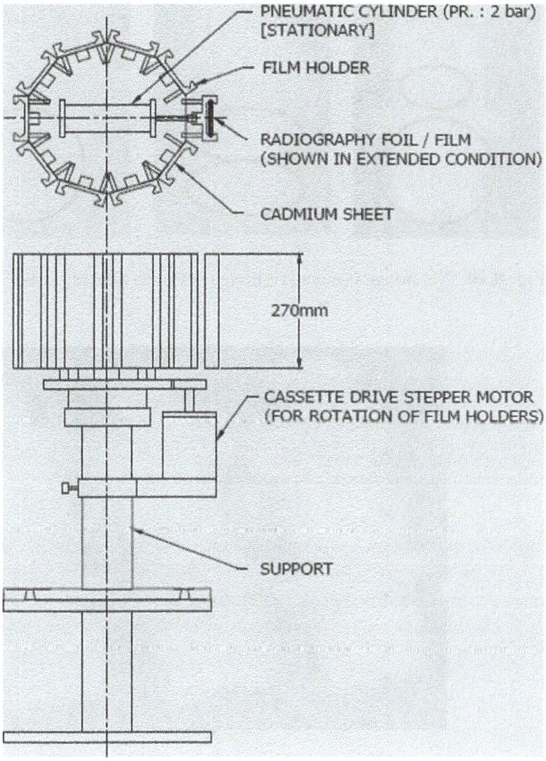

Fig. 8.8 Cassette drive mechanism used in NR facility at IGCAR. Reproduced from [4] with permission from AIP Publishing

Few of the images of inert pyro cartridges tested at KAMINI reactor are shown in Figs. 8.10, 8.11 and 8.12. The encircled regions highlight the presence of explosives (white region).

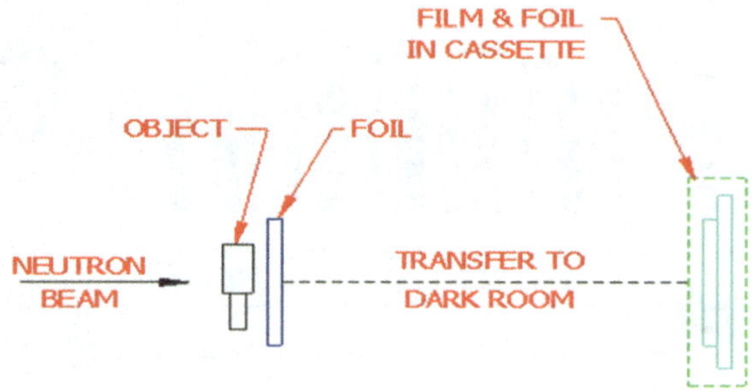

Fig. 8.9 Transfer method for neutron radiography

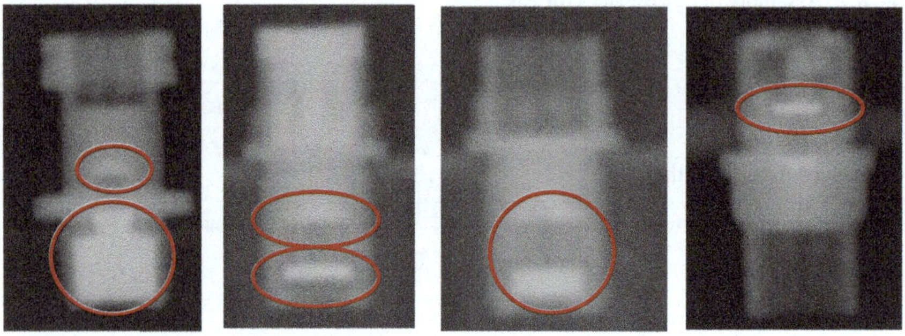

Fig. 8.10 NR images (negative image) of pyro devices at IGCAR Kalpakkam

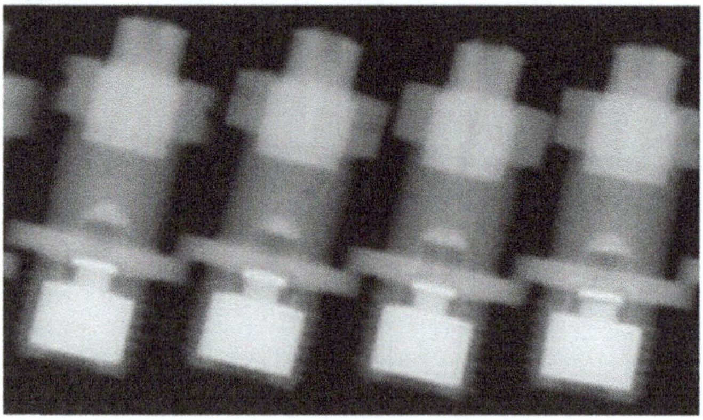

Fig. 8.11 NR images (negative image) of cartridge at IGCAR Kalpakkam

8.4.3 D-T Neutron Source-Based Neutron Imaging

A D-T neutron source-based thermal neutron imaging system was designed and developed by Bhabha Atomic Research Centre (BARC), and the system was installed

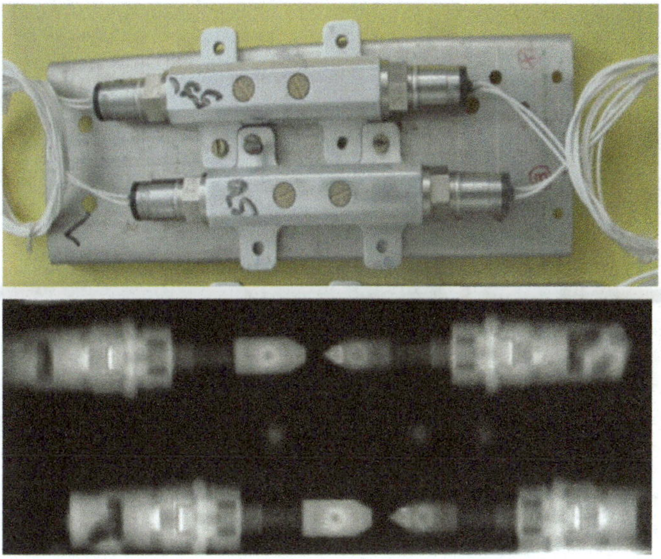

Fig. 8.12 (Top) Photograph and (Bottom) NR image (negative image) of cable cutter with cartridge assembly at IGCAR Kalpakkam

at NR Facility, Vikram Sarabhai Space Centre (VSSC), Thiruvananthapuram, India [3, 5, 6].

A D-T generator is used as neutron source. It emits 14.1 MeV neutrons by means of following nuclear fusion reaction

$$ {}_1^3H + {}_1^2H \rightarrow {}_2^4He + {}_0^1n(14.1\text{MeV}) $$

The fast neutrons produced by the source are moderated and collimated using a moderator–collimator assembly and an output of 1×10^4 n cm^{-2} s^{-1} thermal neutron flux is obtained at the sample plane which is used for neutron radiography. Since the thermal neutron flux output is low, dysprosium foil-based transfer technique cannot be used towards optimum use of the neutron generator. Instead, a 16-bit cooled intensified CCD camera-based neutron imaging system is used for capturing the NR image. It uses a neutron scintillator screen [^{6}LiF: ZnS(Ag) (0.4 mm thick)]. This camera is well suited for capturing the image within few minutes. Figure 8.13 shows the neutron radiography assembly.

The NR beam output is fixed, and hence, a sample manipulator was designed in such a way that the object will be in line with the beam path. A steel base plate moves over an aluminium support structure using rack and pinion mechanism. The plate has provisions to hold multiple fixtures separated by a fixed distance. The support structure is wide enough to provide stability to the plate during motion. Vibration isolator has been provided to the support structure to prevent vibration of the fixtures while in motion. The L-shaped plate to hold fixture containing pyro devices is made

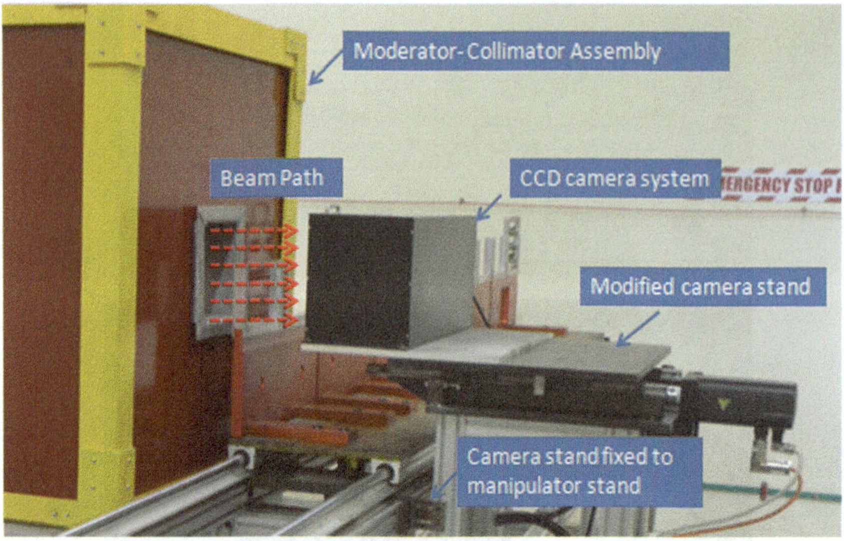

Fig. 8.13 Photograph of neutron radiography assembly at NR facility, VSSC

of aluminium plates of thickness 30 mm. The fixture made of aluminium plates of thickness 3 mm is attached to the top of this L-shaped plate and is designed in such a way that the object is just 3 mm from the imaging plane. The camera base plate is designed to move up to 0.5 mm distance from the fixture. This design has been incorporated to reduce the geometric unsharpness.

The pyro devices are assembled in a cassette which has removable plates containing holes to assemble the devices. 12 such cassettes are fixed to the assembly frames on the manipulator. When the fixture is in motion, it is ensured that no vibration occurs that might cause fall of pyro devices. Nearly 100 components can be mounted over 12 fixture stations and imaged in a single run without regular user intervention to mount samples. Figure 8.14 shows the fixture which is designed to easily assemble the plates containing pyro devices.

The intensified CCD camera is an integral part of the system and is incorporated in the manipulator assembly. This requires a similar base plate movement on guided rails mounted on an aluminium support structure holding the camera. Since L-shaped plate was used to hold fixture, the base plate of camera was designed so as to include the distance that the camera has to move in the forward direction. The system was adjusted in such a way that a gap of just 0.5 mm only will exist in between imaging plane and the object holding fixture. Once the manipulator system is in motion, the camera shall be retracted remotely to prevent camera and the pyro devices from touching each other accidently.

The manipulator is controlled using servo motors which are provided with programmable human–machine interface (HMI) stations in control room and NR hall. They provide manual and automatic movements to the manipulator system and

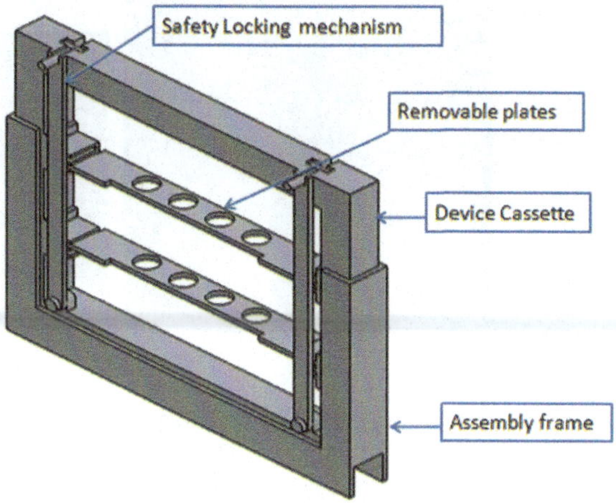

Fig. 8.14 Fixture to hold pyro devices

the camera. The HMI station inside the hall helps in checking the fine gap while system moves, while the HMI station outside helps in controlling the movement of fixtures remotely from the control room. Thus, it is possible to move each station to come in line with the neutron beam and ICCD camera one by one after each exposure time period is over. Each time, image shall be captured, interpreted and archived. Provisions to finely adjust the position of fixtures and to bring back already radiographed station again for radiography purpose so as to clear any doubts in interpretation shall also be done manually from the control room.

Image Quality Requirements

ASTM E 545 describes the image quality requirements in neutron radiography; however, it is for direct NR technique by using gadolinium (Gd) screen. The standard uses a beam purity indicator (BPI) and a sensitivity indicator (SI) as reference standard to classify a neutron radiograph category. As per the standard, the radiograph shall be of category I or II. However, achieving such high-quality radiographs with accelerator-based neutron source is very difficult.

However, para 5.2 of ASTM E 545 states, "The only truly valid sensitivity indicator is a reference standard part". A reference standard part is a material or component that is similar to the object being neutron radiographed except with a known standard discontinuity, inclusion, omission or flaw. The sensitivity indicators were designed to substitute for the reference standard such that achieving/detecting a standard discontinuity is adequate to qualify a neutron radiograph. The sensitivity indicators were designed to substitute for the reference standard and provide qualitative information on hole and gap sensitivity which can be interpreted as if one is able to generate a reference standard with intended defects and the defects thus generated are captured by the NR technique. Then, it is not essential to achieve category I or category II.

The qualification of image is based on detecting the intentionally created defect in the reference standard part. For each item that is to be radiographed at NR facility, a reference standard part is prepared and it is radiographed along with normal part

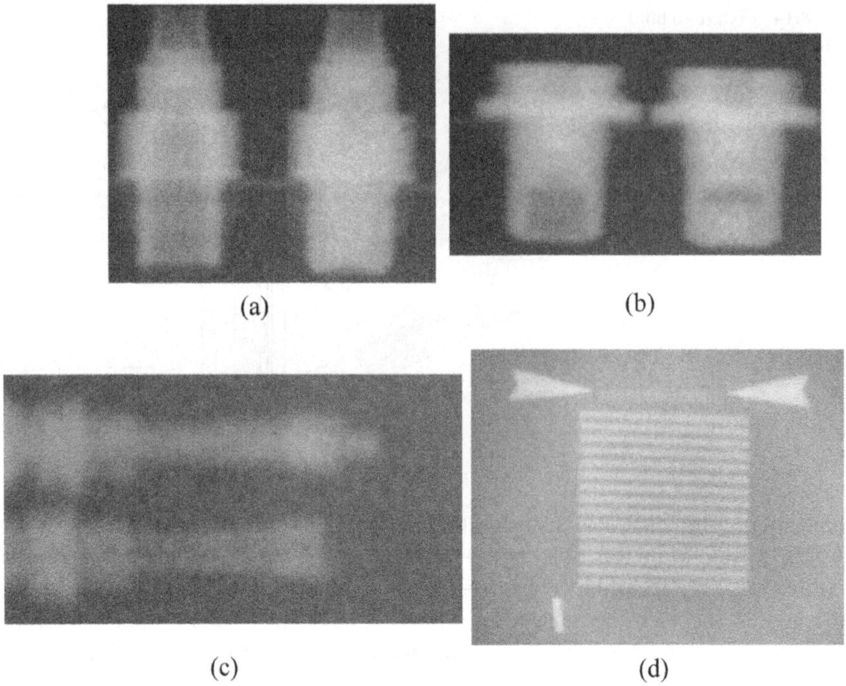

Fig. 8.15 NR images (negative image) of **a** cryo cartridge, **b** bolt cutter, **c** explosive transfer assembly tip, **d** explosive manifold leads

which is identical. All the geometrical, neutron tube and image acquisition parameters were varied to get the optimum image which can clearly discern the intentionally created defect in reference standard part. Some of the images taken at NR facility are shown in Fig. 8.15.

8.5 Comparison of the NR Facilities Used in Indian Space Applications

A comparison of the three neutron radiography facilities used for aerospace applications is shown below.

S. No.	Parameter	IGCAR facility	SHAR facility	VSSC facility
1	Neutron source	Research reactor (KAMINI)	Electron Linac -based photoneutron (15 MeV Linac-Be^9)	D-T neutron generator
2	Source neutron flux available	10^{12} n/cm^2 s	10^{11} n/cm^2 s	10^{10} n/s (14.1 MeV neutron yield)

(continued)

(continued)

S. No.	Parameter	IGCAR facility	SHAR facility	VSSC facility
3	Thermal neutron flux at imaging plane	10^6–10^7 n/cm^2 s	10^5–10^6 n/cm^2 s	1×10^4 n/cm^2 s
4	L/D ratio	160	18	35
5	Imaging area (sq. mm.)	220 × 70	225 × 225	100 × 100
6	Image capturing mode	Indirect method using dysprosium sheet and X-ray film	Indirect method using dysprosium sheet and X-ray film	Direct method using cooled intensified CCD camera (digital image)
7	Exposure time/frame using the detector mentioned in Sl. no. 6	10–15 min	30–90 min	5 min

8.6 Summary

Neutron imaging in aerospace industry has become a major non-destructive technique for testing of aerospace components such as turbine blades and pyro devices. This chapter emphasized the application of neutron imaging in testing of pyro devices used in space launch vehicles. As an example, we have discussed neutron imaging in the Indian space programme where three different facilities have been used over the years. These include reactor-based neutron imaging, electron Linac-based photoneutron imaging and D-T generator-based thermal neutron imaging. This shows the potential of both reactor- and non-reactor-based imaging facilities for such applications.

References

1. https://phoenixwi.com/. Accessed on 01–07–2021
2. Viswanathan K (1999) Neutron radiography in Indian space programme. Nucl Inst Methods Phys Res A 424:113–115
3. Sambamurthy E, Namboodiri GN, Gunasekaran R, Thomas C, Thomas CR (2016) Studies on neutron radiography technique for NDE of pyro devices using a low flux neutron source. NDE (2016)
4. Raghu N, Anandaraj V, Kasiviswanathan KV, Kalyanasundaram P (2008) Neutron radiographic inspection of industrial components using Kamini neutron source facility. AIP Conf Proc 989:202. https://doi.org/10.1063/1.2906066
5. Namboodiri GN, Sambamurthy E, Sai Krupa M, Satheesh PK, Ramlet U, Gunasekaran R, Rajendran L, Thomas C, Thomas CR (2016) Challenges faced in realization of object manipulator for neutron radiography of pyro-devices. NDE (2016)

6. Namboodiri GN, Kumar MCS, Nallaperumal M, Umasankar S, Levin G (2020) Detection and characterization of low dense charges inside metallic devices used in space applications by neutron radiography. J Nondestruct Eval 39. https://doi.org/10.1007/s10921-020-0657-7

Chapter 9
Neutron Imaging for Nuclear Power Industry

Shefali Shukla and Tushar Roy

9.1 Introduction

The ever increasing demand for electric energy owing to industrialization and growing economy necessitates expansion in nuclear power production throughout the world. To increase the safety and improve the efficiency of nuclear power, research and development in this field are constantly progressing. The next-generation reactors are focussed on better fuel utilization, proliferation resistance and enhanced safety. This requires better materials that can sustain these temperatures and increased radiation damage. The components related to operating reactors also require periodic examinations to ensure safe working conditions. Apart from that, the issue of waste disposal in a safe manner is also a challenge. Neutrons being unique probes offer assistance in a wide range of problems associated with reactors. With increased flux at modern neutron sources together with advanced sample environments, it becomes possible to conduct studies of reaction kinetics at reactor operating temperatures. The data generated using neutrons facilitate benchmarking simulations and modelling of crystal structure evolution and thermo-mechanical treatment. Development of advanced neutron detection methods widened the scope of neutron imaging in nuclear industry especially for nuclear materials, where heavy elements (e.g. uranium) are to be imaged together with light elements (e.g. hydrogen, oxygen).

S. Shukla (✉) · T. Roy
Technical Physics Division, Bhabha Atomic Research Centre, Mumbai 400085, India
e-mail: shefali@barc.gov.in

T. Roy
e-mail: tushar@barc.gov.in

D. K. Aswal et al. (eds.), *Neutron Imaging*,
https://doi.org/10.1007/978-981-16-6273-7_9

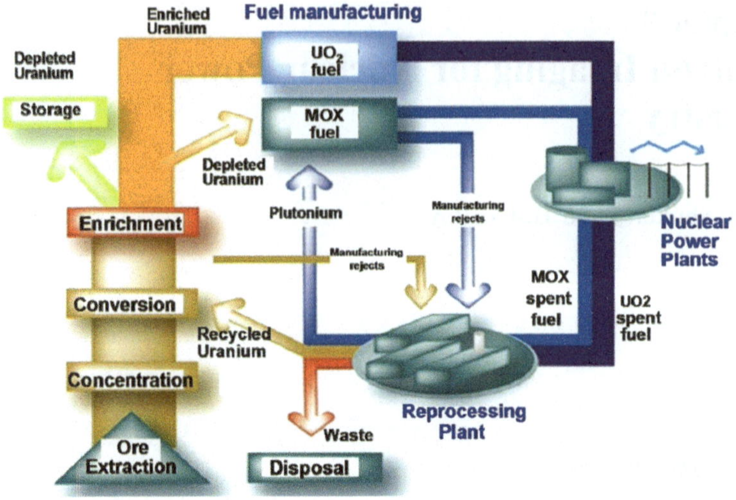

Fig. 9.1 Schematic of nuclear fuel cycle [1]

The nuclear fuel cycle begins with the preparation of fuel, the service period, i.e. the time during which the fuel is being used in the reactor for electricity generation and ends with the management of the spent fuel which can either be reprocessed and reused or diposed off.

Figure 9.1 shows a schematic of a general nuclear fuel cycle. Various stages involved from ore extraction to its disposal have been shown. Nondestructive testing (NDT) is mandatory at every step of nuclear fuel cycle to maintain the safe working and operation of the plants and facilities. Several techniques like radiography (neutron and X-ray), ultrasonic testing, eddy current and magnetic testing have been routinely used for NDT. Of these X-ray and neutron radiography are being explored in the nuclear field providing complementary information. Both these techniques form an indispensable part of NDT studies. Neutron imaging finds immense applications in the nuclear industry; ranging from testing of new fuels and reactor materials to investigation of irradiated failed components to find out the root cause of damage. In the subsequent subsections, we discuss the application of neutron imaging for hydrogen embrittlement studies, postirradiation investigations and testing of new fuels.

9.1.1 Neutron Imaging for Fuel Enrichment Studies

U-235 is the only nuclide existing in nature (in any appreciable amount) which is fissile with thermal neutrons (OECD Nuclear Energy Agency, 2003). Natural uranium has 99.284% U-238 isotope, with the U-235 isotope constituting only about 0.711%. Uranium enrichment is required for light water reactors (LWRs) and for other reactors. Neutron radiography can be used for monitoring of isotopic enrichment in fuel pellets loaded into a fuel pin [2] through attenuation comparison. Frajtag [2] conducted neutron radiography studies on fresh fuel pellets in a fuel pin. With neutron radiography, varying degrees of enrichment also could be studied.

Plutonium recycling in existing thermal reactors requires design of fuel assemblies containing pellets of different enrichments at different positions. A final check of the loading pattern is required to ensure the quality assurance of these fuels. In 1983, Gosh et al. [3] used neutron radiography to monitor plutonium enrichment in mixed oxide fuel pellets inside fuel pins.

Tremsin et al. [4] used neutron resonance absorption imaging and transmission Bragg edge imaging for fuel pellets study at LANSCE pulsed moderated neutron source of Los Alamos National Laboratory. The significant difference in the absorption cross sections of U-235 and U-238 isotopes was used, and very accurate nondestructive spatial mapping of the enrichment level of fuel pellets was carried out. The distribution of other isotopes in the spent fuel elements (Nd, Gd, Pu, etc.) was also demonstrated. This information can then be used for the investigation of fuel burn up rates for fuel elements placed at different rod positions in the reactor core. Tremsin concluded that the contrast between the pellets of different densities depends strongly on the range of neutron energies used. The more thermal part of the beam spectrum reveals the pellet with the lowest density as an object with the highest transmission whilst the coldest part of the neutron spectrum shows the opposite (Ref. Fig. 9.2).

9.1.2 *Quality Assurance of Fabricated Fuels*

As a result of the increasing demand for nuclear power, there has been an emphasis on the reliability of nuclear fuel performance for safety considerations. Fuel assemblies must withstand the operational requirements of the reactor and the irradiated fuels need to be stored in water basins for prolonged periods before reprocessing or disposal. One of the keys to consistent and reliable fuel production is the quality assurance of fresh fuels. Neutrons being transparent to metals can be used to image voids, inclusions, cracks or any other defects in freshly fabricated fuels. One of the first applications of neutron radiography was nuclear fuel characterization in the early 1960s using the film technique. Since then, neutron radiography has frequently been used as a quality control measure for fresh fuels. The fuel pin of PHWR type reactors contains natural uranium dioxide in the pellet form encased in a zircaloy cladding tube and sealed at both ends by welding of end plugs. Figure 9.3 shows the neutron radiography studies conducted on a fresh PHWR fuel pin. The radiography image shows some inclusion in the cladding material. For a detailed examination, tomography studies were conducted. The reconstructed image as well as a slice showing the flaw is shown. The study was carried out at neutron imaging beamline at Dhruva [5].

Neutron imaging can be used for studying any inclusions in fuel assemblies. Figure 9.4 shows cadmium and wax inclusions in PHWR and BWR fuel pin [6].

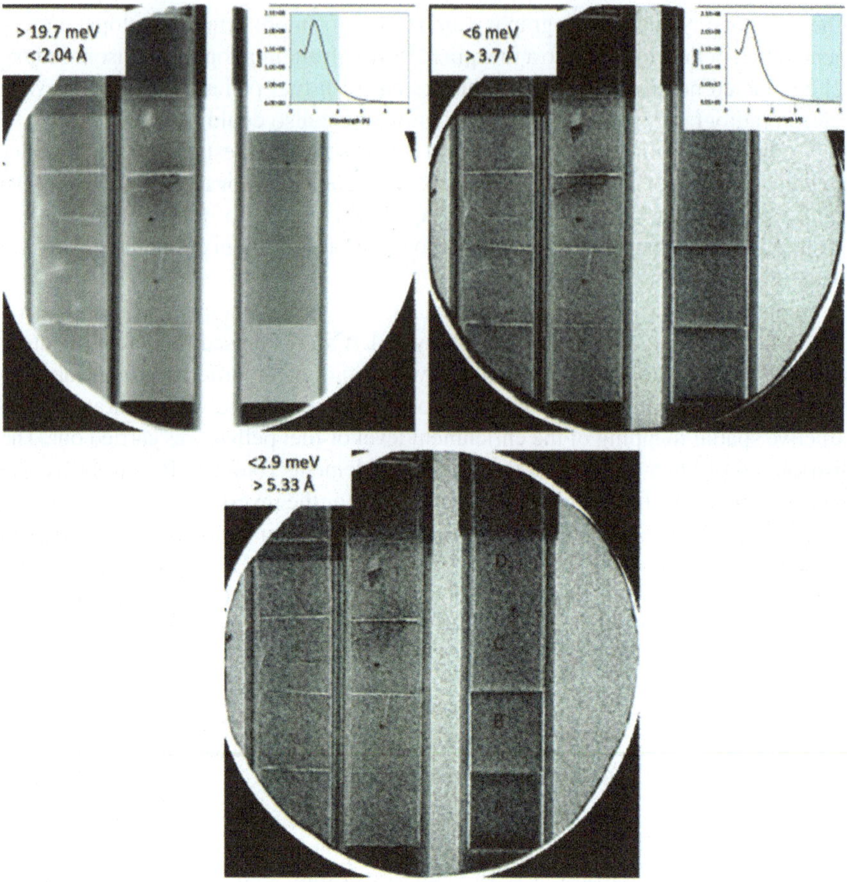

Fig. 9.2 Thermal neutron transmission radiographs of fuel pellets. Different energy ranges highlight the lowest(>19.7 meV) and highest(<6 meV) density object as per the transmission [4]. {Reproduced with permission from Elsevier Copyright Clearance Centre}

The images have been taken by the transfer method. The zirconium hydride formed is also visible owing to the difference in the cross section between zirconium and zirconium hydride. A typical experimental nuclear fuel pin consists of a thin-walled cladding tube filled with MOX(uranium, plutonium and thorium oxides) fuel pellets and hermetically sealed at both ends. The fuel pins are subjected to irradiation testing in research rectors and thus serve the purpose of studying fuel performance. Enriched uranium-aluminium alloy fuel sandwiched between aluminium-clad plates and roll-bonded into fuel plates are used as a core in small research reactors employed for physics research and applications like neutron radiography [7]. To reduce the centre temperature of the fuel pellet during irradiation and thereby reduce the pellet clad mechanical interaction, annular pellets (pellets with a central void) or pellets with

Fig. 9.3 Neuton radiograph of fresh PHWR fuel pin showing fuel pellets and an inclusion

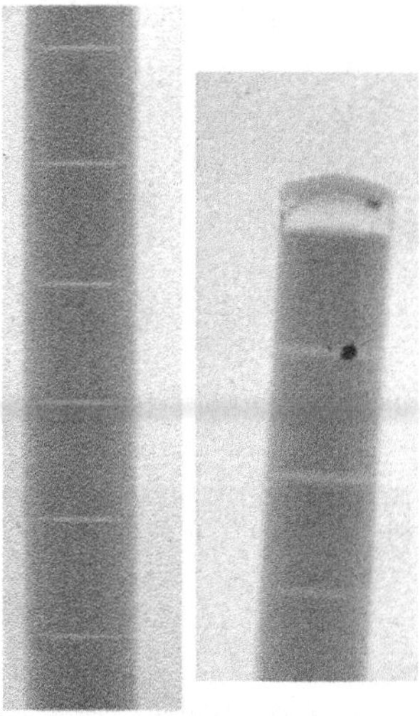

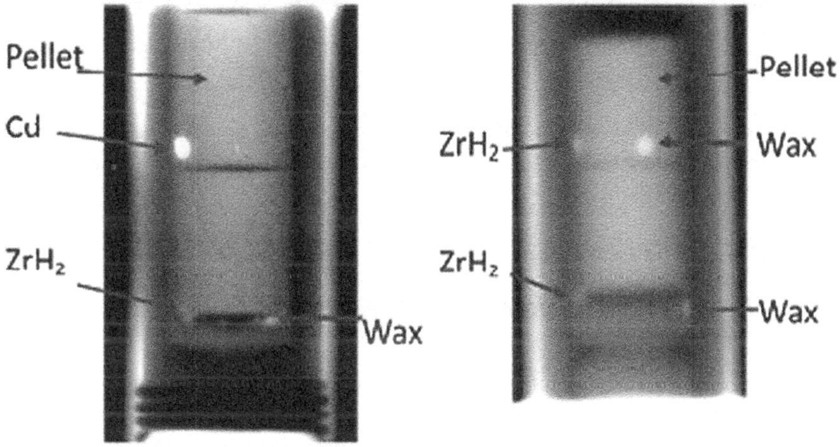

Fig. 9.4 Neutron radiography conducted on fresh PHWR and BWR [6]

depleted UO2 central core are being considered in BWR fuel pins. The deep penetration of neutrons in dense and high Z material is utilized to check the integrity of fuel pellets and also identify the anuular pellets. Water ingress inside a fuel pin because of a breach in the clad can be seen in the neutron radiograph (Ref. Fig. 9.5).

Neutron radiography unlike X-ray can be used for discriminating nearby elements. This unique capability can be utilized to identify UO_2 and PuO_2 -UO_2 mixed oxide fuel pellets inside sealed experimental fuel pins. Whilst mixing of PuO_2 with UO_2 for making MOX pellets, there are chances of PuO_2 agglomerates being present as

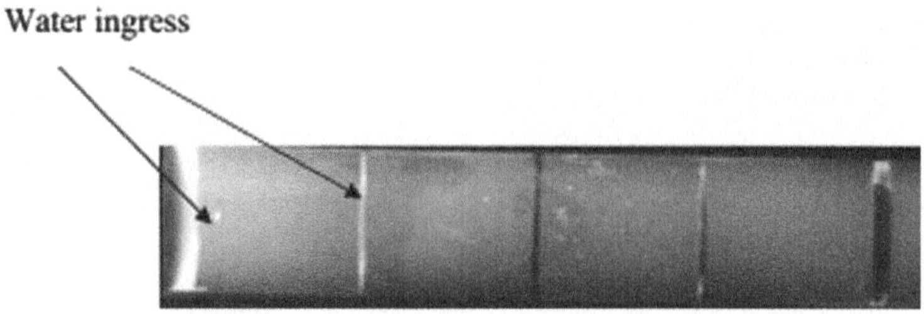

Fig. 9.5 Water ingress inside a PHWR type fuel pin because of a breach in the clad [7]

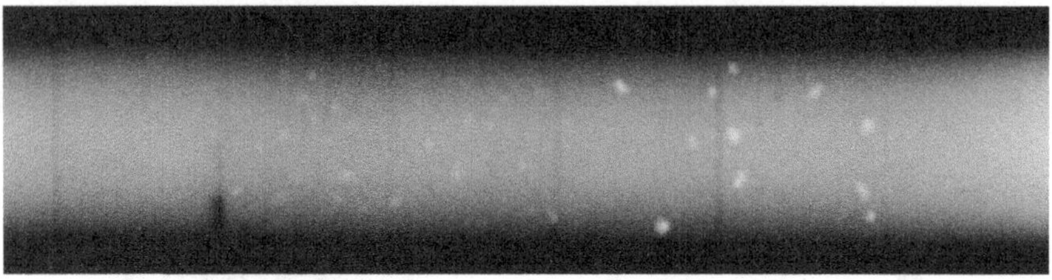

Fig. 9.6 Neutron radiography of plutonium oxide agglomerates inside MOX fuel pellets [7]

inclusions in the MOX matrix. Neutron radiography can be used to check homogeneity of this mixing. Figure 9.6 reveals PuO_2 agglomerates in MOX matrix in the neutron radiograph.

9.1.3 Testing of New Materials for Reactors

The future of nuclear energy depends not only on the availability of nuclear fuels but also on the development of advanced structural and functional materials. These materials are to be used in extreme environments like high temperature/pressure, intense neutron irradiation, presence of magnetic/electric fields, corrosive agents and all that in combination with complex loading states and cyclic operation histories. Further, developments in advanced materials for the existing thermal reactors may improve safety margins, reduce the frequency of component replacement and, ensure higher burn ups thereby improving fuel cycle economics and safety. For advanced future reactors (e.g. fast reactors, high-temperature reactors, fusion reactors, accelerated driven systems), the development of advanced materials is crucial due to the extreme environment and operating conditions associated with these innovative systems [8]. Nuclear technologies rely on the availability of advanced fuel and structural materials. These structural materials should be able to resist extreme conditions like high neutron, proton and gamma fluxes and fluencies, high temperatures, or corrosive and abrasive media and combinations of them. In particular, new concepts for nuclear energy production like fusion or next-generation fission reactors require materials suitable for the application at temperatures up to 1000 °C under strong irradiation resulting in atomic displacements up to 120 displacement per atom (dpa). The very high safety standards valid for nuclear installations involve excluding material failure under all real operating or hypothetical incident accident conditions. To predict the material behaviour even at the end of the usage, the structural processes occurring must be understood. These processes can result in the degradation of the mechanical properties. Examples are thermal ageing, irradiation and hydrogen embrittlement, formation or relaxation of residual stresses, corrosion and erosion or material fatigue. In the development phase of new reactor concepts, the material behaviour during operation and transient conditions has to be understood, often by extrapolation of the known behaviour.

For irradiated materials where the background may be high because of the activation of the materials, neutron studies can be conducted. SINQ is the Swiss neutron source at the Paul Scherrer Institute (PSI) for research purposes, based on spallation in a lead target irradiated by 590 meV proton beam. The target rod consists of lead in a zircaloy tube. Since some of the lower central rods are regularly heated by the proton beam to beyond the melting temperature of lead (327 °C) with the corresponding volume expansion, the tubes have a lead filling of only 90%. Lehmann et al. [9] and the group conducted studies to investigate the targets (virgin and irradited) to be used in SINQ at NEURAP beamline. Figure 9.7 shows radiography for a virgin rod (top) and of an irradiated rod (middle and bottom). This way, an empty horizontal gap is formed in the top region of each rod. Studies were also conducted to see the effect of irradiation on target rods and it was observed that spallation products are formed and also because of cyclic melting and solidification, redistribution of lead also takes place.

Fig. 9.7 Neutron radiography of a virgin SINQ target rod (**top**); Neutron radiography of a target rod after long irradiation (**middle**); Neutron radiography depicting the change in Pb distribution due to irradiation during operation (**bottom**) [9]

9.1.4 *Neutron Radiography for Nuclear Waste Disposal*

Nuclear power is said to be the solution of the energy crisis around the world. It can produce a large amount of energy from relatively small amount of fuel and also the waste generated in the whole process is small. The problem lies in the fact that almost all the waste generated is radioactive and thus requires careful management. Radioactive waste is generated not only in the nuclear fuel cycle but in many other sectors like agriculture, research, minerals exploration, medicine, etc., which use radioisotopes or radioactive particles. The only respite is that the level of hazard of all radioactive waste decreases with time. So the storage of radioactive waste is a crucial aspect of its management cycle. The waste must be stored such that it can be retrieved whenever required, and at the same time, it is completely isolated from the environment. The materials used for storing and transporting nuclear or radioactive waste must be fool-proof to ensure no leakage of radioactivity as it can be hazardous. Nondestructive testing of the casks and packing materials ensures their integrity. Nuclear materials are usually transported in packages which should ensure stringent safety standards like impact resistance, thermal shock, etc. Aluminium foam plays a major role as cushioning material in transport and prevents damage from shock or impact. The advantage of such faoam structure is that it can have the strength of the metal but with comparatively less weight. Neutron radiography allows us to image the entire bulk and calculate pore size pore fraction which can be used for modelling the structure. Figure 9.8 shows the 3D tomogram generated at Dhruva neutron imaging beamline [5].

The nuclear or radioactive waste which cannot be reprocessed immediately or which has to be dumped in radiological repositories needs to be packed in drums.

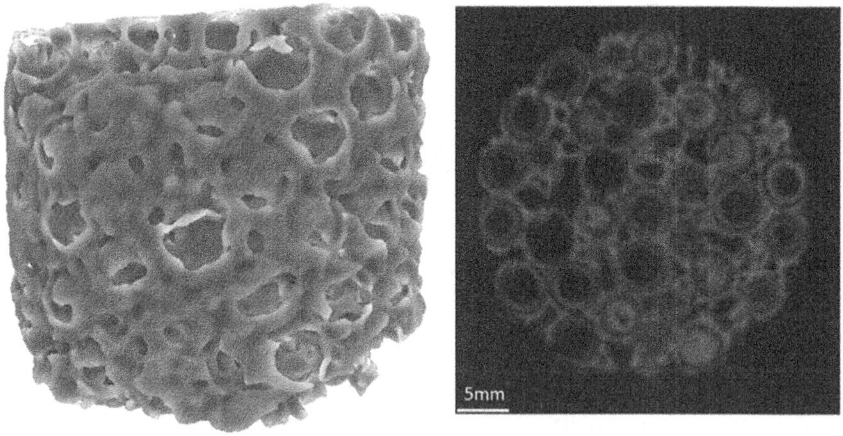

Fig. 9.8 3D reconstruction of aluminium foam used as cushion for nuclear material transport packages. The right picture shows a slice from the reconstructed tomogram. Pore size and density can be calculated from the images [5]. {Reproduced with permission from Elsevier Copyright Clearance Centre}

There are many active and passive methods that are being routinely used for drum scanning. They include segmented gamma ray scanning [10], gamma-transmission measurements (radiography and tomography using radioactive sources [11] or accelerator [12]), neutron emission counting (with time correlation analysis to distinguish between neutrons originating from spontaneous fission or (alpha, *n*) reactions, respectively, [13]) and neutron interrogation techniques (to induce fission events) [14]. Recently, neutron radiography studies have also been attempted for nondestructive analysis of the radioactive waste packages. Fission neutron radiography of 200 L waste drums was demonstrated first time at the NECTAR facility by Bucheral and coworkers [15]. Figure 9.9 shows the tomographic investigations of 200 L and 280 L mock-up waste drums using fission neutrons and cobalt-60 source.

Cementitious material is the most common medium used for safely encasing and disposing of low and intermediate-level radioactive waste. It acts as a barrier to prevent water from contacting the radionuclides and as a cover that prevents the migration of radionuclides. Both these processes depend basically on limiting water transport through the cementitious layer, thereby defining the performance of the cementitious waste form. Neutrons can be an excellent probe when studying water transport especially in high Z materials like cement. Mcglinn [16] studied water transmission through two types one unleached and the other leached for 8 years. Results for the unleached sample are shown in Fig. 9.10.

Fig. 9.9 Radiography of a mock-up waste drum using fission neutrons (**left**) and 60Co source (**right**) [15]. {Reproduced with permission from Elsevier Copyright Clearance Centre}

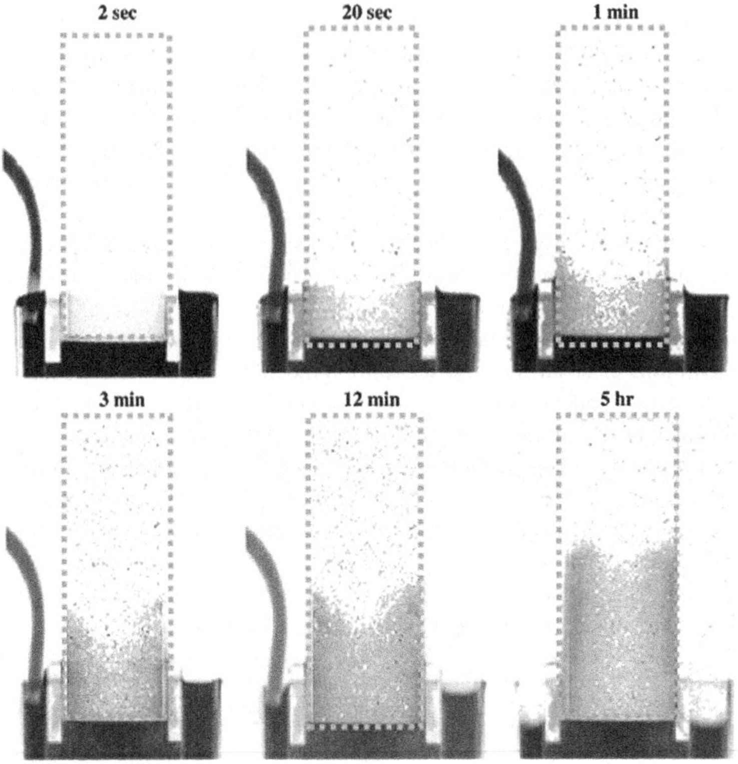

Fig. 9.10 Neutron radiographs taken at differents instant of time to show the movement of water front [16]. {Reproduced with permission from Elsevier Copyright Clearance Centre}

9.1.5 Postirradiation Studies

For the nuclear fuel development programme information about the fuel performance at various stages—pre-irradiated, interim and postirradiated is extremely important. Postirradiation examinations provide valuable information like fuel cracking, swelling, melting, fissile material redistribution, etc. Postirradiated samples cannot be examined using conventional techniques because of their inherent high radiation. However, neutron radiography can help qualitative as well as quantitative information about the fuel and associated structural materials. This feedback helps in improving fuel design. Neutron imaging plays a valuable role in postirradiation examination as it is a nondestructive method and can be used even for highly radioactive samples. Indirect neutron radiography used for such samples is a two-step process involving building radioactivity on a neutron-absorbing screen followed by exposing a film in contact with this screen to generate the image of the test sample.

One of the most common forms of degradation found in metals is hydrogen embrittlement. It is a metal's loss of ductility and reduction of load-bearing capability due to the absorption of hydrogen atoms or molecules by the metal resulting in reduced strength of the material. When talking about the nuclear power industry, the first and foremost material that comes into mind is zirconium. Zirconium alloys are frequently used at nuclear structural materials for water reactors because they low thermal neutron capture cross section, excellent mechanical properties along with possess good resistance to corrosion in high-temperature water [17, 18]. They, however, like other metals are susceptible to hydrogen induced damage: hydrogen embrittlement [19, 20]. It could be manifested in two major forms, namely delayed hydride cracking and hydride blisters. Therefore hydrogen estimation is, necessary to predict the life of in-service components. Though there are several techniques available for the same, because of hydrogen atom's high absorption cross for thermal neutrons, neutron radiography has the advantage of being a nondestructive technique providing bulk information within a very short duration. Once calibrated using standard samples of known hydrogen content, neutron radiography can estimate the exact amount of hydrogen in any sample within experimental errors. Figure 9.11 shows the attenuation coefficient as a function of neutron wavelength. The utility of neutron radiography for studying hydrides is quite evident because of the huge difference between zirconium and its hydride [21].

Figure 9.12 shows the tomography studies on zirconium alloy coupons pre-charged with a known amount of hydrogen carried out at CIRUS imaging beamline, BARC by Ashish et al. [22]. Calibrating the setup with a known amount of hydrogen

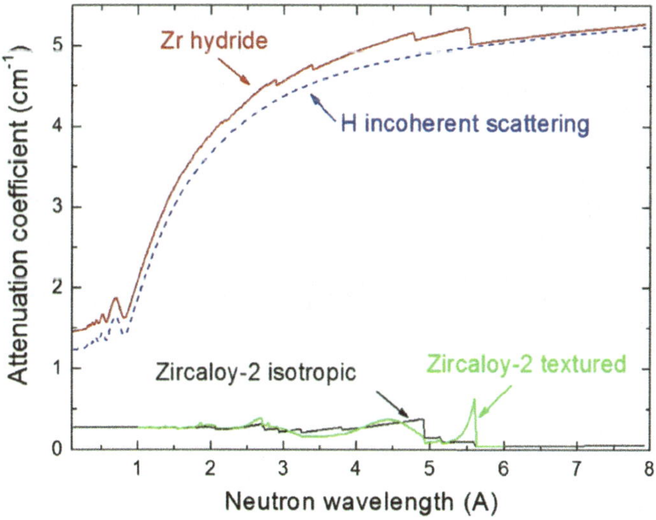

Fig. 9.11 Neutron attenuation vs neuton wavelength graph for zirconium and its hydride. The difference in mean attenuation coefficient of zirconium alloy versus hydride is quite evident [21]. {Reproduced with permission from Elsevier Copyright Clearance Centre}

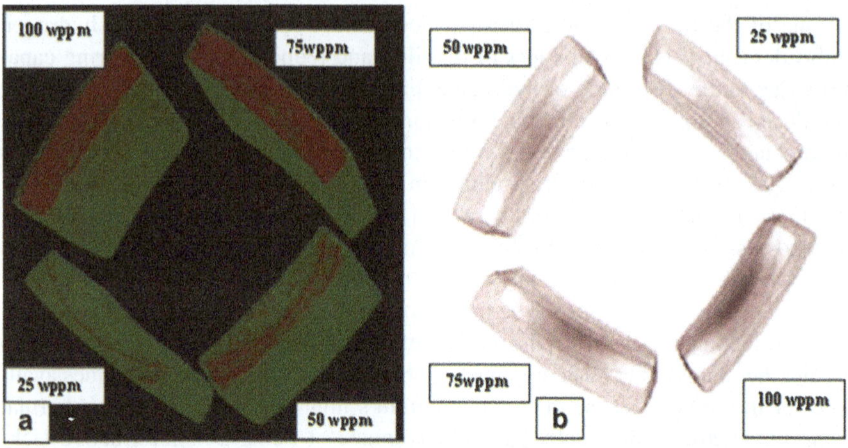

Fig. 9.12 Neutron tomography results on zirconium alloy pre-charged coupons. The green part is the base matrix and red indicates the amount of hydrogen present [22]. {Reproduced with permission from Elsevier Copyright Clearance Centre}

pre-charged coupons, it was also attempted to measure the hydrogen content of an unknown sample.

Figure 9.13 shows the neutron radiography images of the failed fuel element. Studies were carried out to understand the reason for rupture. Hydrogen embrittlement in metals particularly zirconium alloys manifests as blistering and delayed hydride cracking [23–25]. These two processes can occur even if the total hydrogen content in the pressure tube is less than the critical value. This occurs because the hydrogen which enters the zirconium matrix is highly mobile and is quick to respond to any concentration, stress or temperature gradients.

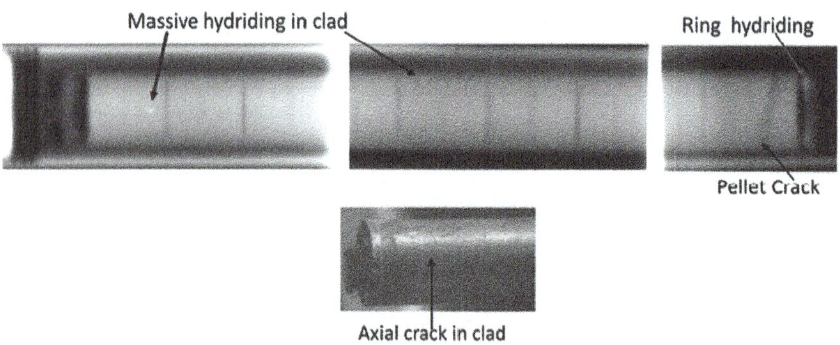

Fig. 9.13 Radiograph showing metal hydriding in fuel. The metal hydride has attenuation cross section 16 times the matrix hence is clearly visible in the radiography [7]

Hydrogen diffusion studies in various zirconium alloys have been attempted by Grosse and Buitrago [21] using neutron radiography. Shukla [26] conducted systematic studies for measuring diffusivity parameters for Zr-2.5%Nb alloy used in Indian PHWRs as pressure tube material. Figure 9.14 shows the neutron radiography and Fig. 9.15 shows hydrogen depth profiles obtained.

Hydrogen present in the zirconium pressure tube as a result of downside corrosion migrates towards regions of higher stress and gets precipitated making that specific

Fig. 9.14 Radiograph showing depth profile of hydrogen in samples. The hydride layer on one extreme end denotes the source of hydrogen [26]. {Reproduced with permission from Elsevier Copyright Clearance Centre}

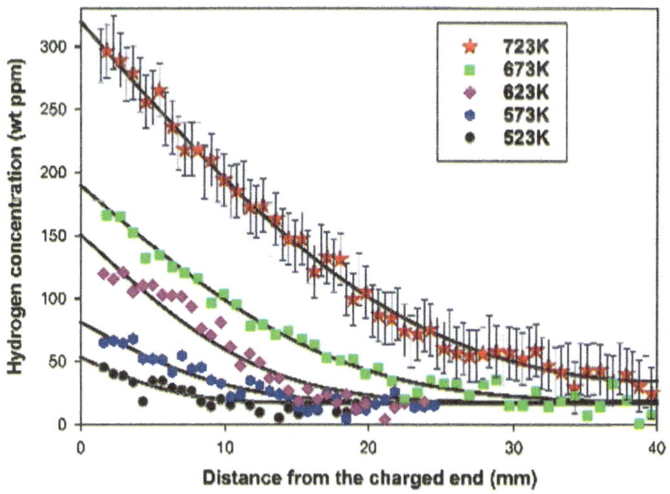

Fig. 9.15 Hydrogen concentration change with depth data is fitted to obtain diffusivity values [26]. {Reproduced with permission from Elsevier Copyright Clearance Centre}

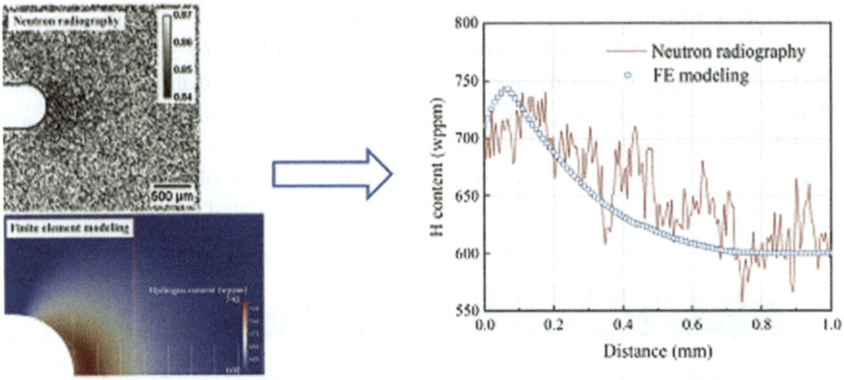

Fig. 9.16 High-resolution radiography showing the hydrogen concentration near a notch [27]. {Reproduced with permission from Elsevier Copyright Clearance Centre}

area brittle. This phenomenon is known as delayed hydride cracking (DHC) and is the major cause of failure of pressure tubes. Studies have been carried out to understand the kinetics of hydrogen movement under stress gradients. Gong [27] carried out hydrogen diffusion studies in Zircaloy nuclear fuel cladding materials under stress using high-resolution (10 μm) neutron radiography. The results obtained were in agreement with finite element modelling. Figure 9.16 shows the high-resolution neutron radiography results and the FEM results for comparison.

Whenever a temperature gradient occurs, the hydrogen migration leads to the formation of blisters. Agarwal [22] studied blistering in a pressure tube spool and PHWR fuel pin. Figure 9.17 shows the data which neutron tomography studies have obtained.

Hydrogen-assisted cracking in electrochemically hydrogen-charged technical iron and supermartensitic steel was studied using neutron tomography. Griesche [28] carried out blister-related studies on technical iron and developed a 3D model. Hydrogen accumalations around cracks and hydrogen gas trapped in cracks could be visualized through neutron tomography. Figure 9.18 shows the reconstructed 3D model of the hydrogen-charged technical iron sample.

Figure 9.19 shows the hydrogen-assisted cracking results obtained in super martensitic steel. Dog bone shaped samples pre-charged with hydrogen were exposed to constant tensile load with subsequent sample rupture. It was observed that the tensile sample ruptured at a load of 2/3 of the theoretical tensile strength value for non-hydrogenous material. An apparent degradation in the mechanical property had occurred and was proved by this experiment.

Nuclear fission has long been explored and widely used in reactors worldwide. The efficiency of nuclear fission is low compared to the energy produced if we opt for nuclear fusion. Also the problem of spent fuel storage and disposal can be tackled. The fission–fusion hybrid energy system may be a practical approach in attaining sustainable development of nuclear energy. Metallic fuel used in subcritical blankets can be promising candidates for a fusion-fission hybrid energy system. U–Zr alloy

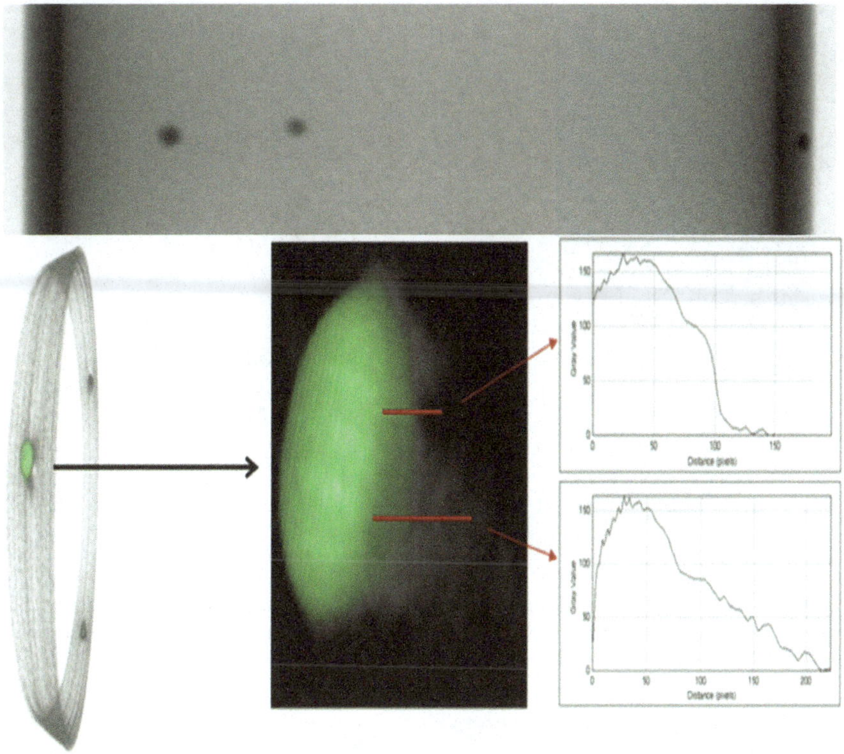

Fig. 9.17 Hydride blister in PHWR pressure tube [22]. {Reproduced with permission from Elsevier Copyright Clearance Centre}

which is used as blanket material might become the most potential material for metallic fuel as it can withstand high temperature, high pressure and radioactive environment. Nondestructive testing, however, is required to ensure reliability. Since the irradiated U–Zr alloy fuel is highly radioactive routine methods of inspection cannot be used. Indirect neutron radiography is used for the same. Yong Sun [29] and group carried out neutron radiography studies to obtain the inner information of irradiated U–10Zr alloy. They studied the swelling behaviour of irradiated U–10Zr alloy fuel at various burn ups using indirect neutron radiography. It was demonstrated that the swelling rate increases with an increase in burn up. Figure 9.20 shows the neutron radiography result for a particular sample, the metallic alloy enclosed in an aluminium powder shell is shown. Swelling is quite evident at position A as shown in Fig. 9.20. Figure 9.20b shows that the grey value on line A is different from B and C indicating some leakage, and uranium has flown out into the aluminium powder shell.

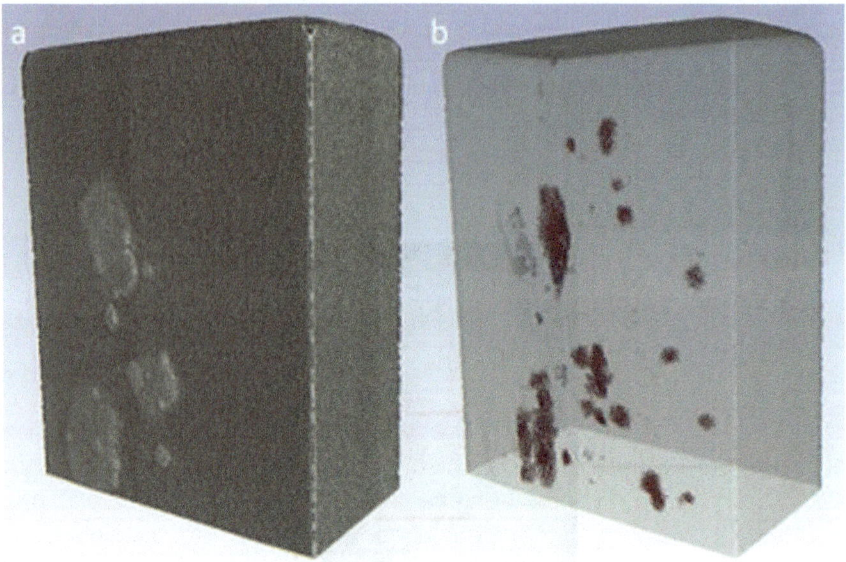

Fig. 9.18 Inclined view on the reconstructed 3D model of a hydrogen-charged iron sample. The rendered surface with blisters is shown, the cracks are shown in blue and the trapped hydrogen in red [28]. {Reproduced with permission from Elsevier Copyright Clearance Centre}

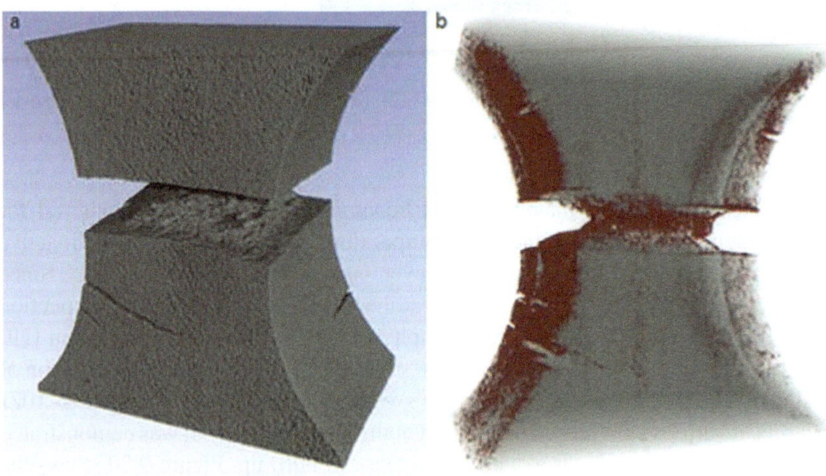

Fig. 9.19 Inclined view on the reconstructed 3D model of a part of a hydrogen-charged super martensitic tensile sample. **a** and **b** depict the rendered surface and the hydrogen distribution in red, respectively [28]. {Reproduced with permission from Elsevier Copyright Clearance Centre}

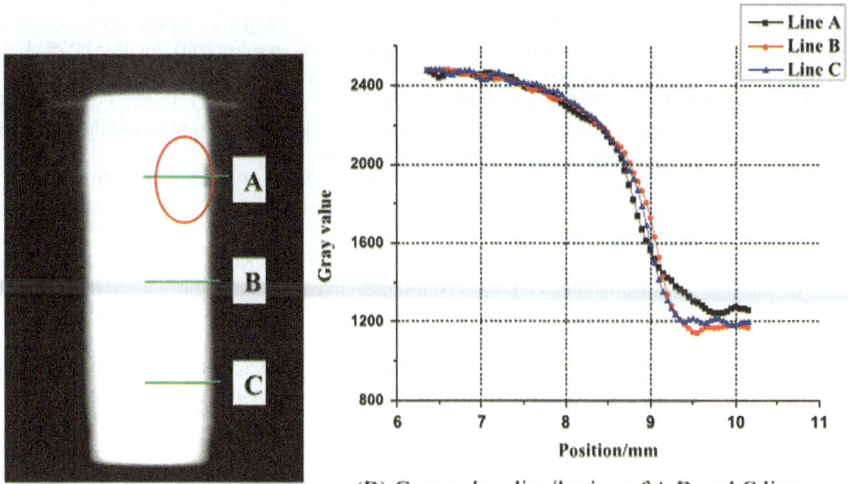

(A) INR result

(B) Gray value distribution of A,B and C line which indicated in (A)

Fig. 9.20 Indirect neutron radiography (INR) result and grey value distribution of A B C line indicated in (left figure) of sample [29]. {Reproduced with permission from Elsevier Copyright Clearance Centre}

9.2 Summary

This chapter deals with one of the most important applications of neutron imaging: the nuclear industry. Neutron radiography and tomography are routinely used to investigate of fresh and irradiated fuels. It also serves the purpose of testing new fuel compositions and designs. Indirect neutron radiography provides the answers especially for irradiated samples where most of the techniques fail to provide information because of the high gamma background. The materials used for reactor structure, fuel and coolant clad need to withstand extreme conditions. Neutron imaging allows the testing and development of such materials. Hydrogen embrittlement is a major issue for all structural materials particularly zirconium and iron. Being sensitive to hydrogen, neutrons can be used to evaluate hydrogen content in different metals as well as to study its diffusion parameters under stress temperature and concentration gradients. Neutron imaging, thus, finds numerous applications linked to the nuclear power industry. This chapter attempts to give a glimpse of it.

References

1. www.tsoturkey.com/page_fields_fuel_cycle.html. Accessed on 01 Aug 2021
2. Frajtag P (nd.) Radiation protection and radiation applications: gamma and neutron radiography. http://moodle.epfl.ch/pluginfile.php/1593971/mod_resource/content/2/RRA-EPFL-FS2014-Week14a.pdf. [Accessed 1 May 2015]

3. Ghosh JK, Panakkal JP, ROY PR (1983) Monitoring plutonium enrichment in mixed-oxide fuel pellets inside sealed nuclear fuel pins by neutron radiography. NDT Int 16(5):275–276, Oct 1983. https://doi.org/10.1016/0308-9126(83)90127-X
4. Tremsin AS, Vogel SC, Mocko M, Bourke MAM, Yuan V, Nelson RO, Brown DW, Feller B (2013) Nondestructive studies of fuel pellets by neutron resonance absorption radiography and thermal neutron radiography. J Nucl Mater 440:633–646
5. Shukla M, Roy T, Kashyap Y, Shukla S, Singh P, Baribaddala TP, Gadkari SC (2018) Development of neutron imaging beamline for NDT applications at Dhruva India. Nucl Instrum Methods-A 889:63–68
6. Singh JL, Mondal NK, Dhotre MP Pandit KM, Bhandekar A, Kumawat N, Ranjan R, Sahu AK, Ramesh N, Anantharaman S (2011) Non-destructive evaluation of irradiated nuclear fuel pins at cirus research reactor by neutron radiography. In: Proceedings of the national seminar and exhibition on Non-destructive evaluation NDE, Dec 2011, pp 8–10
7. Characterisation and testing of materials for nuclear reactors IAEA techdoc-1545 (2007). https://www.iaea.org/publications/7760/characterization-and-testing-of-materials-for-nuclear-reactors
8. Use of neutron beams for materials research relevant to the nuclear energy sector IAEA techdoc-1773 (2015). https://www.iaea.org/publications/10824/use-of-neutron-beams-for-materials-research-relevant-to-the-nuclear-energy-sector
9. Lehmann E, Thomse K, Strobl M, Trtik P, Bertsch J Dai Y (2021) NEURAP—A dedicated neutron-imaging facility for highly radioactive samples. J Imag 7:57. https://doi.org/10.3390/jimaging7030057
10. Bücherl T, Kaciniel E, Lierse Ch (1998) Synopsis of gamma scanning systems, comparison of gamma determining systems and measuring procedures for radioactive waste packages. Report WG-A-01 Sep 1998. www.en-trap.eu/doc/gammasynopsis.pdf
11. Martz HE et al (1996) Application of gamma-ray active and passive computed tomography to nondestructively assay TRU waste. UCRL-JC-123342 23–26 Jan 1996. www.iaea.org/inis/collection/NCLCollectionStore/_Public/27/060/27060283.pdf
12. Rizo Ph et al (2000) Application of transmission tomography to nuclear waste management. In: Proceedings of the 15th World conference on non destructive testing, Toma, Italy
13. Bucherl T et al (2001) Synopsis of neutron assay systems, Report WG-A-02, September 2001, www.em-trap/doc/neutronsynopsis.pdf
14. DeSimone D et al (2010) Review of active interrogation techniques and considerations for their use behind an information Barrier. LA-UR-10- 06958 29 Sep 2010
15. Bucherl T (2017) A feasibility study on reactor based fission neutron radiography of 200-l waste packages. Phys Procedia 88:64–72
16. Mcglinn PJ (2010) Appraisal of a cementitious material for waste disposal: Neutron imaging studies of pore structure and sorptivity. Cement Concr Res 40:1320–1326
17. Northwood D (1985) The development and applications of zirconium alloys. Mater Des 58–70
18. Nikulina AV (2003) Zirconium-Niobium alloys for core elements of pressurized water reactors. Met Sci Heat Treat 45:287–292
19. Singh RN, Kumar N, Kishore R, Roychaudhary S, Sinha TK, Kashyap BP (2002) Delayed hydride cracking in Zr--2.5Nb pressure tube material. J Nucl Mater 304 189–203
20. Northwood D, Kosasih U (1983) Hydrides and delayed hydrogen cracking in zirconium and its alloys. Int Met Rev 28(1):92–121
21. Buitrago NL, Santisteban JR, Tartaglione A, Marin J, Barrow L, Daymond MR, Schulz M, Grosse M, Tremsin A, Lehmann E, Kaestner A, Kelleher J, Kabra S (2018) Determination of very low concentrations of hydrogen in zirconium alloys by neutron imaging. J Nucl Mater 503:98–109
22. Agrawal A, Kashyap Y, Sarkar P, Behera A, Mayank Shukla RN, Singh AS, Chakravartty JK (2011) Study of hydride blisters in Zr-alloy using neutron tomography. J Nucl Mater 421:47. https://doi.org/10.1016/j.jnucmat.2011.10.047
23. Sawatzky A (1960) Hydrogen in zircaloy-2: Its distribution and heat of transport. J Nucl Mater 2:21–28

24. Varias AG, Massih AR (2000) Simulation of hydrogen embrittlement in zirconium alloys under stress and temperature gradients. J Nucl Mater 279:273–285
25. Jovanovic M, Stern A, Kneis H, Weatherly GC, Leger M (1988) Thermal diffusion of hydrogen and hydride precipitation in Zr–Nb pressure tube alloys. Can Metall Q 27(4):323–330
26. Shukla S et al (2021) Investigation of hydrogen diffusivity in Zr-2.5%Nb alloy pressure tube material using metallography and neutron radiography. J Nucl Mater 544:152679
27. Gong W, Trtik P, Valance S, Bertsch J (2018) Hydrogen diffusion under stress in Zircaloy: high-resolution neutron radiography and finite element modeling. J Nucl Mater 508. https://doi.org/10.1016/j.jnucmat.2018.05.079
28. Griesche A (2016) Measuring hydrogen distributions in iron and steel using neutrons. Phys Procedia 69:445–450
29. Sun Y (2016) Heng yong Huo, Yang Wu et al, Swelling behavior detection of irradiated U-10Zr alloy fuel using indirect neutron radiography. NIM 837:23–27

Chapter 10
Neutron Imaging for Material Science and Engineering

P. S. Sarkar and Shefali Shukla

10.1 Introduction

Material science and engineering is an interdisciplinary area concerning the properties of matter and its application to various areas of science and engineering. The basic of all material science involves relating the material's desired properties and relative performances on a specific application to the structure of the atoms and phases in that material through characterization. Neutrons are sought for probing materials because they can provide bulk information, which is difficult to obtain using other techniques. Neutrons are very sensitive to light elements and can be used for their mapping even if they are encapsulated in high Z materials. The attenuation coefficients of various materials for neutrons also depend on the neutron energy, and this property has been exploited for Bragg neutron imaging for providing better contrast.

10.2 Major Applications of Neutron Imaging in Material Science and Engineering

This section deals with the applications of neutron imaging in various fields of material science ranging from flow field and degradation studies of proton exchange membrane (PEM) fuel cells to structural stability of hydrogen storage materials specifically essential for these type of fuel cells. Also like hydrogen, neutrons are very sensitive to lithium, enabling us to visualize lithium in lithium-ion batteries. Various aspects of Li-ion batteries preparation and operation have been covered.

P. S. Sarkar (✉) · S. Shukla
Technical Physics Division, Bhabha Atomic Research Centre, Mumbai 400085, India
e-mail: pss@barc.gov.in

S. Shukla
e-mail: shefali@barc.gov.in

D. K. Aswal et al. (eds.), *Neutron Imaging*,
https://doi.org/10.1007/978-981-16-6273-7_10

An advanced kind of neutron imaging, namely Bragg-edge (already discussed in detail in the previous chapter) can also be used for crystalline imaging phases in materials. Residual stress measurement is of fundamental interest in order to estimate the service life of engineering components and structures subjected to various loading conditions operating in different environments. Neutron imaging application in residual stress measurements has an upper edge over the mechanical relaxation approaches such as the contour method or non-destructive methods such as X-ray diffraction.

10.2.1 Neutron Imaging of PEM Fuel Cells

Proton exchange membrane (PEM) fuel cells, which directly convert chemical energy to electrical energy, have attracted significant attention due to the numerous benefits they offer like high-power density, high energy conversion efficiency, fast start-up, low sensitivity to orientation and environmental friendliness. Figure 10.1 shows a schematic of a PEM fuel cell [1]. The heart of a PEM fuel cell is the membrane electrode assembly (MEA) which includes the membrane, the catalyst layers and the gas diffusion layers (GDL). Apart from MEA few more hardware components are incorporated in a fuel cell. They include gaskets for providing seal around the MEA to avoid leakage of gases and bipolar plates for assembling individual PEM fuel cells into a fuel cell stack and provide channels for the gaseous fuel and air. The membrane present between anode and cathode compartments serves as the electrolyte and helps conduct protons from the anode to the cathode.

One of the keys to PEM fuel cells (PEMFCs) performance and durability is water management. Water management in PEMFC is widely recognized to be a very complex issue, due to the fact that the gaseous reactants share the same pathways as the partly liquid product water. If the water evaporation is very slow, it will flood the membrane and the accumulation of water inside the field flow plate will impede the flow of oxygen into the cell, but if the water evaporates too fast, the membrane will be dry and the resistance across it increases. Both these cases are to be avoided to prevent degraded performance. For addressing the issue of water management, water distribution in the PEMFC must be accurately and efficiently measured for providing data for model validation. A typical PEMFC will be made up of 2 cm thick aluminium plate and about 1 cm thick graphite flow fields. The region of interest where water is present is very small around 5 mm which makes it difficult to measure water using conventional techniques like X-ray and magnetic resonance imaging. Neutron, however, can make the task easier as the all the typical hardware of the fuel cell will be transparent to neutrons and only the hydrogenous part will be visible. Just by analyzing the neutron radiography images, the laminar water flow in an operating PEMFC can be measured.

Neutron imaging, therefore, has been used to investigate different aspects of water management, such as the effect of different flow field geometries, materials, operating

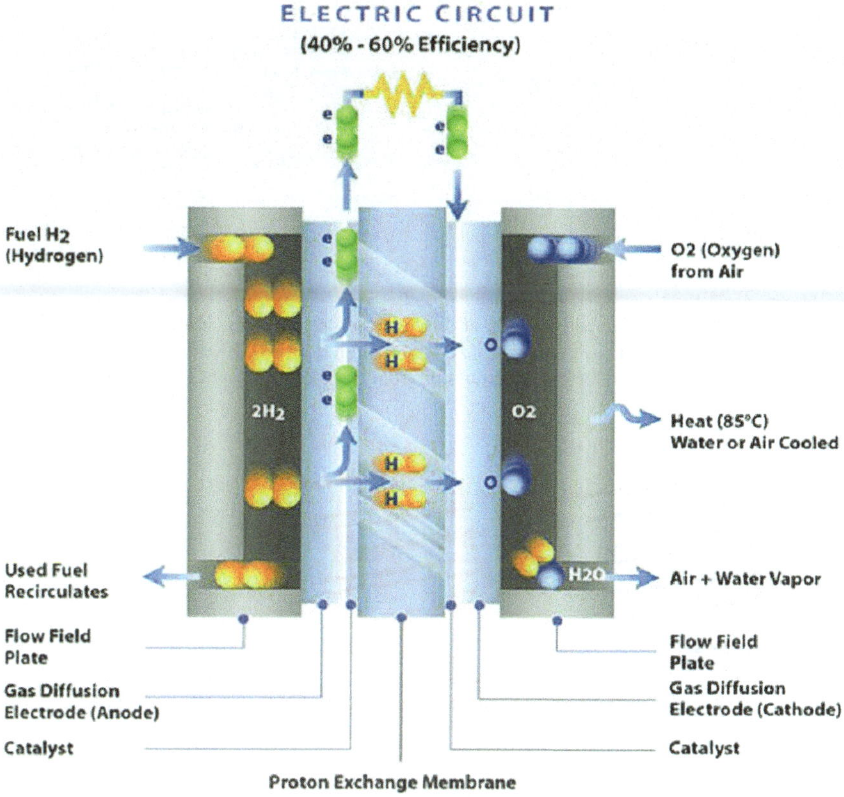

Fig. 10.1 Schematic of a PEM fuel cell [1] (open access)

conditions and degradation. In situ water content visualization for a commercial PEMFC has also been attempted using neutron imaging.

Totzke [2] conducted neutron radiography studies to examine flow field in prepared PEMFCs using a detector with a large FOV and high spatial resolution. The advantage of large FOV was that a complete fuel cell could be imaged in a single shot. Figure 10.2 shows the radiographs depicting the water evolution in the GDL and the flow field channels presented images have been normalized with respect to the reference image (dry cell). As the current density increases, the amount of water formed increases which is quite evident from (A), (B), (C) and (D).

Radiography images taken in trough plane direction were unable to separate the water distributions for the cathode and anode sides. Moreover, for actual analysis, data for fuel cell stacks were required. Since the water distribution may strongly differ between single cell and cell stacks, radiography studies had to be replaced by tomography studies to provide information on the water accumulations in individual cells of the stack. Markotter et al. [3] carried out quasi in situ tomography to study the water content in individual flow field channels of threefold stacks (Fig. 10.3).

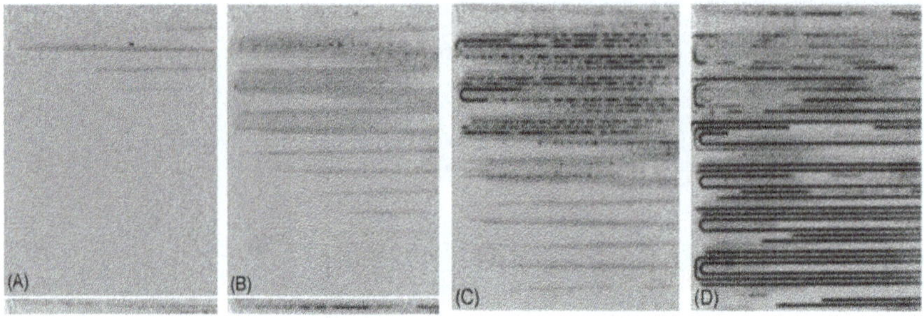

Fig. 10.2 Evolution of water distribution at different current densities, **a** after 30 min, I = 100 mAcm^{-2}; **b** after 40 min, I = 300 mAcm^{-2}; **c** after 65 min, I = 400 mAcm^{-2}; **d** after 105 min, I = 500 mAcm^{-2} [2]. Reproduced with permission from Elsevier Copyright Clearance Centre

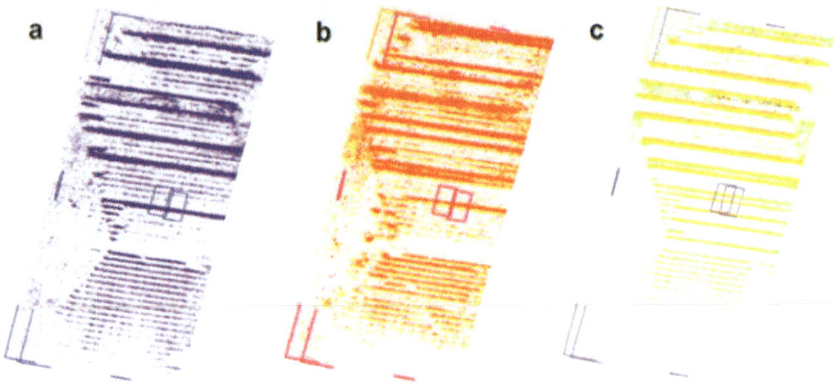

Fig. 10.3 Separated water distribution for each single cell in a threefold stack [3]. Reproduced with permission from Elsevier Copyright Clearance Centre

Fuel cells used for commercial applications have a flow field defined by discrete channel-land structures guiding the reactant gases. Sealing gaskets surround the flow field on its outer perimeter, separated from the flow field by an edge region. Gas and water transport between flow field and edge region strongly depend on the specific design of the fuel cell. Stahl [4] tried imaging water in this area because it could be subject to degradation effects.

Inside a gas diffusion layer (GDL), liquid water circulates under the effect of capillary forces from the catalyst layer (CL) to the flow channel. In the reverse direction, an optimum diffusivity of the reactant gases is required to sustain the electrochemical reaction with minimum mass transport losses. The simultaneous two-phase flow is maintained by the properties of the GDL, such as hydrophobicity, porosity or tortuosity [5]. Truhan et al. [6] studied through-plane water distribution and liquid removal in an actual operating fuel cell for different flow channel surface properties, using high-resolution neutron radiography. Biesdorf et al. [7] also evaluated the impact of varying amounts of PTFE coatings inside the GDLs of polymer electrolyte fuel cells in relation to mass transport losses. Neutron radiography, limiting current density and pulsed gas analysis were used to obtain a detailed characterization of the local mechanisms and water distribution in GDL and CL. Figure 10.4 depicts the in-plane water distribution averaged during the first and last 30 min for the four configurations during continuous operation at full humidification.

Neutron imaging also allows studying the various mechanisms of fuel cell degradation, namely corrosion and ageing of GDL. The corrosion of carbonaceous components in PEMFCs is a concern for long-term durability. This corrosion can be as a result of potential excursions occurring during cycles of operation. Further, the transient power demands and localized water blockages can cause fuel starvation leading to favourable corrosive conditions. Carbon corrosion reduces efficiency of PEMFC through several mechanisms, and efforts are underway to find ways to reduce this phenomenon. Fairweather et al. [8] carried out studies to investigate the effect of cathode corrosion on through-plane water transport.

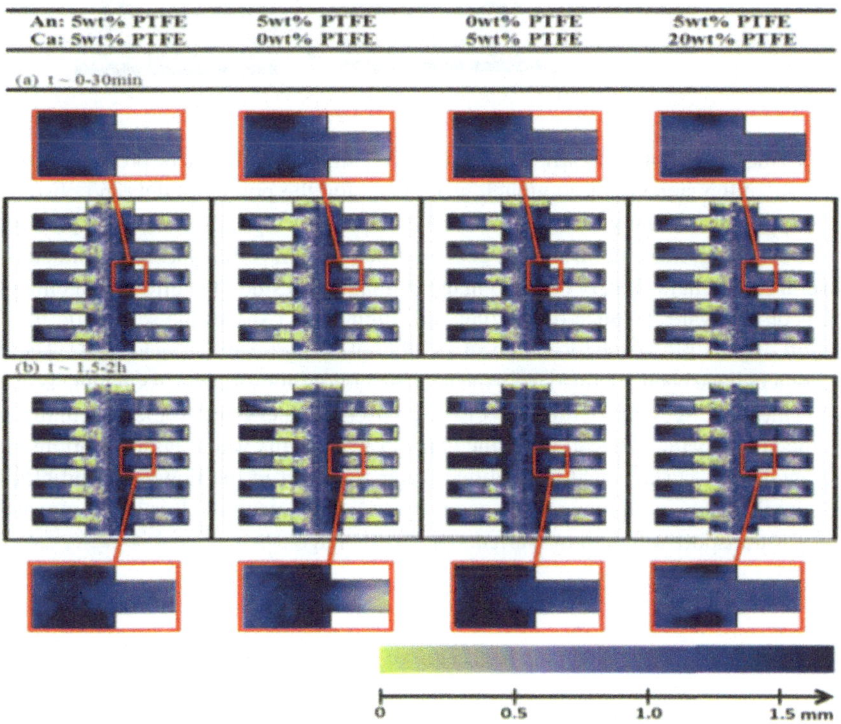

Fig. 10.4 In-plane water distribution averaged during the **a** first and **b** last 30 min for different coatings on anode and cathode during continuous operation at full humidification. The dotted line in the middle depicts the membrane, with its anode on the left and cathode on its right. The five fringes depict the gas channels of the flow field. The amount of water is represented in false colour code (yellow = dry, blue = wet) which corresponds to 0 and 1.7 mm water [7] (open access)

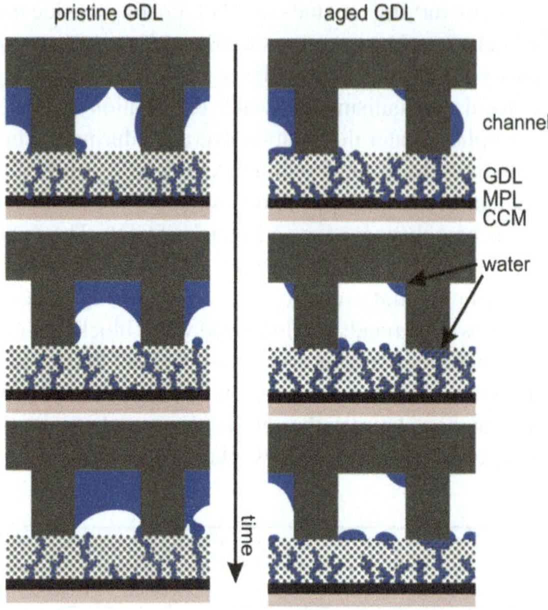

Fig. 10.5 Schematic illustration of possible water formation in a cell containing a new (left) or aged (right) GDL [9]. Reproduced with permission from Elsevier Copyright Clearance Centre

The membrane of the PEMFC must be sufficiently humidified to be proton conductive, but at the same time, too much water can clog the pores of GDL and obstruct the supply of reactant gases into the catalyst layer. During long-term operation due to GDL ageing, the water transport gets modified and this effect has been studied by several researchers. Katzel [9] used neutron imaging to study the effect of ageing of GDL on the water balance of the fuel cell. Figure 10.5 shows the schematic illustration of possible water formation in a cell containing a new or aged GDL. These inferences have been obtained from the neutron radiography results. As evident is that when the GDL ages, the hydrophobicity decreases and the water droplets accumulate on the GDL surface instead of on the channel walls. This water agglomerations on the aged GDL surface lead to modified water transport in the flow field and block the supply gases' path towards the catalyst layer. This results in less efficient water and gas transport which in turn affects the cell voltage.

10.2.2 Neutron Imaging for Hydrogen Storage Materials

Fossil fuel depletion and increase in energy consumption threaten the energy and economic security worldwide. In addition, fossil fuels are not environmentally friendly, so alternatives to fossil fuels are continuously being thought of. Hydrogen energy has long been proposed as hydrogen is the most abundant element on earth, and the by-products of energy production will be just water vapour (hydrogen cycle). Hydrogen also offers the advantage of a high chemical energy per unit mass when used as fuel. The problem, however, is that although hydrogen is the most abundant element but is not available in free form and has to be produced from water or natural gas. Hydrogen production cannot be carried out in situ and thus hydrogen storage is necessary. It is the most challenging task as hydrogen has low density and boiling point. For increasing density, hydrogen can either be compresses to high pressures, liquefied at cryogenic temperatures or stored as a solid in metal hydrides.

Out of all the hydrogen storage methods, metal hydrides are particularly important because they provide a safe and convenient way to store hydrogen. The reason being they offer high volumetric hydrogen storage densities at low hydrogen gas pressures and also metal hydrides for few transition metals (TM) can be formed even at room temperature. TM-based hydrogen-absorbing alloys find applications in the fields of backup power, materials handling (forklifts), marine or submarine, etc.

A further innovation into hydrogen storage is the metal hydride composites (MHC) with expanded natural graphite (ENG). These can be used as a better alternative to metal hydride powders because they offer higher thermal conductivity and reduced porosity. Herbeig et al. [10] carried out neutron imaging studies to compare the structural stability and spatial-temporal hydrogen concentration of a MHC with loose metal hydride powder under cyclic hydrogenation. The results indicated that reversible hydrogen storage can be efficiently done in MHCs as the thermal resistance between MHC and container wall is very less compared to the loose powder bed as the latter tears apart whilst dehydrogenation (refer Fig. 10.6).

Hybrid storage systems which combine high-pressure vessels capability with some alloy-based hydrogen storage materials that absorb hydrogen under elevated pressures are also being considered for solving the issue of hydrogen storage. The internal structure of the hydrogen storage tank determines the performance of the

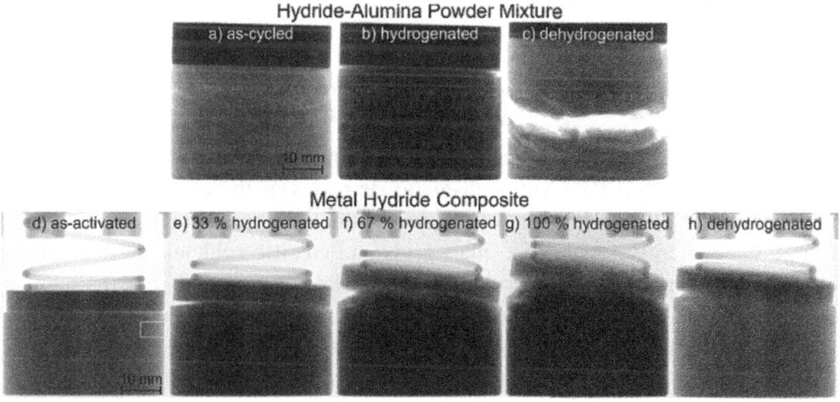

Fig. 10.6 **a–c** In situ neutron images during fast hydrogenation and fast dehydrogenation starting from the as-cycled state of a mixed hydride-alumina powder (HAP) sample. **d–h** In situ neutron images during slow (100 ml-H_2/min STP) hydrogenation and fast dehydrogenation (1st cycle) starting from the as-activated state [10]. Reproduced with permission from Elsevier Copyright Clearance Centre

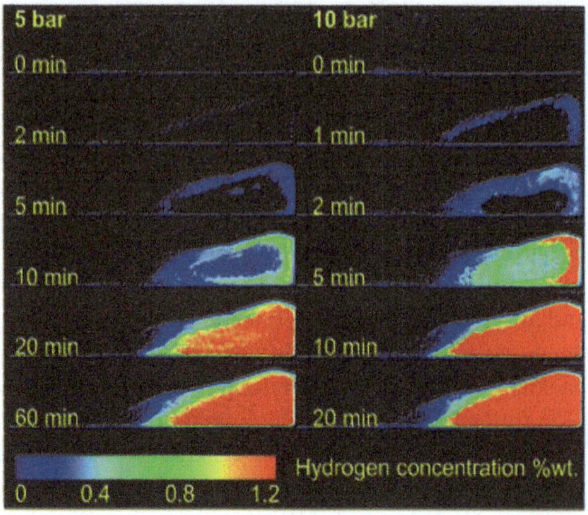

Fig. 10.7 Neutron radiographs showing change in hydrogen concentration highlighting hydrogenating process kinetics at different pressures [11]. Reproduced with permission from Elsevier Copyright Clearance Centre

metal hydride or hybrid storage systems and studies have been carried out on different types of tanks available. Gondek [11] carried out neutron imaging studies on sodium alanate-based hydrogen storage tanks. Neutron radiography measurements were performed for the constant pressure conditions for the time predicted to be sufficient to obtain complete hydrogenation of the active material inside of the container. Radiographs were taken every minute, and results have been summarized as shown in Fig. 10.7. Figure 10.7 depicts the hydrogenation profiles under different pressures and time. It shows that the reaction speed increases by about three upon raising the pressure from 5 to 10 bar.

10.2.3 Neutron Imaging of Lithium-Ion Batteries

Energy storage has been a primary worldwide concern over the last few decades. Lithium-ion batteries (LIBs) with high energy densities are therefore the most sought of energy storage devices for microelectronic products. For meeting the requirements of large scale energy storage like smart grids and electric vehicles, there is a huge demand for producing high performance LIBs. LIBs are essentially electrochemical cells made of electrolytes and electrodes. To enhance their electrochemical performance, an understanding of the various components and processes along with degradation mechanisms of LIBs are important. In recent years, various techniques based on X-ray, electrons and neutron probes have been used to understand the mechanism of LIBs. Usage of neutrons provides complimentary data to X-ray and neutrons being chargeless are offer more penetrability thus are suitable for in situ studies even for commercial LIBs.

Owejan [12] used neutron imaging to get the lithium-ion distribution for an electrochemical cell. The cell used composite graphite as the negative electrode and it was observed that the Li distribution in graphite was dependent on the state of charge. Zhang et al. [13] used neutron tomography to map the Li-ion spatial distribution in a V2O5 electrode of a small coin size cell. Imaging contrast between lithiated/delithiated states of V2O5 electrodes was sufficient to study the Li-ion distribution. Figure 10.8 shows the lithium distribution in a V2O5 cell obtained using

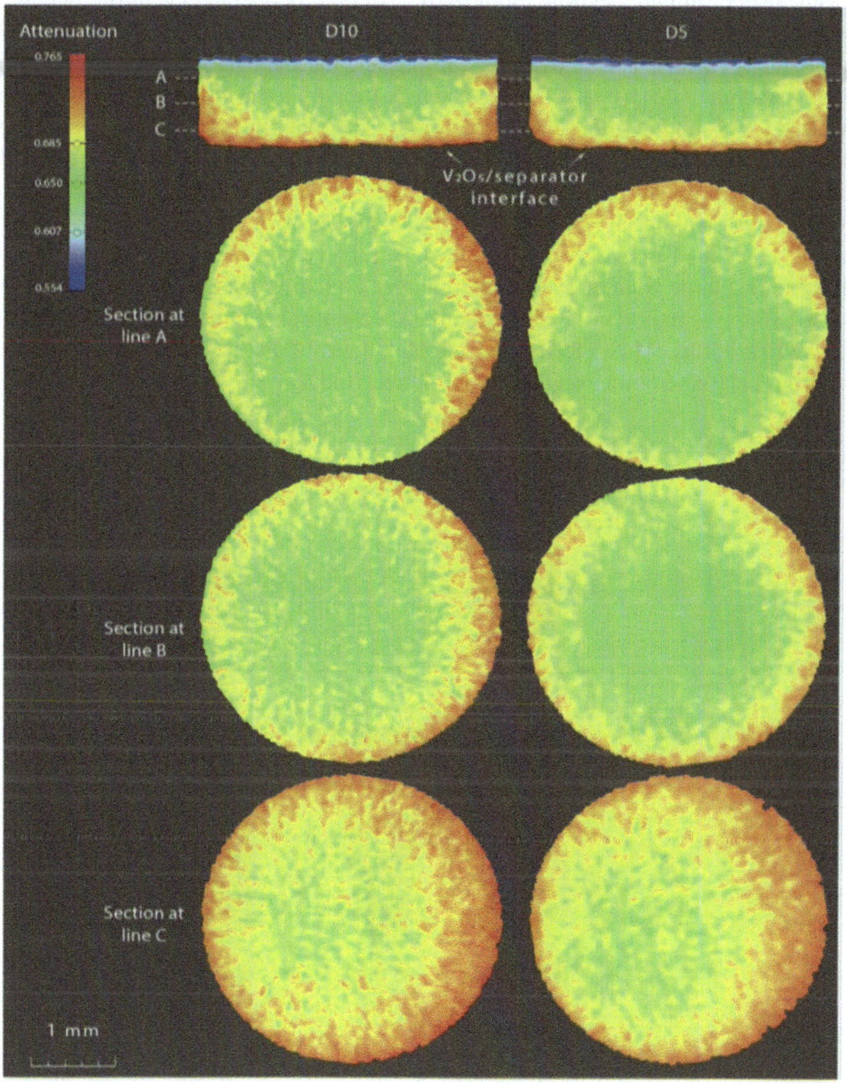

Fig. 10.8 Neutron imaging data showing lithium distribution in V2O5 electrode. In this pseudo-colour plot, higher attenuation represents higher Li concentration [13]. Reproduced with permission from Elsevier Copyright Clearance Centre

neutron tomography. It was observed that the Li spatial distribution is non-uniform within the bulk V2O5 cathodes. The non-uniformity is higher in the electrode where lithiation occurred relatively faster during the discharge (lithiation) [13].

Neutron imaging is an efficient probe for understanding the electrochemical processes inside the electrode. Understanding this can help in improving the batteries' performance both in terms of economic and safety reasons. To study the growth mechanism of lithium dendrite, Song et al. [14] used in operando neutron imaging. Figure 10.9 shows the basic set up used for the same and also the radiography and tomography results. The study helped in provide mechanistic insights with a deep understanding of dendrite Li shorting and redistribution. The information gained can be used for designing safer lithium electrodes. Thermal runaway of LIBs is an issue and the reason for the same is being explored. There are evidences that the thermal runaway begins with the decomposition of the solid electrolyte interphase (SEI) which is passivation layer formed on the electrodes during the first few charging cycles. Once the SEI is decomposed, the electrolyte reacts with electrode materials generating a large amount of heat and thereby exploding the LIBs. For developing safer LIBs, it is important to study the thermochemical stability of electrolyte and its wetting ability. Most electrolytes contain elements like C, H, O and therefore neutrons can serve as an excellent tool for their study.

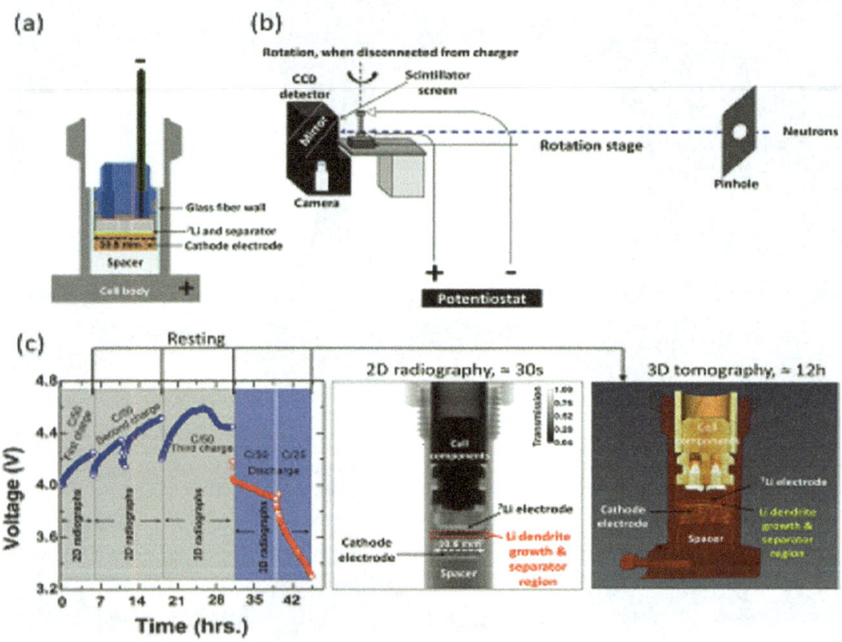

Fig. 10.9 **a** Electrochemical cell, **b** instrument setup at the beamline for the neutron radiography and tomography measurements and **c** charging/discharging profiles of the Li-ion battery during the experiment, highlighting the measurement strategy used for the radiography/tomography (open access)

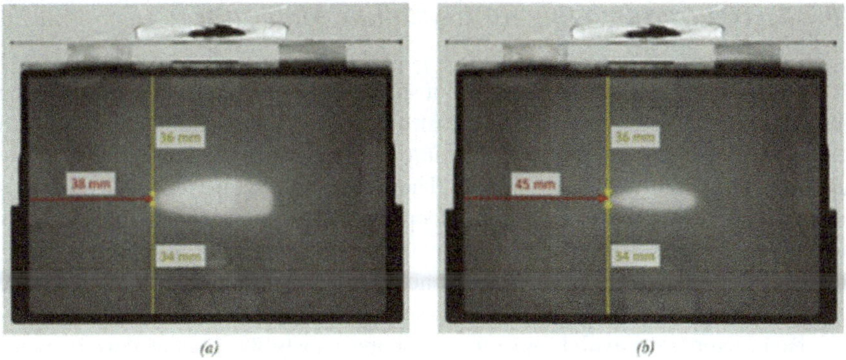

Fig. 10.10 Neutron radiograph of cell filled at vacuum shown **a** 33 **min** and **b** 41 min after the electrolyte filling [15]. Reproduced with permission from Elsevier Copyright Clearance Centre

The first application to electrolyte filling in LIBs visualization using neutron imaging was done by Weynadz [15]. Figure 10.10 shows the neutron radiographs of electrolyte filling in a cell at various instants.

Gases are released in LIB as a result of electrolyte decomposition and lead to the degradation of cells. In situ and in operando experiments are conducted to map the gas release behaviour of electrolyte-based batteries. Since the gases released are mostly light elements, neutron radiography can be used for the online studies. In situ neutron imaging studies have been used to observe the outgassing phenomenon in LIBs by Michalak [16]. Figure 10.11 shows the gassing caused by oxidation and reduction of the electrolyte solvent. It was inferred that the metal dissolution in the electrolyte and decomposition products formed as a result of high potential adversely affect the gas generation. This effect is most predominant in the first charge cycle when the SEI layer is getting formed.

Fig. 10.11 Neutron transmission images of a pouch cell illustrating the different processing steps. **a** After noise reduction, **b** after normalization to the first image and **c** after segmentation, with grey areas representing the generated gas [16] (open access)

10.2.4 Neutron Imaging for Crystalline Materials

Polycrystalline metallic alloys are used for structural support and related engineering applications. Understanding their microstructural properties and state of stress in various loading conditions is of immense importance towards their structural integrity and suitability for the required purposes. The available techniques for such investigations like metallography, electron microscopy and diffraction techniques are towards the surface or near-surface features and require sample preparation. Neutron diffraction has been the mainstay for crystal structure-related information for a long time rather than neutron-based imaging on account of the unavailability of imaging detectors having superior spatial and time resolving capability. Recent developments in transmission-based Bragg-edge imaging which is based upon the recording of transmitted intensity as a function of wavelength have shown promising results in providing scattering information in addition to the transmission image. With the introduction of these new advancements and also better detectors, neutron imaging now is being used as an efficient tool for characterizing crystalline materials.

Welding is an important process in almost every engineering component manufacture. When two dissimilar metals are weld, the weld quality is responsible for the structural integrity of the components. Since strain is developed in the process of welding, evaluation of the residual strain can be used to quantify the weld quality. Residual strain can be characterized by the microstructural features as well as element compositional distribution at the weld location.

Tremsin et al. [17] have reported such studies on Al-steel weld and Ti-steel butt weld with Cu as filler alloy through Time of flight (TOF) technique at the NOBORU beamline of material and experimental facility (MLF) of the J-PARC (Fig. 10.12). Crystallographic phase mapping by Bragg-edge imaging and elemental composition including intermetallic compound formation at the weld has been demonstrated. Nikolay et al. [18] have utilized the comparable lattice spacing length with the cold neutron wavelength to perform wavelength dependent transmission imaging using the time of flight technique with the help of energy resolving detector on weld joint between two austenitic plates. Wavelength dependent attenuation coefficient can be used towards deciphering information related to the microstructure of the specimen and Bragg-edge for its qualitative and quantitative phase composition analysis.

Using energy selective neutron tomography, Robin Woracek and his group [19] have carried out an interesting study on 3D mapping of crystallographic phase distribution in metastable 304 L stainless steel. The study, carried out on five samples, has revealed strain-induced phase transformation from austenite to martensite when subjected to tensile and torsional deformation beyond a certain plastic strain (Fig. 10.13). The transmission imaging results were verified by complimentary neutron diffraction studies. This technique has potential applications for the characterization of natural and advanced materials.

Austempered ductile iron (ADI) is a form of ductile iron that enjoys high strength and ductility as a result of its microstructure controlled through heat treatment. It

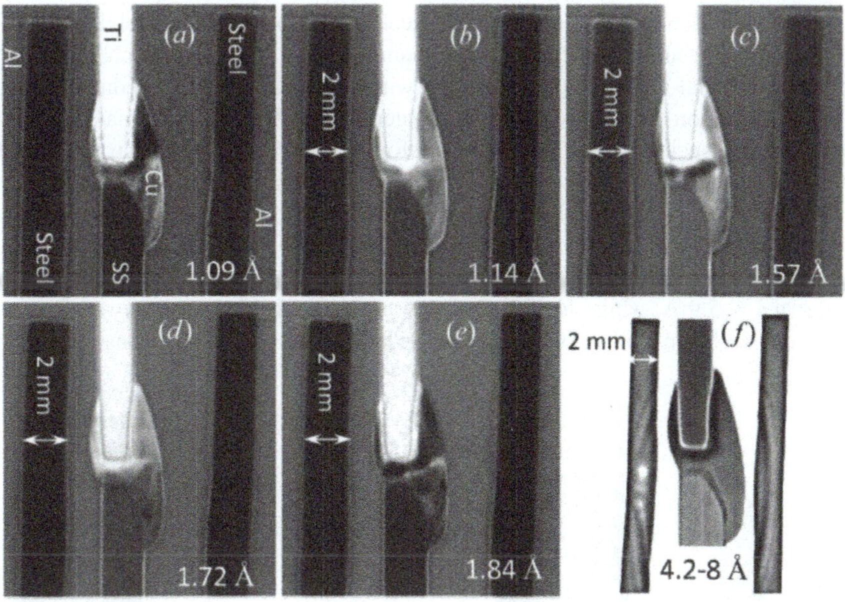

Fig. 10.12 Wavelength dispersive neutron transmission images of weld samples at Bragg-edges of steel and copper [17] (open access)

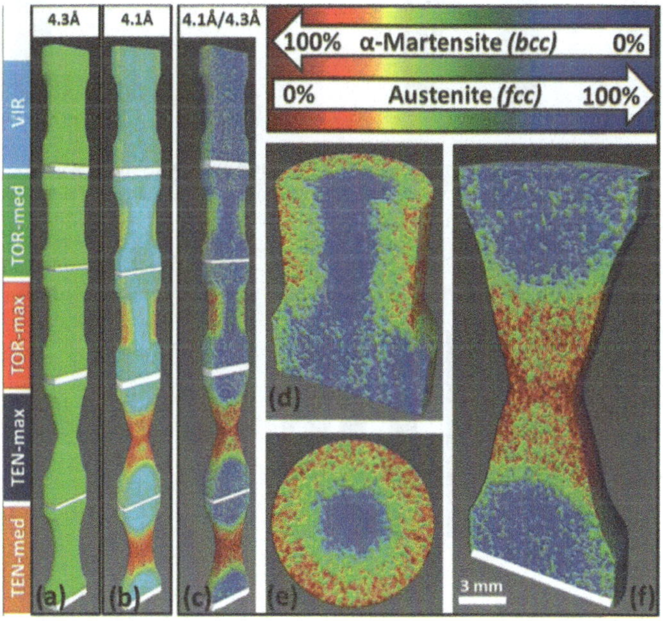

Fig. 10.13 Tomographic reconstruction of martensite and austenite phase fractions in 304 L stainless steel by using simultaneous neutron tomography at wavelengths before (4.1 Å) and after (4.3 Å) the 'Bragg cut-off' corresponding to the austenitic phase [19] (open access)

is a potential candidate to replace cast steel owing to its lighter weight, higher-tensile strength and significantly lower manufacturing cost. ADI is characterized by its spheroidal graphite nodules spaced within the matrix. These nodules reduce microsegregation of solutes within the material. For ADI, the material has been austempered such that the matrix is transformed into ausferrite, or a mixture of acicular ferrite and austenite. Soria et al. [20] investigated the spatial distribution of martensite content in ADI samples under different strain conditions using energy selective neutron radiography. Figures 10.14 and 10.15 show results for this study using two approaches. The first approach included taking images in the wavelength interval 1.6–5.9 Å, whilst for the second approach, only two specific wavelengths were chosen for imaging. For the first method, the martensite content was calculated

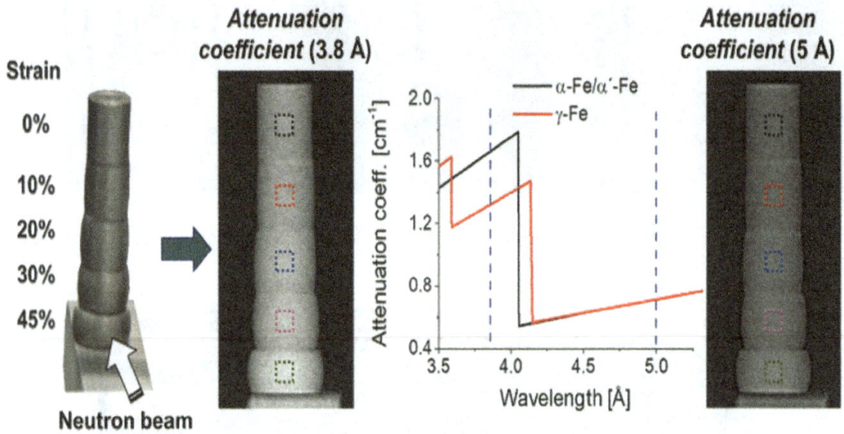

Fig. 10.14 Experimental setup of energy selective neutron radiography along with graph depicting the evolution of theoretical attenuation coefficient with change in wavelength. The change in Bragg-edge is clearly visible in the graph [20]. Reproduced with permission from Elsevier Copyright Clearance Centre

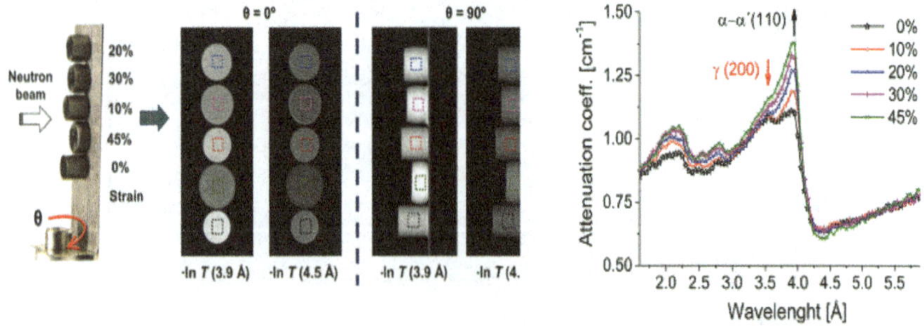

Fig. 10.15 (Left) Experimental setup of energy selective neutron radiography acquired at two wavelengths. (Right) Evolution of the experimental attenuation coefficient spectrum at different deformation level [20]. Reproduced with permission from Elsevier Copyright Clearance Centre

from the height of the selected Bragg-edge, whilst for the second one, the attenuation coefficient spectrum for different deformation levels provides the phase information.

10.3 Dynamic Neutron Imaging for Engineering Applications

Neutron radiography was first widely used in industrial applications in the mid-1950s. But these could only provide images of static samples. For real-time or near real-time events such as flow pattern studies, initial attempts were made using a modified version of medical fluoroscopy system based on image intensifier tube by Haskins [21], Bracher and Garret [22] and Stewart [23]. Berger [24] was the first to demonstrate real-time neutron radiography. Lindsay [25], in 1978, assembled and demonstrated the first successful neutron radioscopy facility at the United States University, University of Missoiri, Columbia. The system comprised of a NE426, li6F/ZnS(Ag) and Trimax, GdO2S, scintillator screens and ISIT camera. Later on with the advent of CCD cameras, imaging systems were developed by coupling them using lens and mirror to the converter screen. The mirror approach was necessary because the silicon imagers would otherwise be damaged by long-term exposure to neutrons and gammas. By the mid-1980s, improvements in digital imaging technology and the availability of advanced computers and image processing softwares enabled neutron radiography or radioscopy community to perform dynamic imaging with better image quality. Cooled CCD-based imaging units are very commonly used nowadays because of their good resolution and low noise. Being costly and also sensitive to radiation, these are properly shielded using B4C and lead. Because of the specific attenuation property of neutrons, they are very much suited for real-time flow pattern studies involving hydrogenous fluids inside metallic clad materials. Lindsay et al. (1992) [26] and Brunner et al. (2002) [27] applied NR for studying internal combustion engines, fuel flow, etc. Takenake et al. (2002) [28] have used neutron radiography to visualize fuel cavitation in a nozzle of a diesel engine. Norris et al. (1996) [29] studied water deposition in aerogel by neutron radioscopy method. Umekawa et al. (1999) [30] reported visualization of large particle movement in fluidized bed, Furui et al. (2002) [31] studied segregation process in fluidized bed, Asano et al. (1999) [32] calculated quantitatively void fraction data in fluid flow patterns using near real-time neutron radiography technique. Sinha et al. (1999) [33] had demonstrated prototype experiments on two-phase flow in metallic pipes using NR. Lots of work related to real-time neutron radiography and their advancements have been carried out by Lehmann [34], Harris [35], Jasti [36].

With the advancement in technology of various components related to neutron imaging, real-time imaging which were not possible earlier is being carried out routinely. The advancement can be linked to the availability of high-yield neutron sources (spallation), better beam quality arrangements (cold neutrons and L/D ~ 500 or more), sensitive as well efficient area detectors and high-end computing power. Real-time imaging with neutron is termed as neutron radioscopy.

10.3.1 Two-phase Flow Studies in Air–water Systems

Thermal hydraulics is an important part of various applications including heat transfer using fluids, immiscible fluid mixing, oil industry, etc. Various flow pattern generating modules have been developed, simulating the actual conditions, with air/water/oil/steam and experiments have been carried out at neutron imaging facilities. Since the flow is dynamic and to have the best result, one needs to capture every moment of the flow pattern. The source intensity, flow speed and the detector sensitivity play inter-dependent and cumulative role in deciding what studies to be carried out at which facility. In case of flow of gas and liquid in a vertical pipe, the liquid and gas phases distribute themselves into several recognizable flow structures. These are described below in brief:

- Bubbly flow—numerous bubbles dispersed in the continuous liquid phase.
- Slug flow—Bigger size bubbles forming specific shapes after coalescing several bubbles. This happens with increasing void fraction.
- Churn flow—With increase in flow speed oscillator motion sets in with net upward flow. This is an unstable flow pattern and is generally avoided.
- Annular flow—In this type of flow, the liquid is expelled from the centre of the flow as a thin film on the wall. There is a clear interface between the high velocity gas and liquid. This pattern is stable and is desired in two-phase flows. In this case, liquid droplets may get entrained in the gas.
- Wispy annular flow—This is formed at higher flow rates when the entrained droplets coalesce and forms cloud like coherent structures in the central vapour core.
- Mist flow—This flow happens at very high flow rates. The annular thin film gets sheared by the gas until it becomes unstable and is destroyed. This is inverse to what happens in the case of bubbly flow.

Since thermal neutrons are attenuated strongly by hydrogenous compounds such as water or oil, their flow inside metallic pipes, which are almost transparent to neutrons, can be visualized with contrast. Initial experimental work on two-phase (air-water) systems can be traced to the work of Mishima and co-workers [37–39] using Japan research reactor 3 M (JRR-3 M) of JAERI. The experimental conditions and components were as follows: thermal neutron flux was ~10^8 n/cm^2/s, L/D was 153 (horizontal) and 176 (vertical), NE 426 scintillator coupled to high speed (1000 fps) video camera system coupled with a compatible image intensifier. Two-phase flow experiments were carried out in aluminium alloy made rectangular ducts. Images obtained from this experiment were processed to extract geometrical information to measure flow regime, velocity of bubbles, wave height, etc., as well as to measure the void profile and the average void fraction.

Advanced heavy water reactor (AHWR) is being developed in India for large scale thorium utilization in the third stage of Indian nuclear power programme. One of the passive safety aspects of this reactor is natural circulation-based heat removal using water. So the flow pattern and their transitions will govern the heat removal from the fuel. In other words, the study of flow pattern transitions can provide insights into the responsible operating parameters causing them. As for example, heat transfer from the fuel to the coolant is maximum in case of whole of the coolant pipe's internal volume is filled with water. But on account of heat, boiling occurs and hence various flow patterns set in such as slug, bubbly, annular and churn. This creates irregular heat transfer and disturbs the efficiency of the system. To perform hydraulic modelling and subsequent design of such a system, two-phase flow visualization for all possible scenarios is a must as such data are not available. A test setup involving a simulated loop made up of stainless steel pipe containing de-mineralized water with operating pressure range 1–70 bar was established at the old APSARA reactor (400 KW swimming pool type reactor which has been decommissioned) in BARC, India. The heating was carried out by a 10 kW heater at the portion that was radiographed. ^{6}LiF-ZnS(Ag) scintillator (NE 426) coupled to cooled intensified CCD camera was the imaging system used for this purpose. The imaging system was shielded with 150 mm thick concrete blocks to reduce background noise in the images. The radiography facility provided thermal neutron flux ~10^6 n/cm^2/s at an L/D of 90 and cadmium ratio 6.3. Different case scenarios were generated by changing heating and water flow rate conditions [40, 41]. The configuration of the facility and imaging system along with storage processing provided 15 frame per second acquisition. Acquired data were processed using exponential and equally weighted schemes and void fraction were calculated for each frame and plotted against the process parameters. Experiments were conducted using 3/8" and 1/2" test sections. Though stable flow could be observed in 1/2" pipe, it was not the case for 3/8" pipe. This had been confirmed using in-house developed codes. Details of the Fig. 10.16 are as follows; (a) schematic of the APSARA reactor with test loop, (b) schematic of the test loop, (c) the imaging system, (d) typical frames showing the water flow in pipes (black represents water and white represents steam/air), (e) the instantaneous void fraction as seen on the monitor screen along with the averaged data followed, (f) exponentially weighted average and (g) equally weighted average plot. Prior to performing actual test loop experiment, feasibility experiments involving compressed air and water flow using neutron radiography were carried out.

Similar study involving air-water two-phase flow test loop has also been carried out at the thermal neutron beam port of IEN/CNEN Argonauta reactor, Rio de Janeiro [42]. The imaging conditions and system were as follows: thermal neutron flux at the exit of the port was 4.46×10^6 n/cm^2/s, L/D:70, NE425 scintillator coupled to CCD-based video camera well shielded by borated paraffin, cadmium and lead. Different flow conditions were generated using a water pump and an air compressor in 6 mm diameter aluminium pipe. The studies carried out were drift flux correlation with different velocities of total two-phase flow and the void fraction.

The cases described above had the gas-liquid flow from bottom to top, i.e. in the upward direction. To understand the dynamics of flow towards efficient heat

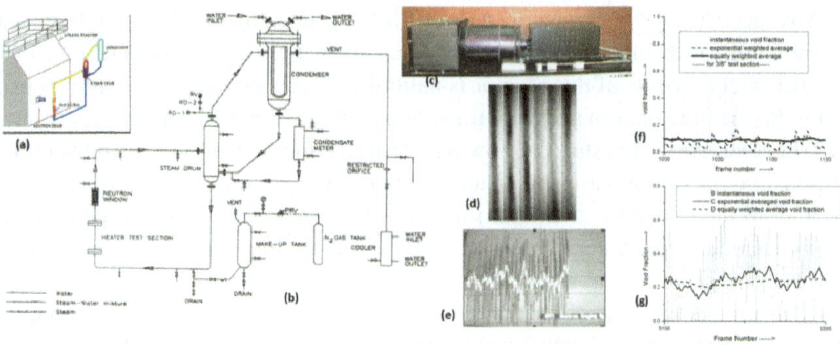

Fig. 10.16 **a** Schematic of the APSARA reactor with test loop, **b** schematic of the test loop, **c** the imaging system, **d** typical frames showing the water flow in pipes (black represents water and white represents steam/air), **e** the instantaneous void fraction as seen on the monitor screen along with the averaged data followed, **f** exponentially weighted average and **g** equally weighted average plot. Reproduced from [40] with permission from RightsLink by CCC

extraction in plate type heat exchangers Hitoshi Asano and co-workers [43] at Kobe University, Japan had carried out real-time neutron radiography towards visualization and void fraction distribution in plate type heat exchanger having downward gas-liquid two-phase flow. Experiments were carried out in the JRR-3 M neutron radiography facility at Japan Atomic Energy Research Institute (JAERI). Differences in flow characteristics between downward and upward flow were observed at low gas velocity conditions. It was also pointed out that liquid distribution into multiple channels strongly depended on the liquid volumetric flux.

10.3.2 Two-phase Flow Studies in Oil-Water Systems

Though heavy oil accounts for two-thirds of the world oil resources, on account of its high viscosity and presence of heavy distillates, its production is poor. In order to reduce oil viscosity, steam injection has been widely used but this has drawbacks such as severe carbon loss to coking causes low recovery efficiency and high energy consumption. In this regard, supercritical water injection technique for heavy oil recovery has been found to be beneficial as it not only reduces oil viscosity but also dissolves heavy distillates to avoid coking [44]. Shoji E and co-workers [45] have reported neutron radiography-based visualization studies regarding behaviour of heavy oil in supercritical water to understand the flow and mixing behaviour towards improving process efficiency related to supercritical water upgrading of heavy oil. M, Balasko and co-workers [46] at the Budapest Neutron Centre, Hungary have employed NR towards visualization and understanding of the segregation process in a closed loop liquid-liquid two-phase mixture involving 3-methylpyridine and heavy water with varying temperature and heavy water concentration. The images in combination with analysis provided insights for development of models towards this type of flow patterns.

10.3.3 Two-phase Flow Studies in Liquid-Solid Systems

In the area of liquid—metal two-phase flow studies, real-time neutron radiography experiments carried out at the JRR-3 M thermal neutron radiography facility in the Japan Atomic Energy Research Institute (JAERI) using lead—bismuth eutectic as a working fluid and gold—cadmium intermetallic compound (AuCd3) as tracer has been reported by Takenaka et al. [47]. Towards performing safety analyzes and understanding of severe accident in nuclear reactors, neutron radiography has always helped by providing real-time visualization and measurements in various experimental situations. Some notable scenarios are as follows: (a) steam explosion due to direct contact of molten metal and sub-cooled water simulating a severe accident of a light water reactor and (b) gas-molten metal mixture in a pool simulating a severe accident of a fast breeder reactor. Mishima et al. [48] have reviewed important aspects related to application of neutron radiography to thermal hydraulics research. Neutron radioscopy technique has been efficiently utilized for basic understanding of liquid–metal two-phase flow visualization and measurements [49] of nitrogen gas–molten lead bismuth two-phase flow in respect to fast reactor research.

In the area of power production using advanced fourth generation reactors, fast breeder reactors offer the opportunity towards effective use (60–70%) of natural uranium resources [50]. Since the cores of fast reactors are compact, huge amount of heat removal is necessary. Liquid metals offer the best solution in such cases. So, understanding the heat transfer mechanism along with the flow characteristics, neutron radiography-based real-time imaging is being carried out across the globe. These sorts of studies are a necessary prerequisite not only for safety analysis of liquid-metal fast breeder reactors (LMFBRs) but also for the development of liquid-metal magneto-hydro-dynamics (MHD) system. High frame-rate neutron radiography of liquid metal two-phase flows in metallic vessels was studied by Saito et al. [51] towards understanding of gas—liquid metal flow characteristics.

10.3.4 Lead Melting and Solidification Studies

As per performance standards specified by IAEA, AERB in India, radioactive materials are transported in steel shells canisters inside lead shielded packages. The design of packaging of nuclear materials for such cases is aimed towards their safety, taking into considerations of all accident scenarios (example fire) such that there should not be any streaming of radiation. Lead casks are subjected to open pool fire test at 800 °C for 30 min amongst other tests to ascertain their performance stability. Study of lead melting towards designing transportation packages for radioactive materials is very important for estimating their performance. Experimental data in such cases are required for the development of benchmarking with high confidence. Numerical modelling becomes impossible when inaccuracies due to strong non-linearity of the governing equations combined with a moving boundary, created by the heating mechanism, sets in. During the melting process, solid-liquid phase transition takes place and the flow patterns are governed by the temperature, heating mechanism and the flow velocity. Knowledge of these parameters is of utmost importance to carry modelling. Thermocouple probe-based techniques have been used by lot of researchers (Emrich 1981) for recording the temperature at various locations in the bulk and get information of the melting or solidifying front with respect to time. Disadvantages with this technique are that they disturb the molten propagating front as they are embedded inside the bulk and also they provide information at a particular point. Basically, this technique is of intrusive type. A number of studies towards melting have been reported by F Wolf and R. Viskanta for tin [52] and by C Beckermann and R Viskanta for gallium [53]. Ultrasound wave-based imaging has been proved to be a promising tool for getting temperature information (Sielschott 1997; Xu et al. 2002). Ultrasound thermometry-based laboratory experiments have been carried by Fife et al. [54] for optical and opaque metallic fluids for understanding of the convection flow patterns. Various techniques towards obtaining temperature profiles at boundaries of the test specimens for understanding the flow pattern transitions have also been employed. These are flow velocity measurements using hot wire anemometry, laser Doppler anemometry or ultra sound anemometry, infra-red thermography (Carlomagno 1993).

All the techniques mentioned above are somewhat indirect in nature towards interpretation of the melting or solidification phenomena related to the process parameters. None provided direct images signifying distinct regions in terms of solid and its molten counterpart. Projection-based radiography technique has the ability to provide two dimensional images with clear demarcation of the solid-liquid phase. For light elements, X-rays can be employed as probes. Observation of melt-solid interfaces, their shape and movement in germanium and lead tin telluride crystals during their growth has been reported by Yin et al. [55], Barber et al. [56] using X-rays. On a similar manner, gamma ray tomography-based studies have been carried out by Chun et al. [57] for online monitoring of solidification front in metal castings. But, X-ray or gamma ray imaging fails for thick and high Z materials such as lead. Neutrons come to the rescue in such cases as they have the unique property of getting highly attenuated by materials such as hydrogen, water, boron, gadolinium and cadmium and much lesser for materials such as aluminium, iron and lead. Though neutron radiography has been used as non-destructive technique for various studies such as flow pattern transition in water-air mixture inside metallic pipes, oil flow inside motor engines [58–61], water ingress in concrete [62] and metallic composite structures [63], there has been limited study on lead melting and solidification process and their interface propagation using it. Some research work in the domain of solid melt studies using NR are Pb-Bi specimens by Takenake et al. [64] and Ogino et al. [65] employing tracer induced and dye injection methods and observation of segregation in Sn-Cd alloy metals by Reijonen et al. [66].

Feasibility experiments on imaging of solid-liquid interface of lead melting and solidification phenomena using non-invasive NR technique have been carried out at neutron imaging beamline, CIRUS reactor (at present decommissioned), India for visualization of the interface during both melting and solidification process and understand the convection driven phenomena with respect to time and heating power.

The neutron imaging beam line characteristics were as follows: L/D ~ 125, bismuth plug at the collimator inlet to reduce gamma background, flux at sample position ~3 × 10^6 neutrons /sec. The neutron imaging system consisted of ^{6}LiF-ZnS converter screen, a front coated aluminium glass mirror and a cooled CCD camera (1250 × 1050 pixels, pixel size 22 μm, Model: DW432N) and provided resolution ≈ 100 μm. For the safety of the CCD camera, it was shielded using lead and borated polythene blocks. The schematic of the imaging system is shown in Fig. 10.17a and the actual photograph of the installed imaging system in Fig. 10.17b.

50 mm cube Pb block (99.9%) was heated from one side using heating coil. Attached thermocouples provided the temperature details. To ascertain the contrast sensitivity across the Pb block, small Pb sheets of thickness 1.5 mm and 3 mm were placed behind it and imaged. They provided 7% and 15% contrast with respect to the Pb block which amounted to a density difference of 3% and 5.6%, respectively. This was better than the required condition of 6% contrast which is the case for 50 mm solid and liquid Pb. Real-time neutron radiography was carried out at heating powers of 150 W, 175 W and 200 W with the individual exposure of 10 s and frame to frame time difference of 3.8 s. This was optimized with experiments to provide sufficient contrast in the images towards visualization and quantitative analysis. Figure 10.18 shows the lead melting front shape and position at different time instances for three heating powers. These experiments provided data related to modelling and benchmarking [67]. In a similar way, solidifications of the molten lead were also carried out for all the heating conditions by putting OFF the heater at the completion of melting phenomena. Typical images during the solidification are shown in Fig. 10.19. Also

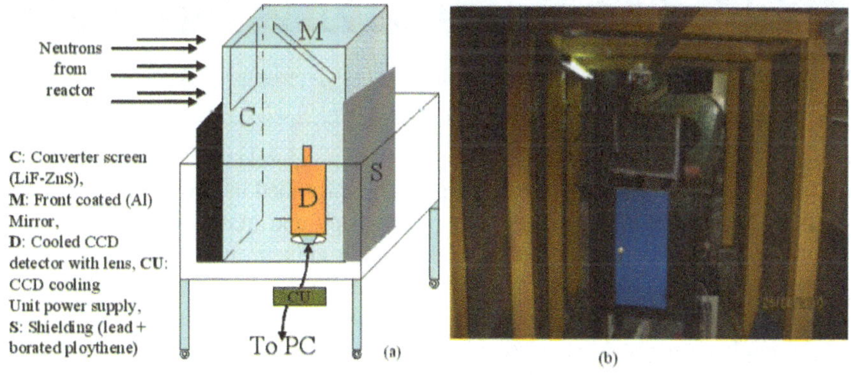

Fig. 10.17 **a** Schematics of the imaging system along with its components and **b** the actual photograph of the imaging system as installed at e-12 beamline, CIRUS Reactor, India. The imaging box is on top of the blue coloured table [68]

Fig. 10.18 Typical neutron radiography images clearly showing the propagation of melting front at various instances for heating powers, **a** 150 W, **b** 175 **W** and **c** 200 W

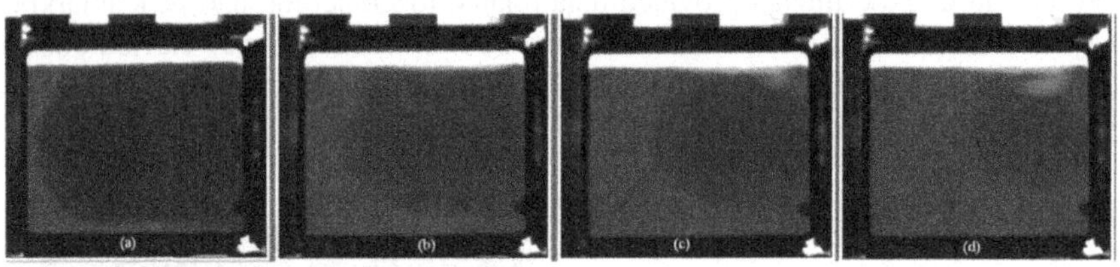

Fig. 10.19 Neutron radiography images of solidifying front propagation at four instants of time

when the study was carried out using impure lead, a separate mushy region was also visualized. Details of the experiment at result are provided in [68].

Using pulsed neutron transmission spectroscopy and nGEM as detector, visualization of solidification phases of lead bismuth eutectic in the energy resolved pulsed neutron imaging beamline at BL22, RADEN facility in J-PARC has been carried out [69]. LBE is under study for its excellent characteristics towards it use as coolant in accelerator driven systems. The radiographs and Bragg-edge images provided transient data during solidification for better understanding through numerical modelling.

10.3.5 Dynamic Imaging of Combustion Engine

Since thermal neutron is sensitive to hydrogen like elements, study of fuel (hydrocarbon-based) behaviour such as calibration or flow in a diesel engine or nozzle design efficiency are possible using neutron as probe. These are important related to fuel injection in diesel engine towards increasing the system efficiency. Cavitation process hampers the supply of fuel to the engine. This sort of study has been reported by Takenaka and co-workers [70, 71] at JAERI, KAERI and Kyoto University Research (KUR) reactor facility. Various configurations such as one injection hole and four injection holes in a nozzle for a truck, a nozzle for a ship and a nozzle for a dynamo were studied for exposure time ranging from 4 to 120 s using cooled CCD camera system utilizing LiF-ZnS(Ag) screen as neutron convertor. Typical images of gas bubbles in the nozzle are shown in Fig. 10.20. Synchronization of the motor with imaging system was carried out through chopper. Owing to this, the signal to noise ratio was poor as neutron flux to the sample was limited.

With the advent of spallation neutron sources, high yield of thermal neutron flux as well as in pulsed mode was possible. Researchers at TUM, Berlin and PSI, SINQ, NEUTRA Switzerland have carried out dynamic radiography of a 125 cc four stroke Yamaha make motorcycle engine using spallation source SINQ at PSI [72]. Synchronized images at various positions of the piston were acquired at source strength ~7×10^6 n/cm^2/s and L/D 350. The imaging system comprised of ^{6}Li loaded neutron scintillator and image intensifier coupled CCD camera images (exposure 0.1 ms) at different rotation speeds up to 1200 rpm were taken and averaged images provided crucial information regarding fuel flow in the running engine.

Similar work on short time stroboscopic neutron imaging and measurement on a rotating engine has been carried out at ILL, Grenoble [73] which is one of the most intense continuous neutron flux in the world. They have examined an electrically driven BMW engine, operated at 1000 rpm with the resolution of 200 μs. For this short time imaging, multichannel plate-based intensified CCD camera, capable of

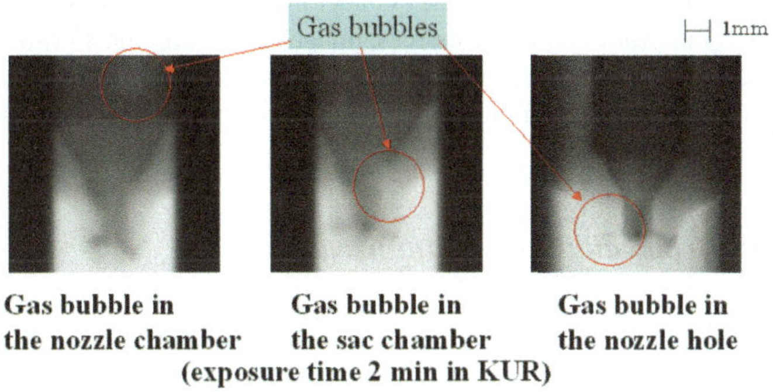

Fig. 10.20 Visualization of gas bubbles in the nozzles using real-time neutron radiography at the KUR. Reproduced from [71] with permission from RightsLink by CCC

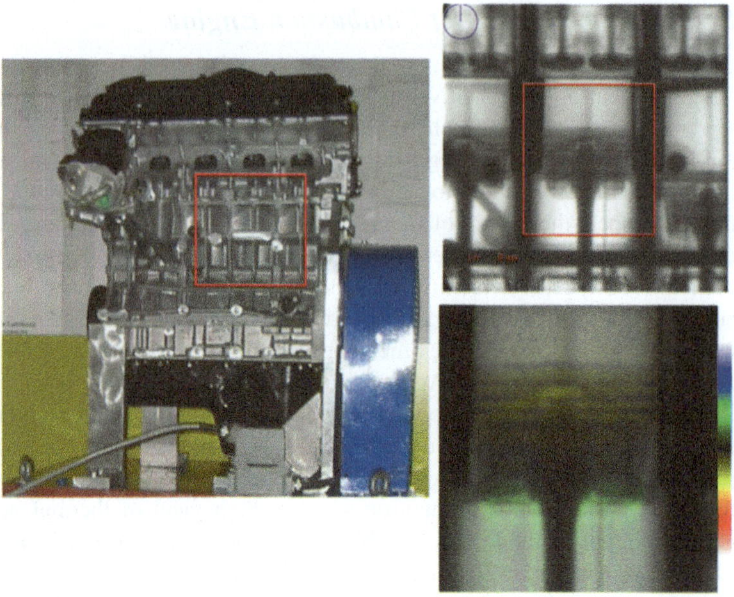

Fig. 10.21 (Left) Photograph of the BMW combustion engine, the field of view for neutron radiography is marked by red rectangle, (right top) a typical NR frame of the running engine and (right bottom) overlapped images, with pseudo colouring, revealing the oil inside the engine. Reproduced from [73] with permission from Author

triggered imaging, was used. Photograph of the BMW combustion engine, a typical NR frame of the running engine and overlapped images, with pseudo colouring, revealing the oil inside the engine is shown in Fig. 10.21.

On the same line of work, PSI researchers have imaged in real time the moving part of a two-stroke chain saw engine running at idle speed (~3000 rpm) and at 8000 rpm, the combustion chamber and the crank shaft [74, 75]. Visualization of fuel and lubricants distributor inside the cylinder in real-time condition was possible. Owing to high flux from spallation sources, image acquisition for short time ~50 μs with a spatial resolution of 100 μm could be carried out (Fig. 10.22). To increase the contrast in the images, gadolinium oxide nanoparticles were dispersed in the fuel.

10.3.6 Neutron Imaging of Fluidized Bed

A fluidized bed is a process where mixing of solid particulate with fluid/gas takes place under appropriate conditions underlying the physical phenomenon of optimum mixing for process enhancement. This phenomenon, where the solid particles behave fluid-like, is known as fluidization. During the fluidization process, fluid drag force and particles gravitational force become equal and the solid particles become

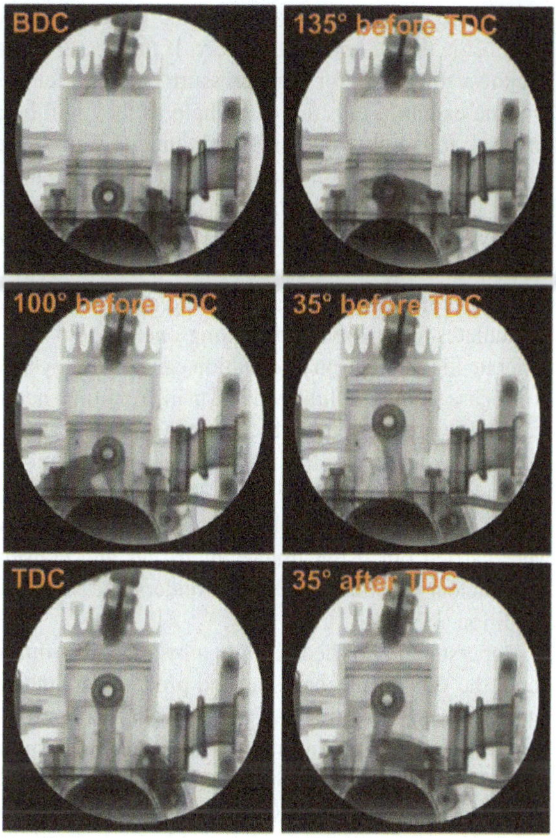

Fig. 10.22 Dynamic neutron radiographs of the prepared two-stroke engine at different crank angle positions between the bottom dead centre (BDC) and the top dead centre (TDC). Reproduced from [74] with permission from RightsLink by CCC

suspended. Fluidized beds are advantageous for the conditions such as area of interaction increases as the particle size goes small which means increase in efficiency, gases/fluids can transfer heat and mass transfer at much higher rates as opposed to solids, operations can be controlled smoothly, rapid mixing enable isothermal conditions throughout the mixing volume can be scaled up, small surface requirement for heat exchangers within the fluidized bed and so on. Though there are some disadvantages, such as very short duration residence time of solid particles for very high flow rate and erosion/abrasion in vessels, the advantages outweigh their use for various applications. Some commercial applications of fluidized bed involving solid/gas/liquid phases are combustion and incineration, roasting of ores, gasification, drying of particles, granulation, etc. To make the reader understand about the fluidized bed concept, let us take a simple example. In case of power generation by burning coal and coal based fuels, the efficiency is ~35 to 37%. Towards increasing the efficiency, formation of a gaseous mixture containing hydrogen and carbon monoxide generated by gasification of coal with air/oxygen and steam at high pressure is beneficial. Pressurized fluidized bed gasification is such a process which increases the coal efficiency to 45%.

To understand the process and the effect of process parameters, one needs to look inside the bed non-intrusively. If the vessel is made up of metal and has some thickness to be safe for high-pressure and high temperature conditions, optical and to some extent X-rays are of no help as they will be attenuated mostly by the clad material. In the case one uses high energy X-rays, there will be almost no contrast between the solid phase and the mixed phase due to very low attenuation in the bed material. Several works related to use of probe method have been cited but they are intrusive type and modifies the bed conditions to some extent. In such cases, neutrons are helpful as they can be used as non-intrusive probe and their attenuation to metals is very less and much higher with fluid/air. In case where mixing of two types of solid particulates are involved, providing one of them with neutron absorbing coating to generate higher neutron attenuation can be employed to obtain high contrast images. Some case studies in fluidized bed using real-time neutron radiography are discussed.

Initial work carried out by Catchen and his team [76] on development of neutron radiography technique for investigating physical models of fluidized bed has laid the way towards this domain of study. They have setup a 15 cm thick sand model and thin-walled aluminium shells simulating spherical voids as a fluidized bed. Experiments were carried out using image intensifier and convention photography techniques in a nuclear reactor beamline to understand various scenarios including dynamic condition.

Understanding of bed particle movement in a fluidized bed provides information about the various mechanisms and process parameters involved. Hisashi Umekawa and his team [77] at the Kansai University, Japan in collaboration with JAERI have carried out neutron radiography of simulated aluminium plate fluidized bed having sand as the bed material. Since both the materials are transparent to neutrons, sand particles were coated with $CdSO_4$, which brought out the contrast. To visualize bulk bed movement, particles made up of B_4C (diameter ~2 mm) were used in the bed as tracer particles. The experiment proved with no doubt that tracer-based NR has high potential towards meaningful data in understand fluidized bed movement in such cases of bed material. Using $GdSO_4$ in place of $CdSO_4$, similar experiments have also been carried out by S. Furui and group [78] towards understanding segregation process in fluidized bed involving binary mixture systems.

Theoretical modelling is a prerequisite in designing any combustion/reaction system based on fluidized bed concept. The relationship between the heat transfer, mixing process and/or diffusion of the reacting gas with bubble or void nature of the bed needs understanding. Real-time NR images with suitable image processing tools have also provided great deal of information regarding the bubble behaviour such as bubble size, bubble frequency, bubble trajectory in vertical tube-banks immersed in a bed [79]. Fluidized bed studies involving different particles have pointed out that segregation phenomena hampers the efficient combustion or reaction, say in a chemical plant. The segregation characteristics depend upon various factors including the particle size distribution, packing ratio and fluidizing gas velocity. Utilizing both X-ray and neutron radiography such experiments have been performed at JAERI [80].

10.4 Summary

This chapter has tried to provide a glimpse on some important applications of neutron imaging related to material science and engineering with the aim to study process parameter, process efficiency, new material development, etc. The know-how obtained through neutron imaging forms the basis for future capacity enhancement towards societal and economic development. To name a few, attenuation-based neutron imaging in the field of PEM-based fuel cell research, hydrogen storage materials, issues related to lithium distribution and thermal runaway in Li-ion batteries have yielded great deal of information towards maximizing their output. Study of material distribution at the weld juncture between two materials using wavelength dispersive imaging has enriched us with crystallographic phase mapping as well as elemental composition including intermetallic compound formation. With the advent of advanced and efficient neutron imagers coupled with the availability of intense sources, dynamic neutron imaging has proved an indispensable probe towards understanding of processes and its controlling parameters. Various flow conditions probed using neutron imaging technique in systems involving air-water, oil-water, liquid-solid, melting-solidification phenomena in lead, combustion engine, fluidized bed have been discussed. The outcome of these sort of experiments provides information related to the processes and helps in improving their utility in their intended applications.

References

1. Nsour W, Taa'mneh T, Ayadi O, Al Asfar J (2019) Design of stand-alone proton exchange membrane fuel cell hybrid system under Amman climate. J Ecol Eng. https://doi.org/10.12911/22998993/111800
2. Tötzke C, Manke I, Hilger A, Choinka G, Kardjilov N, Arlt T, Markötter H, Schröder A, Wippermann K, Stolten D, Hartnig C, Krüger P, Kuhn R, Banhart J (2011) Large area high resolution neutron imaging detector for fuel cell research. J Power Sources 196:4631–4637
3. Markötter H, Manke I, Kuhn R, Arlt T, Kardjilov N, Hentschel MP, Kupsch A, Lange A, Hartnig C, Scholta J, Banhart J (2012) Neutron tomographic investigations of water distributions in polymer electrolyte membrane fuel cell stacks. J Power Sources 219:120–125
4. Stahl P, Biesdorf J, Boillat P, Kraft J, Friedrich KA (2015) Water distribution analysis in the outer perimeter region of technical PEFC based on neutron radiography. J Electrochem Soc 162(7):F677–F685
5. Manke I, Banhart J, Haibel A, Rack A, Zabler S, Kardjilov N, Hilger A, Melzer A, Riesemeier H (2007) In situ investigation of the discharge of alkaline Zn–MnO_2 batteries with synchrotron X-ray and neutron tomographies. Appl Phys Lett 90:214102 (Forner removed)
6. Turhan A, Kim S, Hatzell M, Mench MM (2010) Impact of channel wall hydrophobicity on through-plane water distribution and flooding behavior in a polymer electrolyte fuel cell. Electrochim Acta 55:2734–2745
7. Biesdorf J, Forner-Cuenca A, Schmidt TJ, Boillat P (2015) Impact of hydrophobic coating on mass transport losses in PEFCs. J Electrochem Soc 162(10):F1243–F1252
8. Fairweather JD, Spernjak D, Weber AZ, Harvey D, Wessel S, Hussey DS, Jacobson DL, Artyushkova K, Mukundan R, Borup RL (2013) Effects of cathode corrosion on through-plane water transport in proton exchange membrane fuel cells. J Electrochem Soc 160(9):F980–F993

9. Katzel J, Markotter H, Arlta T, Klages M, Haubmann J, Messerschmidt M, Kardjilov N, Scholta J, Banhart J, Manke J (2016) Effect of ageing of gas diffusion layers on the water distribution in flow field channels of polymer electrolyte membrane fuel cells. J Power Sources 301:386–391
10. Herbrig K, Pohlmann C, Gondek Ł, Figiel H, Kardjilov N, Hilger A, Manke I, Banhart J, Kieback B, Rontzsch L (2015) Investigations of the structural stability of metal hydride composites by in-situ neutron imaging. J Power Sources 293:109–118
11. Gondek L, Selvaraj NB, Czub J, Figiel H, Chapelle D, Kardjilov N, Hilger A, Manke I (2011) Imaging of an operating $LaNi_{4.8}Al_{0.2}$ based hydrogen storage container. Int J Hydrog Energy 36:9751–9757
12. Owejan K, Gagliardo J, Harris S, Wang H, Hussey D, Jackson D (2012) Direct measurement of lithium transport in graphite electrodes using neutrons. Electrochem Acta 66:94–99
13. Zhang Y, Ravi Chandran KS, Bilheux HZ (2018) Imaging of the Li spatial distribution within V2O5 cathode in a coin cell by neutron computed tomography. J Power Sources 376:125–130
14. Song B, Dhiman I, Carothers JC, Veith GM, Liu J, Bilheux HZ, Huq A (2019) Dynamic lithium distribution upon dendrite growth and shorting revealed by operando neutron imaging. ACS Energy Lett 4:2402–2408
15. Weydanz WJ, Reisenwebera H, Gottschalk A, Schulz M, Knoche T, Reinhart G, Masuch M, Franke J, Gilles R (2018) Visualization of electrolyte filling process and influence of vacuum during filling for hard case prismatic lithium ion cells by neutron imaging to optimize the production process. J Power Sources 380:126–134
16. Michalak B, Sommer H, Mannes D, Kaestner A, Brezesinski T, Janek J (2015) Gas evolution in operating lithium-ion batteries studied in situ by neutron imaging. Sci Rep 5:15627. https://doi.org/10.1038/srep15627
17. Tremsin AS, Ganguly S, Meco SM, Pardal GR, Shinohara T, Bruce Feller W (2016) Investigation of metal welds by neutron imaging. J Appl Cryst 49:1130–1140
18. Kardjilov N, Manke I, Hilger A, Williams S, Strobl M, Woracek R, Boin M, Lehmann E, Penumadu D, Banhart J (2012) Neutron Bragg-edge mapping of weld seams. Int J Mater Res 103(2):151–154
19. Woracek R, Penumadu D, Kardjilov N, Hilger A, Boin M, Banhart J, Manke I (2014) 3D mapping of crystallographic phase distribution using energy-selective neutron tomography. Adv Mater 26(24):4069–4073
20. Soria SR, Li XH, Schulz M, Boin M, Hofmann M (2020) Determination of martensite content and mapping phase distribution on Austempered Ductile Iron using energy-selective neutron imaging. Mater Charact 166:110453
21. Haskins JJ (1973) Evaluation of a real time imaging system for Neutron Radiography. General Electric Report, NEDC 12512, 29 May 1973
22. Bracher DA, Garrett DA (1975) Abstract only. Mater Eval 33:47A
23. Stewart PAE (1975) Patent Specification 1 542 860, filed, 23 Dec 1975
24. Berger H (1966) Characteristics of a thermal neutron television imaging system. Mater Eval 24(9):475–481
25. Lindsay JT (1983) Development and characterization of a real time neutron radiographic imaging system. Ph.D. Dissertation, University of Missouri
26. Lindsay JT et al (1992) A summary of neutron radiography and its application to the study of internal combustion engines and fluid flow. In: Proceedings of the fourth world conference on neutron radiography, San Francisco, California, USA, 10–16 May 1992, pp 324–332
27. Brunner J, Lehmann E, Schillinger B (2002) Dynamic neutron radiography of a combustion engine. Neutron Radiography (7), Proceedings of the Seventh World Conference, Roma, Italy, 15–21 Sept 2002, pp 439–446
28. Takenaka N, Kadowaki T, Kawabata Y, Sim CM, Nam KY (2002) Towards the visualization of fuel cavitation in a nozzle of a diesel engine by neutron radiography. Neutron Radiography (7), Proceedings of the seventh world conference, Roma, Italy, 15–21 Sept 2002, pp 453–458
29. Norris PM, Brenizer JS, Raine DA, Bostain DA (1996) Measurement of water deposition in aerogel by neutron radioscopy. In: Fischer CO, Stade J, Bock W (eds) Fifth world conference on neutron radiography, proceedings, 17–20 June 1996, Berlin, Germany, pp 602–609

30. Umekawa H, Ozawa M, Onodera T, Okura Y, Takenaka N, Matsubayashi M (1999) Visualization of large-particle movement in fluidized-bed by neutron radiography. Neutron Radiography (6), Proceedings of the sixth world conference, Osaka, Japan, 12–21 May 1999, pp 507–514
31. Furui S, Umekawa H, Hayashi K, Ozawa M, Takenaka N (2002) Flow visualization of segregation process in a fluidized-bed by neutron radiography. Neutron Radiography (7), Proceedings of the seventh world conference, Roma, Italy, 15–21 Sept 2002, pp 431–438
32. Asano H, Takenaka N, Fujii T, Arakawa T, Matsubayashi M (1999) Quantitative measurement of void fraction distribution. Neutron Radiography (6), Proceedings of the sixth world conference, Osaka, Japan, 12–21 May 1999, pp 515–522
33. Sinha A, Vijyan PK, Shaikh AM (1999) Visualization and analysis of two phase flow by neutron radiography and its applications in the thermal hydraulics studies of advanced reactors. Neutron Radiography (6), Proceedings of the sixth world conference, Osaka, Japan, 12–21 May 1999, pp 545–554
34. Neutron Imaging, E. H. Lehmann (Department Spallation Neutron Source, Paul Scherrer Institute, CH-5232 Villigen PSI, Switzerland), Neutron scattering in novel materials, 22–36 Nov 2000
35. Harris DHC, Seymour WAJ (1987) Applications of real time neutron radiography at Harwell. In: Barton JP, Farny G, Person JL, Röttger H (eds) Neutron radiography. Springer, Dordrecht
36. Jasti JK, Fogler HS (1992) Application of neutron radiography to image flow phenomena in porous media. AIChE J 38:481–488
37. Mishima K, Fujine S, Yoneda K, Yonebayashi K, Kanda K, Nishihara H (1992). In: Jones OC, Michiyoshi I (eds) A study of air-water flow in a narrow rectangular duct using an image processing technique, dynamics of two-phase flow. CRC Press, Boca Raton, pp 141–160
38. Hibiki T, Mishima K, Yoneda K, Fujine S, Kanda K, Nishihara H, Tsuruno A, Matsubayashi M, Sobajima M, Ohtomo S (1994a) Proceedings 4th world conference neutron radiography, San Francisco, CA, 10–16 May, Neutron Radiography, 4th edn. J.P. Barton Gordon and Breach, Yverdon, Switzerland, p 317
39. MishimaT Hibiki K, Nishihara H (1997) Visualization and measurement of two-phase flow by using neutron radiography. Nuclear Eng Des 175(1–2):25–35
40. Sarkar PS, Kashyap Y, Sinha A, Vijyan PK, Prasada Rao GSS (2005) Applications for real-time neutron radiography for convection driven flow pattern transition studies. IEEE Trans Nuclear Sci 52(1)
41. Sinha A, Vijyan PK, Shaikh AM (1999) Visualization and analysis of two phase flow by neutron radiography and its applications in the thermal hydraulics studies of advanced reactors. Presented at the 6th world conference on neutron radiography held at Osaka, Japan, 17–21 May
42. Luiz LC, Crispim VR, Ferreira FJO (2010) Visualization of two-phase flow in metallic pipes using neutron radiographic technique. Physicæ 9:19–23
43. Hitoshi A, Nobuyuki T, Terushige F, Toshiaki W (2005) Visualization and void fraction distribution of downward gas-liquid two-phase flow in a plate heat exchanger by neutron radiography. Nuclear Instrum Methods Phys Res Sect A: Accel Spectrom Detect Assoc Equip 542(1–3):154–160
44. Zhao Q, Guo L, Wang Y, Jin H, Chen L, Huang Z (2020) Enhanced oil recovery and in situ upgrading of heavy oil by supercritical water injection. Energy Fuels 34(1):360–367
45. Shoji E, Kikuchi T, Yamagiwa K, Kubo M, Tsukada T, Takami S, Sugimoto K, Ito D, Saito Y (2020) In-situ visualization of heavy oil behavior in supercritical water using neutron radiography. Chem Eng Sci 225, Article number 115816
46. Balaskó M, Jancsó G, Kőrösi F (2004) Analyzing of segregation in mixtures of 3-methylpyridine and heavy water by dynamic neutron radiography. Appl Radiat Isot 61(4):597–602
47. Takenaka N, Asano H, Fujii T, Motomura Y, Ono A, Matsubayashi M, Tsuruno A (1996) Liquid metal flow measurement by neutron radiography. Nuclear Instrum Methods Phys Res Sect A: Accel Spectrom Detect Assoc Equip 377(1):156–160. ISSN 0168-9002
48. Mishima K, Hibiki T, Saito Y, Nakamura H, Matsubayashi M (1999) The review of the application of neutron radiography to thermal hydraulic research. Nuclear Instrum Methods Phys Res Sect A: Accel Spectrom Detect Assoc Equip 424(1):66–72

49. Saito Y, Mishima K, Tobita Y, Suzuki T, Matsubayashi M (2005) Measurements of liquid–metal two-phase flow by using neutron radiography and electrical conductivity probe. Exp Thermal Fluid Sci 29(3):323–330. ISSN 0894-1777
50. Khodarev E, Liquid metal fast breeder reactors. IAEA Bull 20(6)
51. Saito Y, Mishima K, Tobita Y, Suzuki T, Matsubayashi M, Lim IC, Cha JE (2005) Application of high frame-rate neutron radiography to liquid-metal two-phase flow research. Nuclear Instrum Methods Phys Res Sect A: Accel Spectrom Detect Assoc Equip 542(1–3):168–174. ISSN 0168-9002
52. Wolf F, Viskanta R (1987) Experimental heat transfer, vol 1, pp 17–30; Wang X, Fauelle Y (2009) Int J Heat Mass Transfer 52:5624–5633
53. Beckermann C, Viskanta R (1989) J Heat Transfer 111:416–424; Wolff F, Beckermann C, Viskanta R (1988) Exp Thermal Fluid Sci 1:83–91
54. Fife S et al (2003) Exp Fluids 35:152–158
55. Yin H, Koster JN (2003) J Alloy Compd 352:175–189
56. Barber PG et al (1986) J Crystal Growth 74:228–230
57. Chun J-H, Lanza RC, Saka N, Hytros MM (1995) Ann CIRP 44(1)
58. Mishima K, Fujine S, Yoneda K, Yonebayashi K, Kanda K, Nishihara H (1988) Proc Japan-U.S. Seminar on Two Phase Flow Dynamics, Ohtsu, Japan, pp C.3-1
59. Mishima K, Hibiki T, Nishihara H (1993) Int J Multiphase Flow 1:115
60. Proc Seventh World Conference on Neutron Radiography, Rome, Italy, 15–20 Sept 2002
61. IEEE Trans Nuclear Sci 52(1)
62. Kanematsu M, Maruyama I, Noguchi T, Iikura H, Tsuchiya N (2008) 3rd ACF international conference-ACF/VCA 2008, C 16, pp 925–931
63. Chalovicha TR, Bennett LGI, Lewis WJ, Brenizer JS Jr (2004) Appl Radiat Isot 61:693–700
64. Takenake N, Jujii T, Ono A, Motomura Y, Turuno A (1995) Fusion Eng Des 27:607–613
65. Ogino F, Kamata M (1994) In: Barton JP (ed) Neutron radiography, 4th edn. Kluwer, Dordrecht, pp 339–346
66. Reijonen H, Forsten J (1972) J Cryst Growth 12:61–62
67. Kumar L, Manjunath BS, Patel RJ, Prabhu SV (2014) Experimental investigations on melting of lead in a cuboid with constant heat flux boundary condition at two vertical walls using infra-red thermography. Int J Heat Mass Transfer 68:132–140
68. Sarkar P, Agrawal A, Kashyap Y, Shukla M, Sinha A, Kumar L, Manjunath BS, Prabhu SV, Markandeya SG (2013) Visualization of lead melting and solidification using neutron radiography. Int J Nuclear Energy Sci Eng 3(1)
69. Ito D, Saito Y, Sato H, Shinohara T (2017) Visualization of solidification process in lead-bismuth eutectic. Phys Procedia 88:58–63
70. Takenaka N, Kadowaki T, Kawabata Y, Sim CM, Nam KY (2002) Towards the visualization of fuel cavitation in a nozzle of a diesel engine by neutron radiography. In: Proceedings of the seventh world conference, Rome, Italy, 15–21 Sept 2002, pp 453–458
71. Takenaka N, Kadowaki T, Kawabata Y, Lim IC, Sim CM (2005) Visualization of cavitation phenomena in a Diesel engine fuel injection nozzle by neutron radiography. Nucl Instrum Methods Phys Res, Sect A 542(1–3):129–133
72. Vontobel P, Frei G, Brunner J, Gildemeister AE, Engelhardt M (2005) Dynamic imaging with a triggered and intensified CCD camera system in a high-intensity neutron beam. Nucl Instrum Methods Phys Res, Sect A 542(1–3):148–153
73. Brunner J (2021) Doctoral Thesis, Quantitative time resolved neutron imaging methods at the high flux neutron source FRM-II, https://mediatum.ub.tum.de/doc/603123/603123.pdf, accessed on 25/05/2021
74. Christian G (2013) Progress in industrial applications using modern neutron imaging techniques. 7th International topical meeting on neutron radiography. Phys Procedia 43:231–242
75. Gruenzweig C (2010) Visualization of a fired two-stroke chain saw engine running at idle speed by dynamic neutron radiography. SAE Technical Paper 2010-32-0013
76. Catchen GL, Hughes DE, Kulkarni AK, Ross KM, Ammerman WR (1987) Development of a neutron-radiographic technique for investigating physical models of fluidized beds. Chem Eng Commun 54(1–6)

77. Umekawa H, Ozawa M, Takenaka N, Matsubayashi M (1999) Visualization of bed material movement in a simulated fluidized bed heat exchanger by neutron radiography. Nucl Instrum Methods Phys Res, Sect A 424(1):77–83
78. Furui S, Umekawa H, Hayashi K, Ozawa M, Takenaka N (2002) Flow visualization of segregation process in a fluidized-bed by neutron radiography. Presented at the WCNR 7, Rome
79. Ozawa M, Umekawa H, Furui S, Hayashi K, Takenaka N (2002) Bubble behavior and void fraction fluctuation in vertical tube banks immersed in a gas–solid fluidized-bed model. Exp Thermal Fluid Sci 26(6–7):643–652
80. Umekawa H, Furui S, Oshima Y, Okura M, Ozawa M, Takenaka N (2005) Quantitative measurement of segregation phenomena in a binary-mixture fluidized bed by neutron radiography. Nucl Instrum Methods Phys Res, Sect A 542(1–3):219–225

Chapter 11
Neutron Imaging for Geo-mechanics and Agricultural Applications

Shefali Shukla and Tushar Roy

11.1 Introduction

Neutrons are highly sensitive to light elements especially hydrogen and thus are a valuable probe for investigating water dynamics in bulk materials because neutrons are mostly non-interacting and hence quite penetrative. Geo-mechanics deals with the study of rock and soil mechanics. The primary objective of neutron imaging in this field is studying the fluid flow (mostly water or water-based fluids) in these materials. The flow will depend on the medium's porosity and structure which in turn defines the structural strength of the material. Cement-based materials like concrete and mortars are important because they are the major components of buildings, bridges and other structures. Drying of cement is another application using neutron imaging because moisture can be a life-limiting factor of the reinforced structure inside the mortars due to corrosion. In agriculture, many different aspects have been studied using neutrons. For plants, the transport of water and minerals has been widely studied. Soil water dynamics, the effect of rhizobacteria on water retention, transport of minerals through water are a few applications that have been explored using neutron imaging. Water also constitutes a major part of biological materials so it can be used for studying them; in particular, the dehydration process is generally conducted to increase the shelf life of materials. This chapter gives an overview of applications of neutron imaging in geo-mechanics, agriculture and biology.

S. Shukla (✉) · T. Roy
Technical Physics Division, Bhabha Atomic Research Center, Mumbai 400085, India
e-mail: shefali@barc.gov.in

T. Roy
e-mail: tushar@barc.gov.in

D. K. Aswal et al. (eds.), *Neutron Imaging*,
https://doi.org/10.1007/978-981-16-6273-7_11

11.2 Neutron Imaging for Geo-mechanics

The distribution or flow of fluids in geo-material can be studied using neutron imaging due to the high attenuation of neutrons by hydrogen. Hydrogen and its compounds constitute a major component of fluids like water, heavy water, acids and alcohols and can be easily imaged using neutrons. Another attractive property of neutrons is their isotopic sensitivity, especially exhibited in the case of hydrogen (^{1}H) and deuterium (^{2}H or D). ^{1}H and ^{2}H though chemically similar have vastly different incoherent scattering cross section (80.27 barn and 2.05 barn, respectively). A direct application is in the form of their compounds—normal water (H_2O) and heavy water (H_2O or D_2O)—which though being chemically similar can be distinguished clearly in neutron imaging, making them excellent contrast agents.

One of the first demonstrations of neutron imaging in geo-mechanics was by Gardner and Kirkham [1] who used neutrons as a probe to detect the presence of water in the soil. Subramanian and Burkhart [2], Reijonen and Pihlajavaara [3], Wilson et al., and Lewis and Krinitzsky [4, 5] were probably the first ones who used neutron radiography as a probe to study rocks, concrete and soils, respectively. Jasti et al. [6] carried out neutron radiography of water movement in rock pores and were the first to analyse quantitatively the dynamic processes in radiography in geo-materials. The first neutron tomography of rock was reported in 1990s [7, 8], wherein tuff samples considered of the permanent disposal of radioactive wastes were studied to characterize the water content.

Neutron radiography after that has been routinely used in geo-mechanics. Some of the key applications of neutron imaging for geo-mechanics include study/determination of water distribution in soil [4, 5], rocks [7, 8], concrete [9], bricks [10, 11] and ceramics [12, 13], and the study of porosity in rocks [14–16], water retention in soils, saturation and hydraulic properties of soil [17], water distribution during curing and drying of concrete [9], water transport in porous structures, etc.

11.2.1 Neutron Imaging for Fluid (Flow): Rock, Bricks and Soil

Neutron imaging can detect fluid localizations in porous materials due to their high sensitivity to hydrogen. This technique remained underutilized in geology because of the large acquisition time associated with imaging. Another hitch was the unavailability of experimental devices to simulate the geological relevant conditions like high pressure and temperature. With the recent advances in new beamlines [18–24] which allow faster tomography scans, it has become possible to capture fluid and contaminant flow with high spatial and temporal resolution. Understanding fluid flow in rocks enables quantifying many natural processes like groundwater flow and naturally triggered seismicity. It also provides information on the displacement of

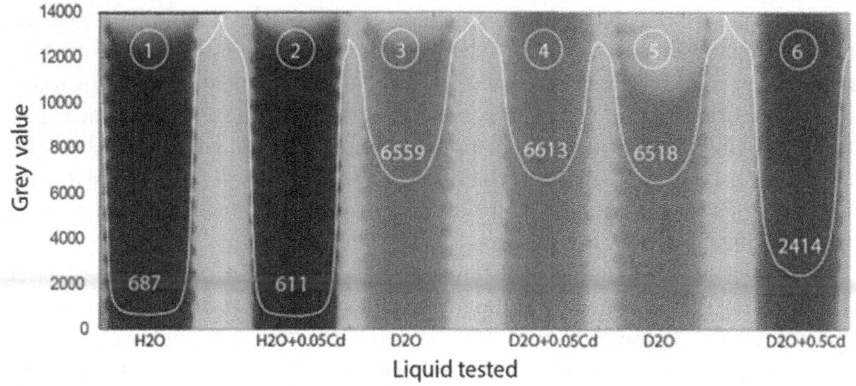

Fig. 11.1 Radiographs and attenuation of injected fluids with different concentrations of Cd, from left to right: see text for details [25]

contaminants, the eligibility of subsurface waste storage, geothermal energy usage, oil and gas recovery and artificially induced seismicity [25].

Cadmium (Cd) is a hazardous compound found in many electronic devices and is a common contaminant in soil and groundwater. Studying the retention and sorption process of Cd and other heavy metals is imperative to developing waste treatment methods. Cd can be used as a tracer because of its high neutron attenuation compared to heavy water. Cordonnier [25] conducted in-situ 2D and 3D studies to study the fluid and pollutant percolation in rocks. Imaging was carried out to study the flow properties such as local permeability and interactions between cadmium and solid rock. Figure 11.1 shows the radiographies and attenuation of injected fluids. 1 shows case of only water, 2 is where 0.05 mol/L cadmium was added to water. It was observed that its challenging to see Cd in the presence of water so for further studies water was replaced by heavy water. 3, 4, 5 and 6 show Cd distribution in the presence of heavy water for different concentrations of Cd. This study demonstrated the use of neutron imaging to provide insights into the subsurface transport of pollutants.

11.2.2 Determination of Moisture Distributions in Porous Building Bricks and Cement

Moisture studies in building construction materials including bricks, cement, etc., are important as the durability of these structures is mediated by water and by biological and chemical agents transported by it. The moisture present in these materials can degrade them over time [10, 26, 27]. Because of its economic consequences, the problem of moisture transport in porous building construction materials has received the attention of physicists and engineers. When in direct contact with water, the hydration products on the concrete surface may get washed out due to leaching. Thus, surface or near-surface of concrete structures are weakened and prone to damage under extreme weather conditions. Many destructive and non-destructive methods are routinely used for the estimation of moisture content in cement-based materials like gravimetric method, electrical conductivity, nuclear magnetic resonance spectroscopy, humidity sensors and radiation attenuation techniques. Neutron imaging can also be used to study moisture distribution and transport in cement-based materials owing to high attenuation of neutron to hydrogen over other components of sand like silica, calcium, iron, sodium.

El Abd et al. [28] studied moisture transport in four different kinds of brick samples using real-time neutron radiography. These samples are cement lightweight brick #1, cement lightweight brick #2, clay brick and sand–lime lightweight brick. The moisture absorption process in these bricks was monitored continuously using neutron radiography (NR). Figure 11.2 shows some selected radiographs taken at different absorption times. Since the neutrons are very sensitive to hydrogen, the water-filled section appears darker in the image. In the image, brick#1 is the rightmost one and the brick#4 the leftmost one. From the radiographs, it can be seen that the rate of water absorption is nearly same for brick#1 #2 and #3 but its quite less for the brick#4. From the data obtained using radiography water diffusivities for the samples were obtained. This work concluded that the diffusion process in the samples investigated was anomalous and does not follow Fick's hypothesis.

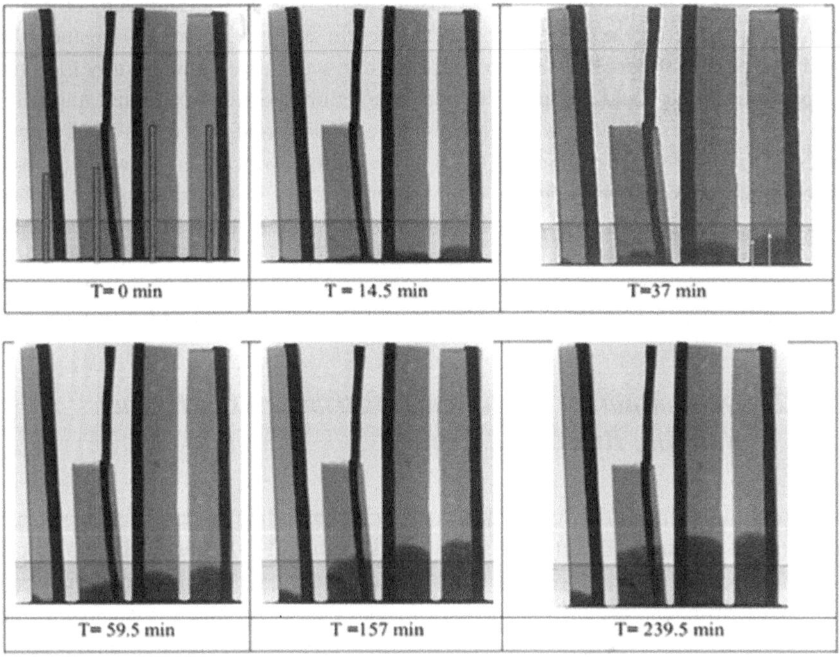

Fig. 11.2 Neutron radiography images for water absorption into brick samples at different absorption times [28]. Reproduced with permission from Elsevier Copyright Clearance Center

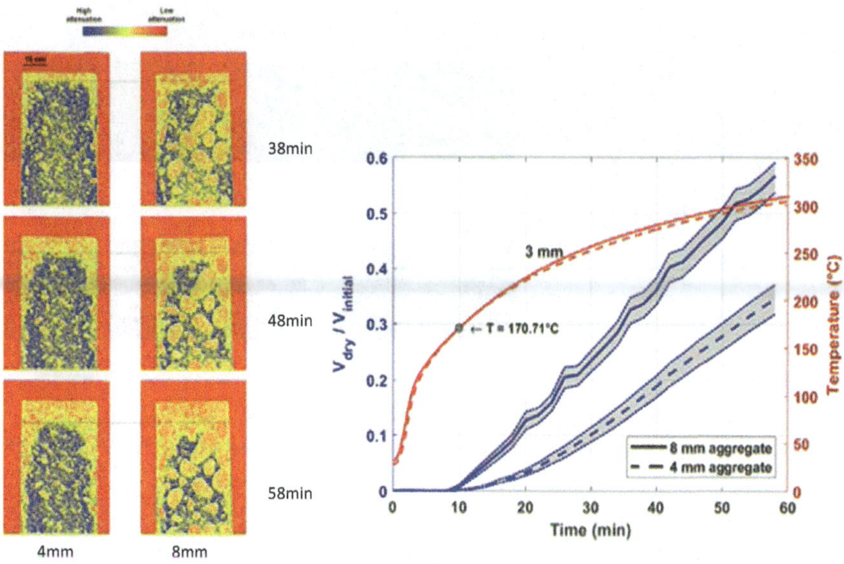

Fig. 11.3 (Left) Vertical slices of rapid (1 min) neutron tomography capturing drying and moisture migration in concrete heated in operand. (Right) Comparison of the effect of aggregate size on temperature (in red) and moisture migration (in blue) measured using neutron tomography [29]. Reproduced with permission from Elsevier Copyright Clearance Center

Studies related to drying of concrete were one of the first applications of neutron radiography. The major focus in early works was the study of cement pastes undergoing environmental drying since this process has a large impact on the mechanical properties of the cement structure. Various procedures like cyclic drying and water absorption and effects of surface treatments on moisture absorption have also been studied. Study of moisture migration in construction materials in the event of fire is crucial as it can result in structural collapse. Since this is high-speed process, many studies have tried to solve this problem in pseudo-2D conditions focusing on either the overall moisture migration or the rapid variation in the crack opening. With the advent of high-speed neutron tomography, Dorjan [29] tested concrete samples heated at 500 °C. The studies revealed a significantly different hydraulic response despite a nearly identical thermal profile. The study also helped identify the moisture accumulation ahead of the heating front which is the possible origin of fire-induced spalling in concrete (refer Fig. 11.3).

11.2.3 Testing of Geo-materials Under Extreme Conditions

Since its early days testing of geo-materials under extreme conditions has been a broad domain in geophysics and geo-mechanics. Imaging allows us to map the

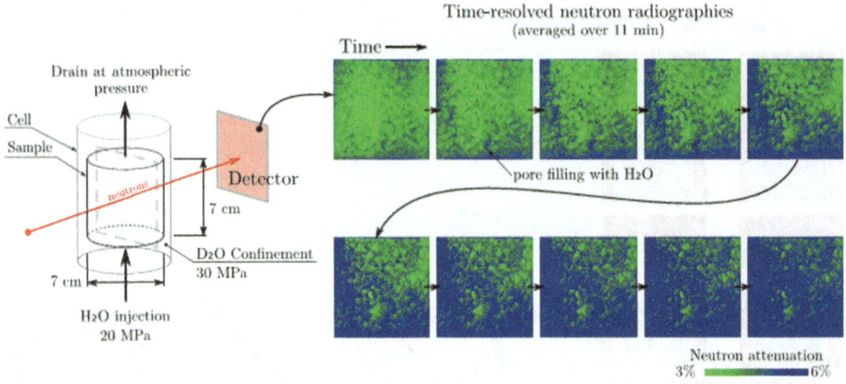

Fig. 11.4 Neutron radiographs were taken at different times after injecting normal water into a concrete sample cured with heavy water [31]. Reproduced with permission from Elsevier Copyright Clearance Center

behaviour in situ and thus is highly beneficial for such studies. However, generating these extreme conditions requires large cells that are difficult to penetrate using conventional X-rays imaging. With powerful X-ray sources, very small samples might be studied but the actual studies of many geo-materials should be conducted on larger samples. These larger samples are studied in thick metallic chambers which can withstand the applied pressures. Neutrons being transparent to metals become an ideal tool for mapping such studies. High pressures are important in many applications, for example hydraulically fracture in rocks, to enhance geothermal energy performances, gas extraction, and to study the effect of fluid on the initiation of earthquakes [30].

A particular example is the study of fluid flow in a low-permeability medium (such as concrete) under high pressure. Neutron imaging studies of geo-materials at high pressure provide a deeper understanding of the processes involved. Studies on fluid flow in concrete at high pressure help in understanding the durability and behaviour of dams and infrastructure. Using the bulk penetration capability of neutrons, studies can be carried out in big confinement vessels over large samples of concrete. Yehya [31] conducted studies using a testing rig capable of withstanding 50Mpa pressure over samples of upto 7 cm in diameter. Water flow in a concrete sample cured in heavy water was imaged using neutrons, and the dominant influence of local heterogeneities was revealed even at such high confinements (Fig. 11.4). The role of reinforcement bars and geometry in 3D has also been explored.

11.3 Neutron Imaging for Agricultural Applications

Neutrons can be one of the best probes for agricultural applications as they are highly sensitive to hydrogen, an element that is in abundance in the plants/roots. On the other hand, since the surrounding background for roots, soil has very little hydrogen and has other elements like silicon for which neutron attenuation is low, high contrast images can be generated. Therefore, neutron imaging has been widely used to study root architecture, root growth and soil and plant water distribution.

11.3.1 Plant Physiology

Plant physiology is the branch of science dealing with the growth of plants and their response to the environment at the cellular and biochemical levels. Various processes associated with plant physiology are metabolism, mineral and water intake along with its process, overall development and reaction to external stimulus or stresses. The soil is the plants nutrient reservoir, and water helps in their transport. Since plants are responsible for feeding the inhabitants of the whole world and providing oxygen for their survival, plant physiology has been under study for centuries to increase their productivity and also to some extent regulate it. Initial studies on understanding mass flow properties were carried out by injecting radioisotopes of ^{3}H and ^{32}P and monitoring their beta emissions [32]. Nutrient transport study involving 'sulphorhodamine G' dye built-up in maize plants from the soil–water mixture to various parts provided means for locating and estimating accumulation of ions present in the soil [33].

Various non-destructive techniques for soil–root system studies such as X-ray imaging [34–36], magnetic resonance imaging [37] neutron imaging [38], ground-penetrating radar are being utilized. But X-rays cannot distinguish between materials with similar physio-chemical properties, and neutrons can compliment this information. Since neutrons are sensitive towards hydrogenous materials, the root and the water inside them can be visualized by transmission imaging technique with contrast in the presence of soil, obviously taking into consideration the optimum thickness of the soil present. Creating simulated samples and performing NR can provide a great deal of information regarding water and nutrient transport [39]. Quasi-dynamic neutron radiography can serve the purpose. Talking about nutrients, boron is essential as it is responsible for maintaining plant cell wall integrity. Plant growth is highly affected by boron imbalance. Since boron has a high absorption cross section for thermal neutrons, determination of boron distribution in plants in simulating conditions using thermal neutrons is possible. Various types of soil systems [40–42] with varied nutrition content have been studied, and the different uptake mechanisms have been proposed towards reaping future benefits.

Cold neutron radiography (CNR) with D_2O tracer has been successfully applied to understand water flow in rose plants from root via stem to leaf or flower [43]. Analysis of the acquired images provided water flow velocities at various locations indicating correlations between water sink and source strength in plant tissues and/or organs. These data might help in the irrigation of plants under deficit conditions. D_2O is the best choice for tracer studies as its thermal attenuation cross section is about one-sixth of that of normal water (H_2O).

Neutron radiography investigations have effectively monitored root functionality and water movements in soil and roots. Warren et al. [45] conducted such studies in *Zea mays L.* (maize) and *Panicum virgatum L.* (switchgrass) seedlings and provided insights regarding root water uptake timing, magnitude and the involved mechanisms critical towards plant biogeochemical cycle. Neutron radiography with D_2O as a tracer was carried out for 12 h period, and suitable arrangements were put in place to ensure proper regulation of root water uptake and transport for photosynthesis mechanism. Nakanishi et al. [44] studied the growth of soybean plants in soil embedded in a small aluminium container. The neutron imaging studies revealed information about the water movement in soil and the root systems morphological development. The experiment was conducted on a soybean plant grown in an aluminium box having soil that has 15% water. The study was done after 8 and 15 days, and root structure was mapped. Since the roots have a high water concentration w.r.t soil a good contrast was achieved, Figure 11.5 shows the neutron image of the soybean root system after 8 and 15 days.

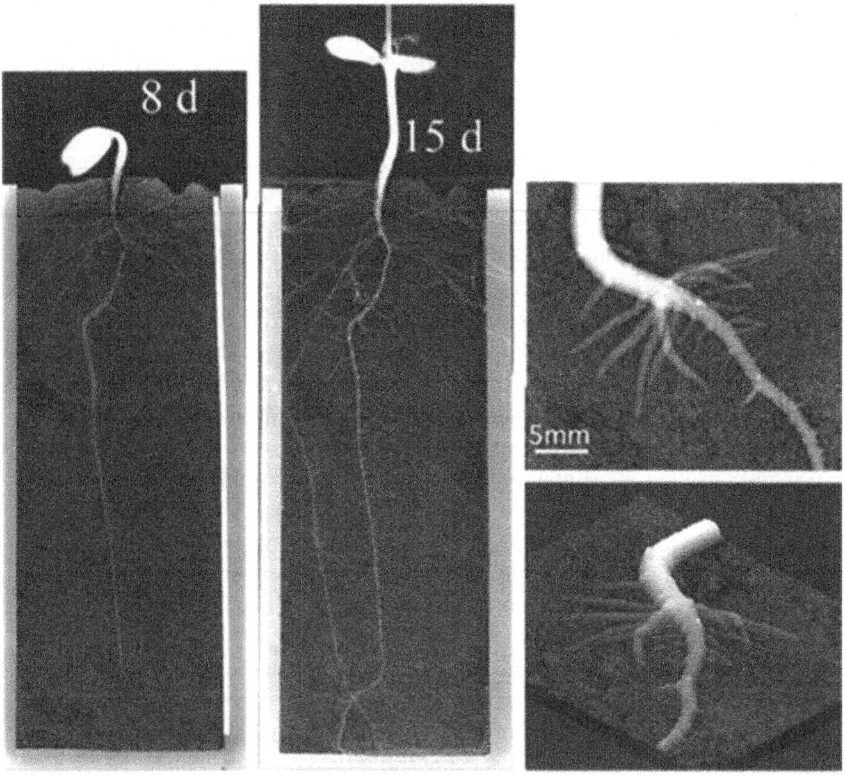

Fig. 11.5 Neutron image of soybean root embedded in soil; **(Upper right)** magnification of the root when side root emerged to develop; **(Lower right)** conversion of the upper image to three-dimensional one, where the water amount was employed as height [44]

Fig. 11.6 Time-resolved neutron tomography of the root system of a lupine after injection of deuterated water through the bottom. From the tomographies conducted at different instants the movement of waterfront can be visualized [46] (Open access)

The Development of ultrafast neutron tomography, where full volume set data acquisition time ~10 s, provides much more information than 2D imaging towards understanding and modelling of water uptake by plants. The experiments [42, 46] carried out using the CONRAD II cold neutron imaging instrument at the neutron source BER II of Helmholtz Centre Berlin provided time-resolved 3D water distribution at 10 s instances. Figure 11.6 demonstrates time-resolved neutron tomography images of the root system, in its 3D perspective, of a lupine after the injection of 4 ml deuterated water (D_2O) through the bottom [46].

In the same direction of research, D_2O-based tracer-induced 3D neutron tomography with fast data acquisition capability has shown promising results towards understanding water uptake by roots and subsequent transport in a spatiotemporal manner [47]. Varying tomogram acquisition times were implemented, shorter after injecting D_2O in the soil and larger afterwards, to understand the hydraulic water uptake methodology by roots. Because of temporally separated 3D data availabilty, sectional views provide specific information related to initial water (H_2O) content, and total water (H_2O $_+$ D_2O) content dynamics in the imaged portion.

Neutron imaging has also been used to study water distribution in veins and lamina, three-dimensional venation architecture and sap flow dynamics in leaves. Leaves play an essential role in water transport in a plant as most of the water loss by transpiration occurs in the leaves. Defraeye et al. [48] investigated venation network architecture and sap flow dynamics of detached leaves using neutron radiography and neutron tomography with cold neutrons. D2O was used as a contrast liquid. The experiment could successfully quantify leaf water content and also the hierarchical structure of water transport pathways.

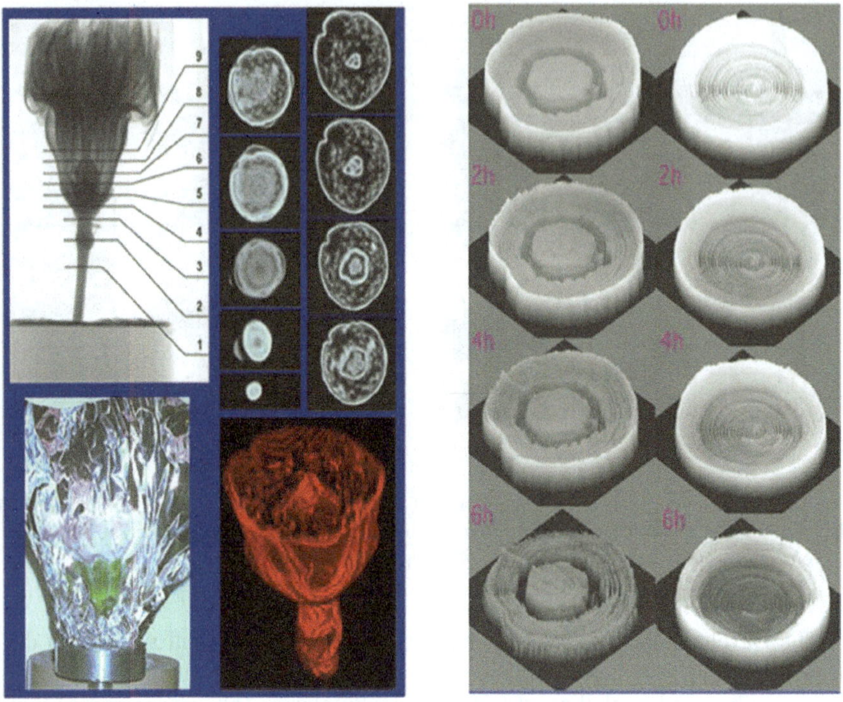

Fig. 11.7 (Left) Three-dimensional image of carnation flower showing internal distribution of water in the flower (Right) Cross-sectional image of wood disc of cedar tree showing internal distribution of water during drying [46] (Open access)

In the flower industry, cut flowers are required to be kept fresh and moist for long durations. Nakanishi et al. [44] showed the effect of changing viscosity of water supplied to cut flowers to keep them fresh. A carnation flower was wrapped with aluminium foil, and neutron tomography was performed to study the water distribution in the flower (Fig. 11.7). From the reconstructed data, water distribution was mapped across various transverse sections of the flower. The neutron imaging data suggested that there are two types of tissues in the stem below the flower, which allows water absorption after drought conditions. Similarly, water distribution during drying of wood disc of cedar tree has also been studied.

11.3.2 Food Science and Food Preservation

Food consists of an organic material having significant hydrogen content; therefore, neutrons that are highly sensitive to this light element become an excellent probe for studying food samples. Also since neutrons are highly penetrating, it becomes easier to investigate within packing and complex chambers mostly of metal. As neutron radiography and tomography can locate even small amounts of water, they are suitable for evaluating water diffusion or water uptake in samples.

In food science, extrusion, drying and heat processing is commonly used for increasing shelf life. Amongs them, drying is the most common thermal processing operation conducted to preserve fresh fruits and vegetables. The resulting extended shelf life and prolonged storage capability guarantees off-season availability of this nutrition source and helps reduce food wastage. The drying of food is still a process being studied. The major objective of drying is to decrease the development of microorganisms or stop some chemical degradation by reducing moisture content. The most common techniques for studying drying are nuclear magnetic resonance (NMR) [49], X-ray radiography/tomography [50] and neutron radiography/tomography [51, 52]. These techniques differ in their spatial and temporal resolution, their ability to detect water content qualitatively and quantitatively and their accessibility and ease of use. Neutron imaging is advantageous in the sense that it provides bulk transmission, and also, water quantification is easier. Defraeye [53] conducted neutron radiography studies on apple fruit slices during convective drying. Figure 11.8 shows the water distribution changes in apple fruit slices during various drying by quantitative neutron imaging [53]. These studies demonstrated that quantitative neutron imaging can successfully monitor the internal moisture distribution inside fruits during various convective drying processes.

The process of drying of fish when studied using neutron radiography and tomography provides valuable information about the moisture content and homogeneity. Tanoi [54] used neutron radiography to analyse the water loss during drying on

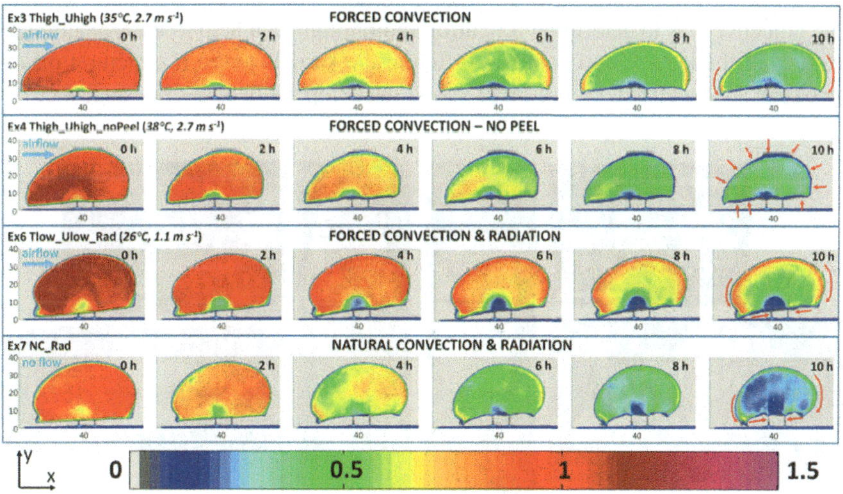

Fig. 11.8 Internal distributions of scaled water content for apple fruit slices from neutron radiographs at different times for four different experiments. The same scaling factor is used for all images. The red arrows illustrate the deformation trend at the end of the drying process. (Dimensions are in mm) [53]

different types of fish. It was observed that the repartition of remaining water was more homogeneous for the freeze-dried fish than the one which was conventionally dried. Additionally, it was seen that water content in few organs like head was high. Tomographic investigations followed which provided better insight into the spatial distribution of water during drying. Few results from that work have been shown in Fig. 11.9.

Radiography of dried fish was also carried at the neutron imaging facility at Dhruva reactor, India [55]. The internal features are easily visible because of reduced moisture content (refer Fig. 11.10).

Cooking is an operation that modifies meat structure by applying heat to enhance taste, aroma and texture. The physico-chemical transformations occurring during the cooking of meat are directly linked to the configuration changes in proteins that constitute the meat: myosin, heam-protein and collagen. Using the advantage of neutron imaging which can be carried out even in sophisticated sample surroundings attempts have been made to study the in-situ cooking process. Scussat [56] attempted to study the impact of cooking on beef meat. The heating was carried out using a semi-professional oven, and radiography was done during the cooking process to determine the key temperature, wherein morphological changes occur due to contraction of meat fibres because of cooking and migration of juices. The main inference of this work was that the main microscopic shrinkage of the meat happens after 70 °C (after myosin and collagen denaturation). Crucial attenuation changes have been observed at higher temperatures (around 80 °C) and are related to juice release following the protein denaturation.

Similar to neutron tomography studies of mechanical objects Cleveland [57] attempted neutron tomography for the structural analysis of dried corn kernels. The objective was to determine if neutron radiography can help to view anatomical features of the kernels and compare different the kernels to see if structural differences can be visualized. Figure 11.11 shows the comparison of (A) physical slices of

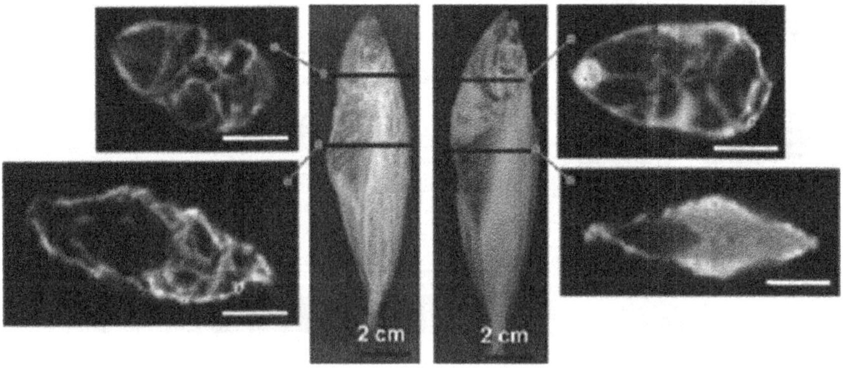

Fig. 11.9 Neutron radiographs of fish (centre) and reconstructed tomographic slices by freeze-drying (left) and conventional drying (right). The position of the slice has been marked in radiograph [54]. Reproduced with permission from Elsevier Copyright Clearance Center}

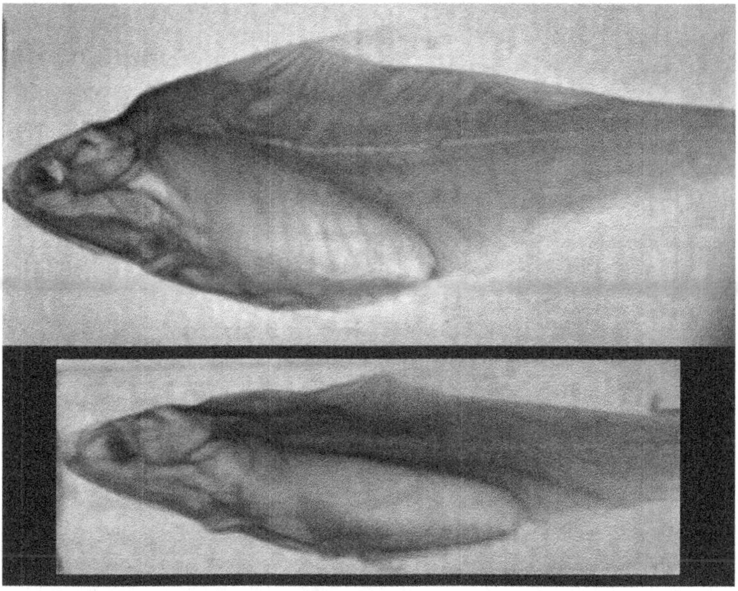

Fig. 11.10 Neutron radiograph of dried fish

kernel (C) stained physical slices of kernel and (B) reconstructed corn kernel slices. This study proved that neutron tomography can identify the well-known internal structure of corn kernel.

11.4 Summary

Neutron imaging has various applications in geo-mechanics and agricultural applications as well as in the food preservation industry. The high attenuation of hydrogen in water is exploited in neutron imaging and is a common theme for most of the above applications where water flow/transport, water content, moisture or dehydration/drying are studied. Moreover, the ability of neutrons to study in-situ processes can be used to probe many dynamic processes as neutrons are unaffected by surrounding material.

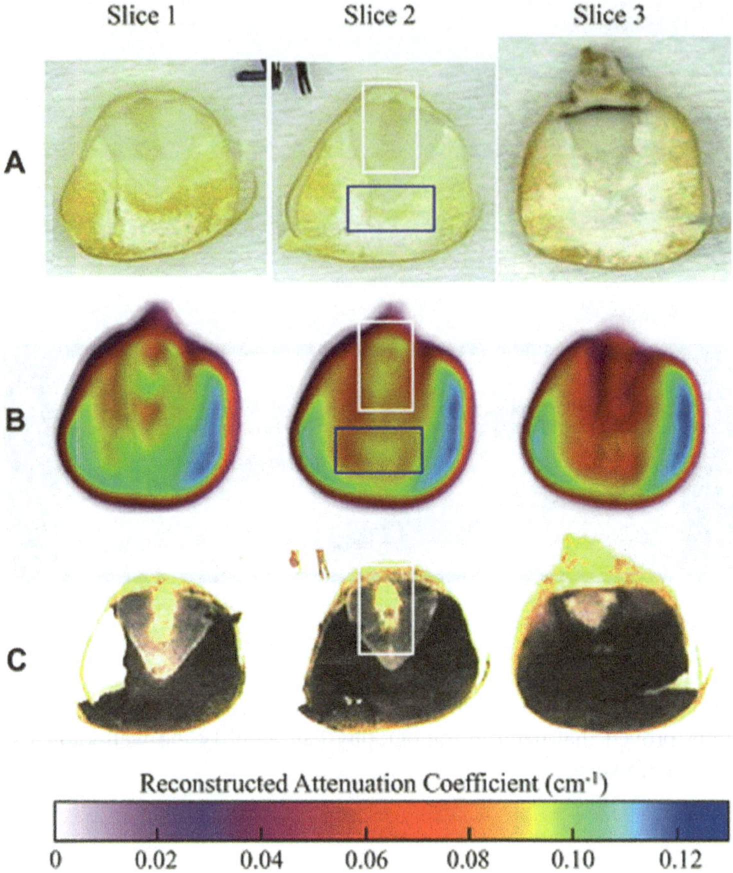

Fig. 11.11 Comparison of (**a**) physical slices of kernel (**c**) stained physical slices of kernel and (**b**) reconstructed corn kernel slices. The colour bar at the bottom gives the neutron attenuation coefficients for the reconstructed slices in **b** [57]. Reproduced with permission from Elsevier Copyright Clearance Center

References

1. Wilford G, Don K (1952) Determination of soil moisture by neutron scattering. Soil Sci 73(5):391–402
2. Subramanian RV, David B (1973) Determination by neutron radiography of the location of polymeric resins injected in rock fissures. Nucl Technol 17(2):184–188
3. Reijonen H, Pihlajavaara SE (1972) On the determination by neutron radiography of the thickness of the carbonated layer of concrete based upon changes in water content. Cement Concr Res 2(5):607–615
4. Wilson NE, Harms AA, Emery JJ (1975) A neutron radiographic in vitro examination of soils. Can Geotech J 12(1):152–156
5. Lewis JT, Krinitzsky EL (1976) Neutron radiation in the study of soil and rock. In: Practical applications of neutron radiography and gaging. ASTM International

6. Jasti JK, Lindsay JT, Fogler HS et al (1987) Flow imaging in porous media using neutron radiography. In: SPE annual technical conference and exhibition. Society of Petroleum Engineers
7. Kupperman DS, Hitterman RL, Rhodes E (1990) Energy dependent neutron imaging. Argonne National Lab, Technical Report, IL (United States)
8. Rhodes E, Kupperman DS, Hitterman RL (1992) Dual-energy neutron tomography of water in rock using the Argonne IPNS. Technical Report, Argonne National Lab
9. Howdyshell PA (1977) A comparative evaluation of the neutron/gamma and kelly-vail techniques for determining water and cement content of fresh concrete. final report. Technical Repoer. Army Construction Engineering Research Lab, Champaign, IL (USA)
10. Pleinert H, Sadouki H, Wittmann FH (1998) Determination of moisture distributions in porous building materials by neutron transmission analysis. Mater Struct 31(4):218–224
11. Nemec T, Rant J, Apih V, Glumac B (1999) Study of building materials impregnation processes by quasi-real-time neutron radiography. Nucl Instrum Methods Phys Res A 424(1):242–247
12. Josef P, Jan T, František P, Tomáš Š (1990) Description of transport of liquid in porous media—a study based on neutron radiography data. Int J Heat Mass Transfer. 33(6):1105–1120
13. Konstantinos K, Christopher H, Bentz Dale P, Nicola M, Wilson Moira A (2009) Relationship between engineering properties, mineralogy, and microstructure in cement-based hydroceramic materials cured at 200°–35 °C. J Am Ceram Soc 92(3):694–701
14. Degueldre C, Pleinert H, Maguire P et al (1996) Porosity and pathway determination in crystalline rock by positron emission tomography and neutron radiography. Earth Planet Sci Lett 140(1–4):213–225
15. De Beer FC, Middleton MF (2006) Neutron radiography imaging, porosity and permeability in porous rocks. South Afr J Geol 109(4):541–550
16. De Beer Frikkie C, Middleton Mike F, Jodie H (2004) Neutron radiography of porous rocks and iron ore. Appl Radiat Isot 61(4):487–495
17. Lopes RT, Bessa AP, Braz D, De Jesus EFO (1999) Neutron computerized tomography in compacted soil. Appl Radiat Isot 50(2):451–458
18. Kaestner AP, Trtik P, Zarebanadkouki M, Kazantsev D, Snehota M, Dobson KJ et al (2016) Recent developments in neutron imaging with applications for porous media research. Solid Earth 7:1281–1292. https://doi.org/10.5194/se-7-1281-2016
19. Kaestner A, Hassanein R, Vontobel P, Lehmann P, Schaap J, Lehmann E et al (2007) Mapping the 3D water dynamics in heterogeneous sands using thermal neutrons. Chem Eng J 130:79–85. https://doi.org/10.1016/j.cej.2006.06.013
20. Lehmann EH, Frei G, Kühne G, Boillat P (2007) The microsetup for neutron imaging: a major step forward to improve the spatial resolution. Nuclear Instrum Methods Phys Res A 576:389–396. https://doi.org/10.1016/j.nima.2007.03.017
21. Masschaele B, Dierick M, Cnudde V, Hoorebeke L, Delputte S, Gildemeister A et al (2004) High-speed thermal neutron tomography for the visualization of water repellents, consolidants and water uptake in sand and lime stones. Radiat Phys Chem 71:807–808. https://doi.org/10.1016/j.radphyschem.2004.04.102
22. Tötzke C, Manke I, Hilger A, Choinka G, Kardjilov N, Arlt T et al (2011) Large area high resolution neutron imaging detector for fuel cell research. J Power Sour 196:4631–4637. https://doi.org/10.1016/j.jpowsour.2011.01.049
23. Tremsin AS, McPhate JB, Vallerga JV, Siegmund OHW, Feller WB, Lehmann E et al (2011) Improved efficiency of high resolution thermal and cold neutron imaging. Nuclear Instrum Methods Phys Res A 628:415–418. https://doi.org/10.1016/j.nima.2010.07.014
24. Tudisco E, Etxegarai M, Hall SA, Charalampidou E-M, Couples GD, Lewis H et al (2019) Fast 4-d imaging of fluid flow in rock by highspeed neutron tomography. J Geophys Res Solid Earth 124:3557–3569. https://doi.org/10.1029/2018JB016522
25. Cordonnier B, Pluymakers A, Tengattini A, Marti S, Kaestner A, Fusseis F, Renard F (2019) Neutron imaging of cadmium sorption and transport in porous rocks. Front Earth Sci. https://doi.org/10.3389/feart.2019.00306

26. Nemec T, Rant J, Apih V, Glumac B (1999) Study of building materials impregnation processes by quasi-real-time neutron radiography. Nucl Instrum Methods Phys Res A 24:242–247
27. Janz M (2002) Moisture diffusivities evaluated at high moisture levels from a series of water absorption tests. Mater Struct 35:141–148
28. El Abd A, Kichanov SE, Taman M, Nazarov KM, Kozlenko DP, Badawy WM (2019) Determination of moisture distributions in porous building bricks by neutron radiography. Appl Radiat Isot. https://doi.org/10.1016/j.apradiso.2019.108970
29. Dorjan D, Alessandro T, Stefano DP, Nikolajs T, Matthieu B, Benedikt W (2018) Analysis of moisture migration in concrete at high temperature through in-situ neutron tomography. Cem Concr Res 111:41–55
30. Roshankhah S, Marshall JP, Tengattini A et al (2018) Neutron imaging: a new possibility for laboratory observation of hydraulic fractures in shale? Géotech Lett 8(4):316–323
31. Yehya M, Edward A (2018) Fluid flow measurements in low permeability media with high pressure gradients using neutron imaging: application to concrete. NIM-A 890:35–42
32. Waring RH, Roberts JM (1979) Estimating water flux through stems of scots pine with tritiated water and phosphorus-32. J Exp Botany 30(3):459–471. https://doi.org/10.1093/jxb/30.3.459
33. Varney GT, Canny MJ (1993) Rates of water uptake into the mature root system of maize plants. New Phytol 123:775–786
34. Warner GS, Nieber JL, Moore ID, Geise RA (1989) Characterizing macropores in soil by computed tomography. Soil Sci Soc Am J 53:653–660
35. Tracy SR, Black CR, Roberts JA, Sturrock C, Mairhofer S, Craigon J, Mooney SJ (2012) Quantifying the impact of soil compaction on root system architecture in tomato (Solanum lycopersicum) by X-ray micro-computed tomography. Ann Bot 110:511–519
36. Garbout A, Munkholm LJ, Hansen SB (2013) Temporal dynamics for soil aggregates determined using X-ray CT scanning. Geoderma 204–205:15–22
37. Meyer M, Buchmann C, Schaumann GE (2018) Determination of quantitative pore-size distribution of soils with 1H NMR relaxometry. Eur J Soil Sci 69:393–406
38. Tumlinson LG, Liu H, Silk WK, Hopmans JW (2008) Thermal neutron computed tomography of soil water and plant roots. Soil Sci Soc Am J 72:1234–1242
39. Hammouda B (2021) Practical neutron applications. https://www.ncnr.nist.gov/staff/hammouda/practical_neutron_applications/practical_neutron_applications.pdf. Accessed date: 20/05/2021
40. Moradi AB, Oswald SE, Menon M, Carminati A, Lehmann E, Hopmans JW (2013) Applications of neutron imaging in soil–water–root systems, (Chap. 6). In: Anderson SH, Hopmans JW (eds) Soil–water–root processes: advances in tomography and imaging, vol 61, 10 June 2013. https://doi.org/10.2136/sssaspecpub61.c6
41. Ogilvie CM, Ashiq W, Vasava HB, Biswas A (2021) Quantifying root-soil interactions in cover crop systems: a review. Agriculture 11:218. https://doi.org/10.3390/agriculture11030218
42. Kardjilov N, Manke I, Woracek R, Hilger A, Banhart J (2018) Advances in neutron imaging. Mater Today 21(6):652–672
43. Matsushima U, Herppich WB, Kardjilov N, Graf W, Hilger A, Manke I (2009) Estimation of water flow velocity in small plants using cold neutron imaging with D2O tracer. Nucl Instrum Methods Phys Res Sect A Accel Spectrometers Detectors Assoc Equip 605(1–2):146–149. ISSN 0168-9002. https://doi.org/10.1016/j.nima.2009.01.187
44. Nakanishi TM (2017) Research with radiation and radioisotopes to better understand plant physiology and agricultural consequences of radioactive contamination from the Fukushima Daiichi nuclear accident. J Radioanal Nucl Chem 311:947–971
45. Warren JM, Bilheux H, Kang M et al (2013) Neutron imaging reveals internal plant water dynamics. Plant Soil 366:683–693. https://doi.org/10.1007/s11104-012-1579-7
46. Tötzke C, Kardjilov N, Manke I, Oswald SE (2017) Capturing 3D water flow in rooted soil by ultra-fast neutron tomography. Sci Rep 7:6192. https://doi.org/10.1038/s41598-017-06046-w
47. Tötzke C, Kardjilov N, Hilger A, Mohr NR, Manke I, Oswald SE (2021) Three dimensional in vivo analysis of water uptake and translocation in maize roots by fast neutron tomography. Sci Rep 11:10578. https://doi.org/10.1038/s41598-021-90062-4

48. Defraeye T, Dereome D et al (2014) Quantitative neutron imaging of water distribution, venation network and sap flow in leaves. Planta 240 (423–236) (2014)
49. Van As H, van Duynhoven J (2013) MRI of plants and foods. J Magn Reson 229:25–34
50. Léonard A, Blacher S, Nimmol C, Devahastin S (2008) Effect of far-infrared radiation assisted drying on microstructure of banana slices: An illustrative use of X-ray microtomography in microstructural evaluation of a food product. J Food Eng 85:154–162
51. Aregawi W, Defraeye T, Saneinejad S, Vontobel P, Lehmann E, Carmeliet J, Derome D, Verboven P, Nicolai B (2013) Dehydration of apple tissue: Intercomparison of neutron tomography with numerical modelling. Int J Heat Mass Transf 67:173–182
52. Defraeye T, Aregawi W, Saneinejad S, Vontobel P, Lehmann E, Carmeliet J, Verboven P, Derome D, Nicolai B (2012) Novel application of neutron radiography to forced convective drying of fruit tissue. Food Bioprocess Technol 6:3353–3367
53. Defraeye T, Nicolai B et al (2016) Probing inside fruit slices during convective drying by quantitative neutron imaging. J Food Eng 178:198–202
54. Tanoi K, Hamada Y, Seyama S, Saito T, Likura H, Nakanishi TM (2009) Nucl Instrum Methods Phys Res A 605:179
55. Shukla M, Roy T, Kashyap Y, Shukla S, Singh P, Baribaddala TP, Gadkari SC (2018) Development of neutron imaging beamline for NDT applications at Dhruva India. Nucl Instrum Methods A 889:63–68
56. Scussat S, Ott F, Hélary A, Desert S, Cayot P, Loupiac C (2016) Food Biophys 11:207
57. Cleveland TE IV, Hussey DS, Chen Z-Y, Jacobson DL, Brown RL, CarterWientjes C, Cleveland TE, Arif M (2008) J Cereal Sci 48:517

Chapter 12
Fast Neutron Imaging Techniques

Saroj Bishnoi

12.1 Introduction

Imaging with neutron or photon is a widely used inspection tool in nondestructive evaluation. It has been applied and adopted in the field of research, medical, industrial and nuclear inspections [1] and many more application. Depending on the types of radiography sources, different sorts of knowledge about structure of the inspected sample can be acquired from their radiographs. In general, X-ray sources are in use for visualizing elements of mid-high atomic number like Al, Fe, mild steel, stainless steel, etc. encapsulated by materials of low atomic number (such as graphite, water, plastic). On the other hand, thermal neutron imaging is used for imaging low atomic numbered elements outlined with high Z materials. High energy gamma and epithermal neutron-based other forms of imaging is mainly required for greater penetrability in bulk samples, while inspecting similar types of materials as discussed above. Imaging using thermal (or epithermal or cold neutron) neutrons has been widely explored and successfully applied for numerous scientific, industrial, medical and technical pursuits [1, 2] as discussed in Chaps. 1 and 2. However, the use of thermal neutron is basically limited by their penetrability in thick objects. For instance, thermal/epithermal neutron imaging is not effective in imaging of plastics, lubricants materials and chemical explosives particularly if these are shielded with thick layers of light material (high-density polyethylene, boron etc.) or in case when these are masked with some specific dense and high Z materials (Cd, Lead, Gd). This is because thermals neutrons are highly attenuated or absorbed by such complex configuration. For imaging of such object configurations, fast neutrons with neutron energy of ~1 MeV or higher, play significant role. That is due to their strong penetration capability through high Z materials and much stronger interaction (higher attenuation) in low Z materials [3, 4] as compared to X-rays. Thus,

S. Bishnoi (✉)
Technical Physics Division, Bhabha Atomic Research Center, Mumbai 400085, India
e-mail: saroj@barc.gov.in

D. K. Aswal et al. (eds.), *Neutron Imaging*,
https://doi.org/10.1007/978-981-16-6273-7_12

imaging with, high energy neutrons (i.e., fission neutron ~1 MeV, 14 MeV D-T neutrons, 2.45 MeV D-D neutrons etc.), could show both types of objects clearly. This makes them unique nondestructive imaging probe for inspection of large-scale objects particularly composed of light and heavy elements.

Over the last few decades, FNI has emerged as one of the most promising technique for various applications and has opened up wide range of possibilities for nondestructive testing (NDT). During 1990s, Hall et al. [5] and scientists at Lawrence Livermore and Los Alamos National laboratories (LLNL) have applied FNI on very large objects of nuclear warhead components and demonstrated related developments towards stockpile stewardship. Later on, proof-of-concept experimental studies performed at the Los Alamos Neutron Science Center (LANSCE). LANSCE was a facility that could produce neutrons of very high energy up to 600 MeV, much higher than those used by industries at that time. Encouraged by the successful initial results, Hall et al. [6] further modeled FNI using relatively lower-energy (<15 MeV) neutrons with the logic that these lower-energy neutrons can effectively penetrate high Z materials and can also strongly interact with low Z materials. Additionally, generation of these lower-energy neutrons cost much less. On the top of that, these lower-energy neutrons are easier to detect with direct use of plastic scintillator type detectors.

With this, FNI has started its journey and progressed over the years and have achieved various milestones in security (inspection for illicit materials such as weapons, explosives, drugs, etc.), flow of gas/liquid or their mixing, radiography or tomography of low Z compound materials masked with heavy shielding. Even though, FNI results were found quite promising for a range of applications, initially it had slow growth and limited use mainly because of two reasons; first of them was unavailability of high intensity fast neutron sources and second was lack of fast neutron sensitive imaging/detection systems. Detection of fast neutron itself is a challenging task. Lack of efficient fast neutron detector offers relatively poor spatial resolution and also results usually in long exposures to obtain reasonable quality image.

Extensive research and development over the years has provided some options to both of these obstacles to a great extent. Chapters 2 and 4 have provided relevant information about electrostatic accelerators-driven neutron sources of high intensity neutron beam and fusion reaction-based neutron generator (NG) capable of producing 10^{11}–10^{13} of 14 MeV/2.45 MeV neutrons per second [7, 8] and compact neutron sources such as sealed tube and plasma-focus type NG, capable of producing over 10^8- 10^{11} n/sec or higher in continuous or in pulsed mode operation [9, 10]. FNI has been demonstrated to be of immense help for national security towards detection of heavily shielded explosives or contraband [11–13]. Researchers from Japan have explored the fast neutron imaging application for qualification of aging-based deterioration of the large-scale concrete structures [14, 15] successfully. The deteriorated social infrastructures (such as tower building, bridges, roads, tunnels etc.,), mostly made of concrete and iron/steel, are subjects of concern worldwide, due to their aging over a long-time span. It has also been used as a probing tool complementing conventional X-ray techniques for imaging of thick cultural heritage objects [16] or for industrial applications like looking into glued wooden boards [17]. In the same direction, fast neutron radiography/imaging study was also performed at radio frequency quadrupole-based NECSA accelerator facility [18], towards preservation of rare samples of cultural heritage, anthropology and paleontology.

Other than the accelerator-driven fast neutrons, there has been renewed interest in the use of reactor source for imaging with high energy neutron. NECTOR is a dedicated imaging beam line developed at FRM II research reactor [19] to utilize the fast neutron components of reactor neutron spectrum. The imaging experiments performed using fast neutron imaging beam line of Budapest Neutron Center [20, 21] developed at 10 MW research reactor, and the fast neutron imaging at YAYOI reactor (JAPAN), have shown reactor as potential source for fast neutron imaging and have been successfully used for industrial applications investigating two-phase flows in nuclear fuel bundles [22–25], under potentially harsh environment.

The increased interest in FNI prompted development of efficient fast neutron imaging detectors. A highly sensitive photo-luminescent novel imaging plates (Chap. 3) in combination with converter screens of threshold activation detectors in transfer imaging technique was suggested and used by various researchers such as Rant, Mikerov and co-workers [26–28]. In this direction, Seki et al. [14] have developed a fast neutron scintillator detector of pixel type with semiconductor photon sensor for transmission imaging of bulk and thick samples. It could identify a metallic sample of steel bar, a void substituting hole feature as well as water in 30-cm-thick concrete blocks via FNI performed at RIKEN, accelerator-driven compact neutron source (RANS) [29]. Makowska et al. [30] have been provided a detail overview of fast neutron imaging detectors and scintillator developments, along with detail performance of commercially available scintillator screen polypropylene (PP)/ZnS type produced by RC Tritec AG [31]. On the other side, such promising applications have opened the area of research in developing very high intensity sources of high energy neutron suitable for imaging. A high intensity neutron source of neutron strength 10^{10}–10^{11} n/s/sr developed at Lawrence Livermore National Laboratory [32], with energies (quasi-monoenergetic) of 7 MeV or 10 MeV suitable for FNI, and has provided fast neutron images of sub mm spatial resolution with good contrast in low Z samples enclosed with thick layer of high Z with an exposure time of 10 -15 min. Capabilities of the FNI, other than fast neutron radiography (FNR), that has attracted attention towards national security are fast neutron resonance radiograph (FNRR) and associated particle imaging (API). FNRR is a variant of FNR that utilizes the characteristic resonance structure of the total scattering cross-section exhibited by the different elements for neutrons in the energy range of mainly 1–10 MeV. The measured transmitted neutron spectra (or intensity) show the variations as per the resonance cross section of the elements present in the inspected sample. Evaluation and analysis of these transmitted neutron spectra reveal the identity of the low Z elements (C, O, N) present in the investigated bulk and determines/image their density distribution. Low Z elements: particularly carbon, oxygen, nitrogen and hydrogen, are of interest because these are primary composition of potential explosives such as RDX and TNT. Explosives can be identified with high fractions of nitrogen and oxygen and relatively low fractions of carbon and hydrogen compared to that in benign materials [13]. Due to the specific resonance's, the spatial distribution of these elements in the image of inspected object (e.g., in air cargo containers or luggage) can be enhanced by FNRR technique. Basic prerequisite for the implementation of FNRR is to have two-dimensional neutron transmission images at selected energies corresponding to the resonance cross-sectional structure of different elements. This can be achieved either in a sequence of several images with quasi-monoenergetic neutron beams [33–35] or, alternatively, with a pulsed broad energy neutron beam and the application of time-of-flight (TOF) methods for energy selection [36].

The high cost, maintenance and complex design of a nano-second pulsed accelerator-based system required for FNRR are one of the major drawbacks, which hinders the implementation of FNRR for field applications. In the area of fast neutron, interaction-based detection-driven imaging technique named as associated particle imaging (API) has shown extremely promising result for explosive detection [37, 38] with the use of a small size compact or sealed D-T NG. API is a 3D imaging method, where no nano-second pulse or collimated neutron beam is needed. Rather than characteristic cross-sectional structures of C, N or O, it utilizes the 14 MeV neutron-induced characteristic prompt gamma rays as elemental signatures for their identification and also material composition of the bulk. It is based on kinematics of D-T fusion reaction, where ^{3}H (^{2}H, n)^{4}He fusion reaction produces a neutron of energy 14 MeV and an alpha particle of 3.45 MeV energy, almost back-to-back. Neutron emission time and its travel direction can be determined by detecting associated alpha particle with pixelated particle detector, and neutron time of flight is determined by the alpha–gamma coincidence time spectrum. The information of neutron direction and time of flight allows localization of specific volume inside the inspected bulk, say, large container from where the neutron-induced prompt gamma ray spectrum is recorded. The characteristic gamma energies from the recorded energy spectrum provide the information of the elements present in the inspected volume, and their intensities provides the amount of the respective elements. So, the relative proportions of the carbon, oxygen and nitrogen determined, and it can differentiate material of illicit nature such as explosives or narcotics from the innocuous materials such as fertilizer. This method is hybrid imaging type in the sense that detection of prompt gamma rays provides location-specific information about the presence of elements in the bulk through TOF technique.

First part of this chapter provides an introduction on basics of fast neutron imaging, its major components including neutron source, collimator, detection system, shielding and their associated technical challenges followed by an overview of the FNI systems/beamlines which either has been developed or under development. Second part of the chapter is devoted to FNRR and API, two unique imaging modalities for explosive or illicit material detection application, their basic methodologies, key components and an overview of the existing systems reflecting the progress so far.

12.2 Basic Principle of the Transmission-Based Fast Neutron Radiography Imaging

The basic principle of the transmission-based fast neutron radiography [39] is similar to the conventional thermal neutron radiography (Please refer to Chaps. 1 and 5 for more details).

It is a 2D projection image of a sample under inspection, obtained on transmission of high energy neutrons through sample, according to the neutron interaction cross section. The 2D transmission image of the sample can provide the first-hand information's at macroscopic scale about internal structure of the sample. For example, for a sample of thickness t and linear attenuation coefficient, say μ (cm^{-1}) the transmitted neutron intensity (I) recorded on passing through a sample can be expressed in mathematical form (Beer–Lambert law)

$$I = I_0 \mathrm{e}^{-\mu t} \tag{12.1}$$

where I_0 is the incident neutron intensity recorded by the detector in absence of sample. Now, let us consider a configuration, where in this same sample, there is a structure made of different material of thickness $t1$ and linear attenuation coefficient μ' (cm^{-1}), then the recorded transmitted intensity by the detector will be (Eq. 12.2).

$$I' = I_0 \mathrm{e}^{-\mu(t-t1)+\mu' t1} \tag{12.2}$$

The transmitted neutron intensity I' recorded by the detector will show a different pattern at a position in line with the location of inhomogeneity inside the sample with respect to the rest of the image.

For practical realization of FNR system, major components are different in their nature than those used for thermal radiography. Saroj et al. [40] have demonstrated the methodology of FNR via modeling and its potential applications using 14 MeV neutrons for broad range of material from low Z, metallic and to high Z materials. It has also explored the capability of FNR towards imaging of low Z sample encapsulated inside a thick layer of the high Z (say, lead) material successfully. However, to develop a fast neutron imaging system and to image large sample or a complex sample containing low and high Z material with good contrast, various practical factors like fast neutron source intensity, source spot size, background of high energy scattered neutron and efficient fast neutron detection system play important roles, as discussed in the next part.

12.3 Major Components of FNR and Challenges

Major components of the FNR system are a high intensity neutron source, a collimator to guide neutrons from source to inspected object, and a detection/imaging system placed after object. Despite many important applications of fast neutron imaging, it is still not a commonly used nondestructive evaluation method compared to thermal neutron imaging which is far matured technique and well adopted across the world. Reason being various issues related to the FNR is still unresolved and is under research. Requirement of high efficiency fast neutron detector, high intensity fast energy neutron source and the issue of the high contribution of scattered neutron makes fast neutron imaging more difficult and challenging as discussed below.

12.3.1 High Intensity High Energy Neutron Sources

Similar to thermal neutron radiography (TNR), basic requirement for fast neutron imaging is high intensity neutron source with a very small neutron emitting source point. Higher the source strength, better the collimation and better is the performance resulting in lower acquisition time and high image quality. Different types of neutron sources existing are radio isotopic neutron emitters, accelerator-based and reactors; each of them produces neutrons via different mechanism as discussed in the Chap. 2. In principle, any of these neutron source producing high energy ~MeV neutron can be used for imaging; however, each of them has their own pros and cons towards imaging of thick and large objects as described here.

12.3.1.1 Radio Isotopic Neutron Source

Examples of radioisotope-based neutron sources are ^{252}Cf, Pu-Be, Am-Be [39, 41]. Chapter 2 has detailed information about such sources. These types of neutron sources generally have relatively low intensity and constant decreasing flux due to an effective half-life (e.g., ^{152}Cf has life 2.645 years [39]). Even though, radioisotope-based sources are easier to operate and portable, the neutron flux achievable with such sources falls short of the requirements for FNR applications. Secondly, their nature of continuously emitting neutrons put a basic requirement of proper storage and a permanent shielding structure.

12.3.1.2 Particle Accelerator or Neutron Generators

Accelerator-based neutron source generates fast neutrons utilizing different nuclear reactions such as ^{3}H (D, n), ^{2}H (d, n), ^{2}H (p, n), ^{15} N (p, n) and ^{15} N (d, n) etc., which could provide the neutron beam of white and monoenergetic neutron energy of range 0.5–24 MeV as discussed in Chap. 2. Particle accelerator-based neutron sources have relatively low gamma energy contents in the neutron beam compared to reactor sources. Neutron generators based on D-D and D-T fusion reactions are most common accelerators-based neutron sources and have been known as a primary choice of source for FNR. D-D NG produces monoenergetic neutrons of typical ~2.5 MeV, and typical yield is in the order of 10^8–10^{10} n/sec, whereas D-T NG can provide monoenergetic neutrons of 14.1 MeV with yield ~10^9–10^{12} n-s^{-1}. This is comparatively lower with respect to that of spallation or reactor neutron source, where neutron fluxes of 10^8 n s^{-1} per cm^{-2} and higher are available for imaging (please refer to Chap. 4 for more details). However, neutron generators are preferred for field applications and compact FNR system development, due to their small size and electronic switching control and need no shielding in idle condition. In case of D-D reaction, neutron energy is nearly monoenergetic to a given emission angle, and it is peaked slightly in forward direction. It is like beam is naturally collimated due to beam kinematics, and it is advantageous to perform FNR without collimator, while in D-T reaction, neutrons emitted isotropically, in 4 pi direction. So, around the source, it is possible to make several radiographic stations but on the other hand, it gives rise to additional background. Well-designed effective shielding around the detector system becomes necessary to reduce the background and to protect the detector and associated electronics from damage due to radiations. The shielding requirement is of more importance in the case of (D-T) compared to that of D-D neutrons because of the higher energy and 2 orders more yield of the former. Also, use of tritium target in lab-based D-T sources is a practical challenge because tritium is a radioactive element, and special safety precautions must be adhered to while working with D-T NG. In this direction, there are sealed type D-T neutron generators available commercially and are safer to handle.

12.3.1.3 Reactor-Based Neutron Sources for FNI

Research reactors produce fast neutrons from fission reactions at their cores. Nuclear fissions are highly probable at thermal neutron energies and the neutrons emitted follow the Watt energy distribution, with peak neutron energy at ~2 MeV and average neutron energy at ~0.7 MeV. A neutron beam from a reactor core is not monoenergetic but contains a distribution of thermal, epithermal, and fast neutrons. From the mixed beam, it is possible to filter thermal neutrons with the use of high thermal neutron absorption cross-sectional material such as Gd, Cd, borated or Lithiated material. Gamma rays produced as a result of fission reactions may be filtered out of the beam with the use of gamma-absorbing materials in the beam collimator such as bismuth or lead. Higher neutron flux, better beam collimation with a relatively large L/D and optimized image acquisition time, prevents image blur and provides better image quality. However, large and effective shielding to highly penetrating fast neutrons to achieve safe and acceptable dose rates around the beam facility poses a significant challenge. Despite these hurdles, scientists across the world could develop reactor based FNI beam lines successfully. NECTOR- a beam line for fast neutron imaging at

FRM II research reactor [19], fast neutron radiography testing setup at YAYOI reactor [20], fast neutron radiography/tomography beamline at 10 MW research reactor of the Budapest Neutron Centre [21] and beam line at Ohio State University Research Reactor's (OSURR) [42] are some examples of reactor based FNI facilities. These facilities are discussed more in detail under Sects. 12.4.

12.3.2 Collimator for Fast Neutron

Collimator is one of the critical components of the neutron imaging system as it defines the beam characteristics and hence directly affects the beam quality (please refer to Chap. 2 for more details). The collimator ratio (L/D) determines the image quality of fast neutron imaging in the same manner as in the case with thermal neutron, where D is the collimator input aperture diameter and L is the collimator length. In case of fast neutrons, their collimation is not straight forward as in the case of thermal neutrons. That is because of lack of efficient fast neutron absorbing materials, which are required to construct walls and aperture of collimator. Realization of imaging system with D-D/D-T neutron generator with collimator having large L/D is not feasible due to their limited neutron source strength. Additionally, neutron scattering may arise from collimator material and can reach at image detector (discussed in the next Sect. 12.3.4). Various collimator designs have been proposed towards optimized utilization. Fantidis et al. [43] have modeled a collimator design for D-D and D-T neutron with L/D ratio ~100 to 300 having source neutron fraction as high as 97–98% and minimum scattered neutron contribution at detection system. Imaging using collimators of such large L/D ratio will definitely provide images of high quality, however with commercially available D-T/D-D NG practically difficult to be realized due to their limited neutron yield ~10^8–10^{11} ns^{-1}. Bishnoi et al. [44] have reported fast neutron imaging results using very small size collimator (L/D = 43) made of iron and borated high-density polyethylene coupled to a D-T NG. Sabo-Napadensky et al. [45] have developed a collimator for a D-T NG based fast neutron tomography setup. It uses massive shield with mixture of borax ($Na_2B_4.10H_2O$) and water as moderator material. The collimator efficiently optimizes scattered neutron in the primary beam that arrives at the imaging plane. Another approach in place of collimator for FNR with NG is to reduce the neutron source size as small as possible (less than ~5 mm), even though it is the challenging task as it reduces the total neutron yield of the NG. However, advanced technological development in the field of sealed or compact neutron generator has made it possible to develop NG with very small neutron source size ~2 to 5 mm. Cremer et al. [46] have used D-D NG of yield 10^9 n/s having source size ~3 to 5 mm to image hydrogen bearing materials. Adam et al. [47] have developed a D-D neutron generator, emitting fast neutrons from source size of 2.2 mm, sufficient to produce radiograph/tomograph with resolution of ~1.5 mm. Recently, Williams [48] reported high resolution fast neutron imaging using a D-T neutron generator of portable size emitting neutron of yield ~3×10^9 n/s and 1.5 mm source size. In case of reactor-based neutron sources, FNR or fast neutron tomography (FNT) beam lines can be designed with a relatively large L/D and a small source size due to high flux availability. However, high background of gamma rays and thermal neutron contributions as well as scattered fast neutrons in the experimental area are to be taken care while designing imaging system for such cases. Design of the beam with use of optimized beam collimators, beam shutters, beam stopper, thermal neutron and gamma filters can reduce/minimize the contribution of the gamma, thermal and scattered high energy neutron.

12.3.3 Efficient Fast Neutron Detection System

Suitable imaging detectors for fast neutrons with high efficiency play major role in proving potential and effectiveness of the FNR/FNT technique for various applications. One of the limiting factors in FNI is low detection efficiency on account of high penetrating power of fast neutrons and very low fast neutron interaction cross section. This imposes long exposure and acquisition time to obtain good image quality in FNI; therefore, efficient fast neutron imaging detector development becomes mandatory to achieve high-quality fast neutron images. Modern neutron imaging detectors are mostly based on digital camera systems with high sensitivity for optical light from a scintillation converter. Therefore, the scintillator is the key component for the detection process, defining the efficiency and the spatial resolution and both of these need optimizations. Since nearly all nuclear processes are accompanied by gamma radiation emission, a scintillator with low gamma sensitivity and more sensitive to fast neutron is a necessity for FNR. Chapter 3 has discussed in detail on various type of fast neutron detection techniques and the different types of the fast neutron imaging detector systems. Some of them are mentioned here. Makowska et al. [49] have presented an overview of scintillator detector developments for fast neutron imaging applications. To perform imaging with high energy neutron in 10–15 MeV range, Dietrich et al. [50] used thick (4 cm) plastic scintillator (BC-408), which showed detection efficiency of ~30% for 14 MeV neutrons. Such thick organic scintillator screens are good for higher detection efficiency for high energy neutrons; however, a thicker screen has an adverse effect on image spatial resolution, and plastic scintillators also have high gamma sensitivity. To improve the overall spatial resolution, new type of detectors, consisting of polypropylene with some amount of silver or copper-activated ZnS, developed were found as more suitable material for efficient measurement of fast neutrons. Polypropylene (PP)/ZnS type scintillator screens have shown better performance compared to that of thick plastic thick slab and are commercially available [produced by RC Tritec AG (Tritec 2017)] [51, 52].

12.3.4 *Scattered Neutrons and Shielding of the Imaging System*

Quantitative analysis on the effects of the scattered neutrons in the fast neutron image formation are quite difficult [53, 54]. Scattered neutron is one of the major concerns to be taken care while designing any FNI system. Other than the scattered neutrons, fast neutron sources have inevitable presence of gamma radiation. Scattering of neutrons could be from surrounding structure or from thick sample itself arising via multiple scattering within the sample. For example, accelerator-based fast neutron radiography system developed at laboratory contains various apparatus may act as source of scattered neutrons. These include accelerator tube, various vacuum pumps, gauges, HV dome, target used for generating the fast neutrons, walls and ceiling of laboratory room which are in general composed of concrete bricks of low or high density. Role and amount of the scattered neutron contribution in image formation via fast neutron-based system is of complex nature and difficult to exclude completely or evaluate experimentally. One possible way to reduce the scattered neutron contribution is by using effective collimator design to shield the detection system from neutrons coming from random directions and allowing neutron from desired direction. However, there may be possibility that the collimator might introduce scattered neutrons into the imaging detection system from the collimator wall materials. On the other side, the large and bulky collimators require large foot print and might reduce the possibility of the portable and compact system design. Combined layers of shielding materials such as hydrogenous material, iron, cadmium and lead need to be put around the detection system to reduce neutron as well as gamma radiation, which on the other hand could introduces scattering effects at the image plane. Hence, an optimized shielding must be designed.

12.4 Review of FNI Experimental Setup/Facilities

Experimental facilities/systems of fast neutron imaging across the world employ different types neutron sources, neutron detectors, beam collimation assemblies and different sample configurations for performance/characterization of the system. Some of the system and facilities are reviewed below.

12.4.1 *FNR Using Accelerator-Driven Source*

FNR using D-D/D-T neutron source

The idea of using fast neutrons for imaging was explored in early seventies with experiments performed towards evaluation its usefulness in nondestructive examination of bulk objects [4, 5]. Proof-of-concept experimental studies carried out at LANSCE facility; LANL has demonstrated the fast neutron radiography as a viable technique for imaging of much thicker samples for applications such as stockpile stewardship in 1997. Frank Dietrich had suggested [7] a conceptual design of fast neutron imaging system with 10–15 MeV energy range neutrons for thick sample imaging and performed experiments in the above-mentioned area. For such applications, neutron source requirement estimated was order of ~10^{12} n/s/sr in the sample direction with neutron source spot size ~1 mm. Further, to minimize the scattered neutron background due to multiple neutrons scattering within the object itself, samples were placed at halfway of the distance between source and imaging plane achieving 2:1 magnification.

Cremer et al. [46] followed the approach established by James Hall and used a small laboratory-based D-D neutron generator of ~10^9 ns^{-1} neutron yield with a large area plastic scintillator detector (thick slab of 5 mm) to image hydrogen bearing materials. Good quality images were successfully acquired by placing objects at a distance of around 1.5–3.2 m far from neutron source and close to the detection system. With exposures time of 5–60 min, overall image resolution ~2.5 mm was determined by a knife-edge measurement. Result emphasized that combination of D-D neutron generator, electronic CCD camera with moderate thickness plastic scintillator can be a suitable choice for development of a low-cost small size portable imaging system to image hydrogen-rich materials embedded in high-density matrix materials. A half-filled wine bottle distinctly showing the liquid level and a concealed cell phone in a purse imaged with this system are shown in Fig. 12.1. In summary, Cremer et al. concluded from its knife edge and imaging experiments that the imaging resolution parameter mainly dominated by neutron scatter than by the neutron divergence angles.

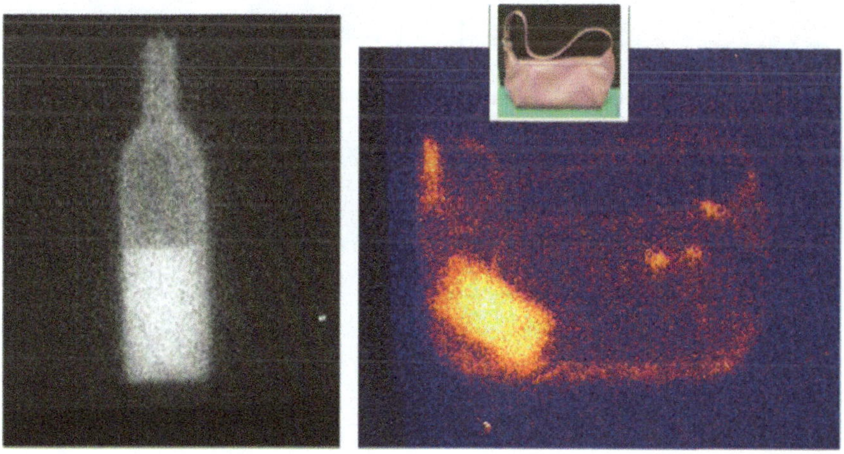

Fig. 12.1 (left) Fast 2.5 MeV neutron radiographic image of half-filled wine bottle acquired in 5.5 min exposure (right) Fast neutron image of a purse with concealed cell phone acquired in 5.5 min exposure (Reproduced with permission from [46])

In the same line of Hall and Cremer, Saroj et al. [44] have developed a small size experimental FNR setup with D-T neutron generator. Saroj et al. have used a neutron source of strength ~2×10^9 n/s emitting from a target of 20 mm size. As source was located inside a small laboratory, so a high contribution of the scattered neutrons was a major issue. To reduce scattered neutron contribution, a metallic collimator having length of 530 mm, consisting of two parts, was used. One part was placed around the neutron target with 10 mm hole at one side and the second part allows neutron to travel towards the sample resulting in fast neutron flux of ~2.1×10^4 n cm^{-2} s^{-1}, at sample position with effective L/D ~43. Fast neutrons were imaged using a 4-cm-thick plastic scintillator and electron multiplying charge-coupled device (EMCCD) camera. Various test objects of low, medium and high Z materials (polyethylene, mild steel and lead) were fabricated in different configurations like step wedge, hole features in step wedge and a polyethylene sample masked with lead. Photographs of the experimental facility and images of HDPe masked with different thicknesses of lead are shown in Fig. 12.2. In images the 5 mm hole and interface region of polyethylene-lead are clearly visible with good contrast in both cases of with and without lead block. With increased thickness of the lead ~100 mm, the contrast of the hole gets reduced and is barely visible; however, polyethylene material presence is

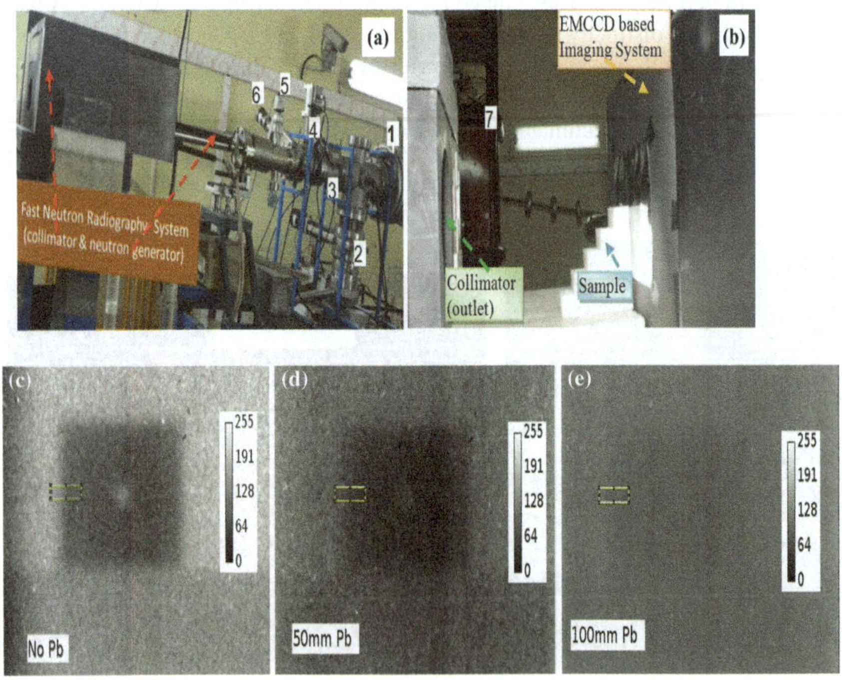

Fig. 12.2 **a, b** D-T NG-based FNR system with iron collimator and images of 5 **mm** hole in HDPE masked with lead of different thickness, **c** no lead, **d** 50 mm, **e** 100 mm (Reproduced with permission from [44])

still observed. Reasonable contrast images obtained showed the imaging capability and limitations of the FNR system.

To overcome the challenges of fast neutron detection efficiency and to have FNR images with improved spatial resolution, development of Zinc sulfide-based thin scintillation screen has been reported by Tang et al. [55] towards proving their worth. Yoshii and Miya K (https://doi.org/10.1016/0168-9002(94)90711-0) through their work on luminescence in fast neutron converters have shown that among different scintillators explored for fast neutron imaging, zinc sulfide-based thin scintillator screen has proven to be of highest quantum efficiency. These screens are developed with silver (PP/ZnS:Ag) or copper (PP/ZnS:Cu)-activated zinc sulfide particles suspended in a rich hydrogenous material. In this type of detector, the hydrogenous material act as a source of recoil protons and also as a light guide for the scintillations. Makowska [51] has performed experiments at the NECTAR facility using silver and copper-activated zinc scintillator screens which are commercially available in different thicknesses. A comparison study of various parameters like light output for fast neutron, gamma sensitivity and spatial resolution for both type of screen was carried out. The result showed that light output is almost same in both compositions but sensitivity for gamma is observed significantly higher in PP/ZnS:Cu compared to that in PP/ZnS:Ag.

Lehmann et al. [56], in collaboration with Chalmers University of Technology, Göteborg, Sweden, have reported FNR with 14 MeV produced from a D-T neutron generator. Fast neutron images were acquired by a system consisting of a scintillator screen (polypropylene, together with certain amount of ZnS), a mirror inside a light proof box and a highly sensitive CCD camera, which was cooled by liquid nitrogen to suppress the camera dark current. Attenuation behavior studied on a steel step-wedge test sample and the calculated cross section was obtained much lower than that expected/given in the data libraries of total cross section. The reason was attributed to scattered neutrons detected at the imaging screen.

Fast neutron radiography using RFQ-based accelerator system

Fast neutron radiography work was initiated at Necsa, a radio frequency quadrupole (RFQ) accelerator facility, in South Africa [18], with the aim of studying cultural heritage, anthropology and paleontology related samples. Imaging of such sample becomes important particularly for preservation of rare samples which symbolizes the ancient culture or civilizations. Necsa produces white neutron beam via accelerating 4–5 MeV deuteron beam (~50 mA) to a solid B_4C target. In order to have longer neutron travel path and improved beam focus, a high energy beam transport (HEBT) section has been developed and implemented with the aim of achieving neutron flux of ~3×10^7 s^{-1} cm^{-2} neutrons at a far distance of 4 m. Fast neutron images of a skull sample obtained using Necsa RFQ facility and an imaging system made of neutron sensitive bicron fiber scintillator viewed by a CCD camera (512×512 with pixel size of 16 μm) is shown in Fig. 12.3.

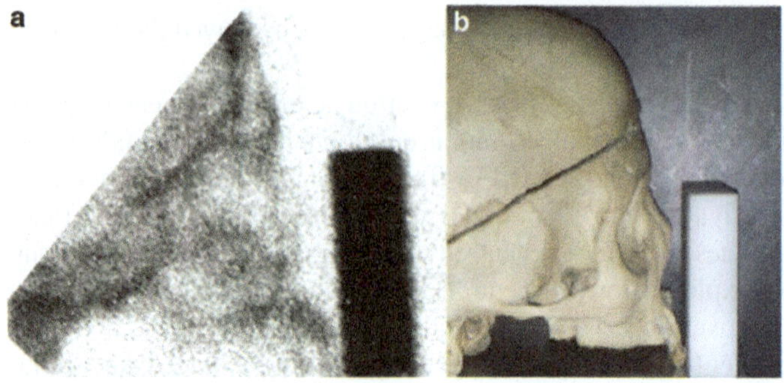

Fig. 12.3 Radiograph of skull sample imaged at the RFQ facility, using a fast neutron imaging system with a 512 × 512-pixel CCD camera and a neutron sensitive bicron fiber scintillator. 4 MeV D + beam irradiating a B4C target was used in acquiring the neutron transmission image with acquisition time ~5 min (Reproduced with permission from [18])

Fast Neutron Tomography (FNCT) at PIAF accelerator facility

Babai [57] has performed fast neutron tomography (FNT) of bulk made of low Z materials masked by thick layers of high Z materials showing its superior imaging capability, compared to high energy X-rays. To demonstrate this, specific phantoms were designed, made up of various plastic objects surrounded by different thickness of lead cylinders and used in experiment. FNT experiments were planned and carried out using PIAF accelerator facility of Physicalisch-Technische Bundesanstalt (PTB), Germany [58, 59]. To produce high intensity high energy neutron beam, PIAF accelerator accelerates deuterons ion beam to 11.5 MeV and strike it on a 5-mm-thick Be target. The produced neutron beam has broad energy spectra with ~5.5 MeV average neutron energy. The focused beam spot size was of 3–5 mm in diameter, and for tomography/radiography, the emitted broad energy neutron beam was collimated either in cone beam or in fan beam geometry using collimators. The collimators were made of composite B_4C loaded polyethylene and steel at 2-m distance from the target the neutron flux was ~10^6 n-cm^{-2} s^{-1}. Tomography setup and sample images taken at PIAF with collimated fan beam and cone beam are shown in Figs. 12.4 and 12.5. The sample of polyethylene cylinder was assembled with drilled holes of different diameters in its surface and inner parts. To mask/cover it with thick high Z material, a polyethylene phantom was inserted into lead cylinders of different thicknesses. This sample configuration was imaged using fast neutron detection/imaging system comprised of a plastic scintillator viewed by image intensifier and a CCD camera. The images obtained were of high contrast as shown in Fig. 12.5.

The results have proven the potential of accelerator-based neutron source for FNT with imaging of low Z material when shielded with thick high Z material. Also, it has shown that the quality of image favors the fan beam configuration over the cone beam configuration, but imaging with fan beam configuration is more time consuming.

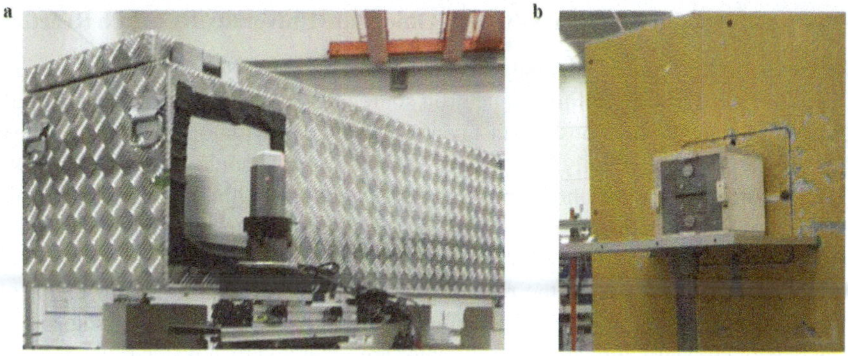

Fig. 12.4 Tomography system at PIAF; **a** the imaged object, a plastic phantom embedded in a cylindrical lead shield placed on a rotation and translation table in front of the detector box. **b** The target station, inside a massive beam collimator, here equipped with a fan beam collimator (Reproduced with permission from [57])

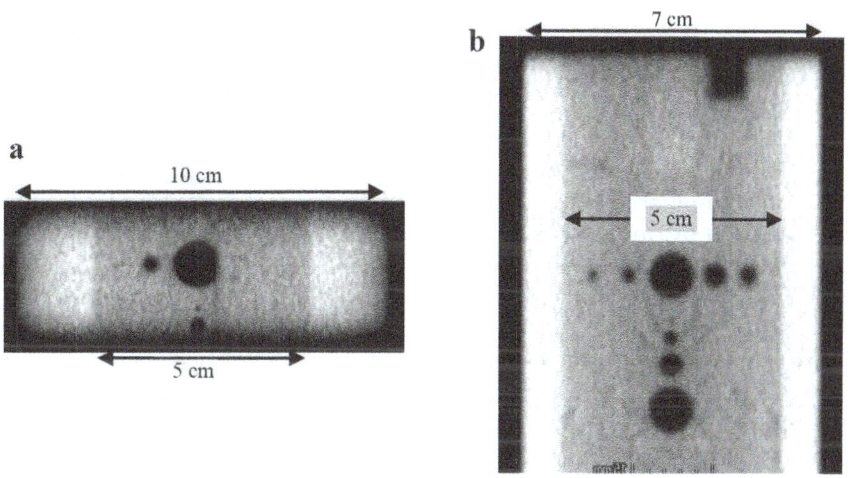

Fig. 12.5 Vertical view of the PE phantom inside Pb sleeve for **a** fan beam geometry, and **b** cone beam geometry. The fan image shows only small part of the phantom while the cone image shows the whole phantom

Anderson [60] has also explored the application of FNT for assessment of steam-water distribution in two-phase flow using a mobile neutron generator-based prototype neutron tomography system "FANTOM." In fast neutron imaging, the scattered neutron background is always a big concern and challenge. Towards reducing its impact, an array of plastic scintillator detector was used. Each detector event recorded using a high-speed digitizer, allowing energy discrimination on event-to-event basis and hence minimizes the scattered neutron background.

Fast neutron beam facility at OSURR (Ohio State University Research Reactor)

Research at Ohio State University Research Reactor (OSURR) has resulted in the development of a well-collimated neutron source of high energy neutrons and high flux for imaging applications and fast neutron radiography/tomography studies have been performed successfully [61]. It consists of three main components, namely two beam collimators, a neutron/gamma shutter between the collimators and a movable beam stop. The beam facility provides beam diameter of 3.2 cm at geometric L/D ~62. The first collimator, a thick graphite cylinder having 3.2 cm aperture and 107 cm long is positioned closer to the reactor core, a solid cylinder (10.16 cm thick) of polycrystalline bismuth is placed just after the aperture for filtering out gamma rays. The reactor core structure and the collimator are separated by a water-filled gap and an aluminum (Al) cap, both 1.6 mm thick, which helps in reducing further thermalization of neutrons before the collimator entrance. Both the first and the second collimators contain disks of lead (Pb), Al, borated Al, and SWX-277z-5 castable shielding material of various thicknesses to further improve the beam collimation.

12.4.2 Fast Neutron Imaging Facilities Using Reactor-Based Source

Neutron Computerized Tomography And Radiography (NECTAR) facility: NECTAR is a reactor neutron source-based FNI facility and has been developed and installed at the Forschungsreaktor München-II (FRM-II) [62, 63]. Schematic layout [19] of the NECTOR is shown in Fig. 12.6. It uses fission neutrons (virtually unmoderated fission neutron spectrum of mean energy ~1.7 MeV). The fission neutrons are generated by highly enriched uranium (93% ^{235}U) used in the form of two plate setup, placed in a D_2O moderator. The thermal or moderated neutrons from compact reactor core region having main fuel elements can induce nuclear fission in highly enriched uranium plates. Isotropically emitted fission neutrons of ~MeV energy range are allowed to travel a horizontal beam tube without moderation. At the given SR10-beam, two measuring stations are installed. One of them is fast neutron imaging setup for fission neutron-based radiography and tomography experimentations (Fig. 12.6). This beam line has various components including beam shutters, borated carbon (B_4C) filter to suppress thermal neutrons, and a lead filter to reduce the gamma radiation is used. Further for beam manipulation both spectrally and geometrically, different moveable filters of lead and polyethylene as well as diaphragms and collimators are installed. The mean neutron energy was 1.9 MeV for uncollimated and 1.3 MeV for the collimated case. The detector system consists of a 2.4-mm-thick sheet of ZnS-scintillator embedded polypropylene of 30 cm × 30 cm detection area coupled to a 1024 × 1024 pixel CCD camera. Some of the experiments performed at this facility had source-to-detector distance of ~L = 935 cm with 2.1-degree beam

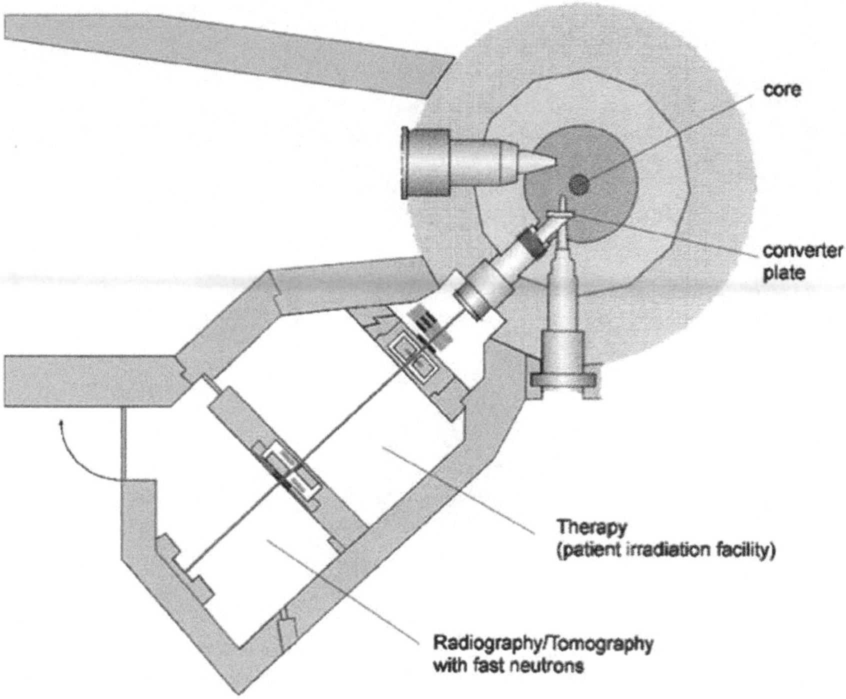

Fig. 12.6 Sketch of the layout of the fast neutron radiography/tomography facility at FRM-II (Reproduced with permission from [19])

divergence and average neutron energy of 1.9 MeV. At sample position, measured neutron flux was 5.4×10^5 n-cm^{-2} s^{-1}, and the L/D ratio is 233.

Fast neutron imaging beamline at 10 MW research reactor of Budapest Neutron Center (BNC)

A thermal neutron imaging beam line at 10 MW research reactor of Budapest Neutron Center (BNC) is routinely in use for thermal neutron imaging [21]. This thermal neutron imaging beamline (RAD, radial channels number 2) has significant fraction of high energy (>2.5 MeV) neutron nearly ~2–3 $\times$ 10^7 n cm^{-2} s^{-1} and was planned towards utilizing it for fast neutron imaging (Fig. 12.7). In this direction, to suppress the thermal neutron component and gamma, a filter made of borated rubber (10 mm) and a thick lead shielding of 300 mm was respectively. Further to reduce the scattered neutron contribution and secondary gamma, additional layers of borated paraffin and lead were placed around these filters. Finally, beamline with 2.8 cm primary aperture diameter and L/D of 165 enabling a quasi-parallel beam imaging geometry, where imaging detector can be placed at a distance of 463 cm from the primary aperture. Radiography and tomography experiments were performed with fast neutron for bulk

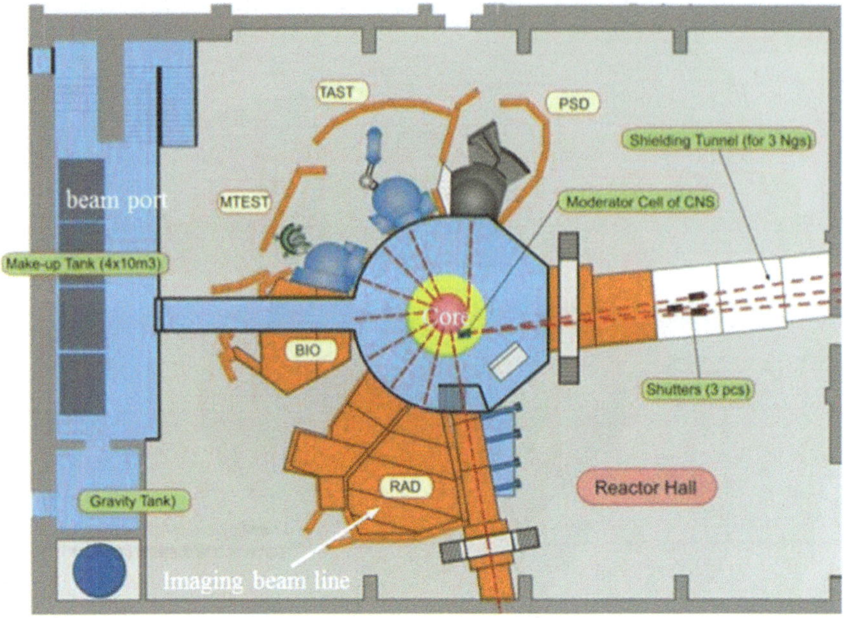

Fig. 12.7 Layout of the reactor hall of the 10 MW research reactor of the Budapest Neutron Center showing some of the horizontal neutron beam facilities including the RAD neutron imaging beamline (Reproduced with permission from [21])

samples having high Z as well as low Z materials simultaneously, using plastic scintillator screen as a neutron detection system coupled with a CCD camera. Typically, with 10 min exposures, reasonable quality radiographic images could be obtained providing spatial resolution of around 1.3 mm. Tomography image of the rectangular lattice fuel bundle mockups is shown in Fig. 12.8.

Neutron radiography at YAYOI reactor, Japan

Fast neutron radiography study using beams of fast neutron at YAYOI—a fast neutron source reactor of the Tokyo University has been started since 1986 [64]. There are five experimental beam holes in this reactor for FNR. Figure 12.9 shows the sectional view of the YAYOI facility. The high energy neutron flux and average energy measured at exit of the reactor beam holes were ~10^6 to 10^7 n .cm^{-2} .s^{-1} and 1.3 MeV, respectively. Major characteristics of different neutron beams are presented in Table 12.1.

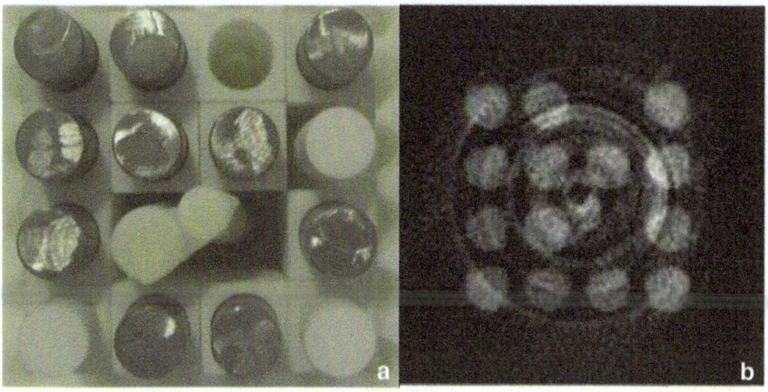

Fig. 12.8 (Left) Photo of fuel assembly mock-up and (right) a typical slice of its tomographic reconstruction at 10 MW research reactor of the Budapest Neutron Center (Reproduced with permission from [21])

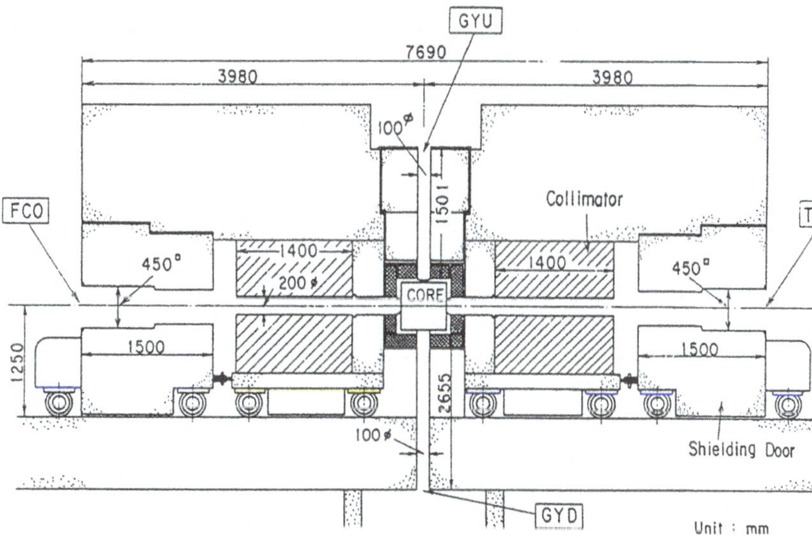

Fig. 12.9 Sectional view of the YAYOI facility (Reproduced with permission from [64])

12.5 Fast Neutron-Based Imaging for National Security

Detection of threat materials like explosives and contraband concealed in cargo (air or sea), truck or air-baggage is a big national security challenge. Presently, inspection systems are based on commercially available X-ray/gamma ray imaging. These photon-based systems have high throughput and are ideally suited for detection of metallic objects with distinctive shape, size and densities such as gun, ammunition, wires etc. At best, these systems provide limited gross information about

Table 12.1 Characteristics of the beam holes for the FNR [64]

Beam hole	Fast neutron flux	Gamma ray dose	Field size	L/*D*	*D* (mm)	L (mm)	Remarks (detector)
TCO	4.8×10^6 ($ncm^{-2}\ s^{-1}$)	0.16 (Gy/h)	450 × 450 mm	74	50	3703	NTV (CCD)
TC	1.8×10^7	1.0	200 mm in dia	39	50	1950	NTV (SIT)
FCO	2.0×10^6	0.13	450 × 450 mm	37	100	3703	
GYU	3.4×10^6	0.23	100 mm in dia	133	20	1501	
GYD	1.2×10^6	0.06	100 mm in dia	53	20	2655	
GZD	6.0×10^5	0.03	100 mm in dia	39	50	2655	

TCO the horizontal thermal neutron column; *TC* the inside of the TCO; *FCO* the horizontal fast neutron column; *GYU* the upper part of the vertical glory hole; *GYD* the lower part of the vertical glory hole; *GZD* the vertical glazing hole near GYD, not face to the center of the reactor core. The fast neutron flux was measured by an activation foil method. The gamma ray dose was measured by BeO TLDs

elemental contents of the inspected object (e.g., low Z vs high Z) and have limitation on revealing material specificity. Unfortunately, most of the current potential threat materials such as chemical explosive and narcotics do not have specific shape, size and are mostly composed of low Z elements (i.e., C, H, O, N, Cl etc.) and X-ray-based systems fail to differentiate them from innocuous materials. On the contrary, neutron-based techniques having ability of simultaneously determining identity and density distribution of low Z elements present in inspected object hold promise towards distinguishing between explosives and benign materials. In this direction, fast neutron imaging techniques named fast neutron resonance radiography (FNRR) and associated particle imaging (API) methods have proven to be most suitable.

12.5.1 FNRR—Fast Neutron Resonance Radiography

FNRR is an advanced modality of the transmission-based fast neutron imaging method for elemental imaging. The technique utilizes the unique resonance structure of the total neutron cross section of different elements, particularly in $E_n =$ 1–10 MeV neutron energy range. The variation of the total neutron cross sections of low Z elements (C, O, N and H) versus neutron energy is shown Fig. 12.10. Sharp energy fluctuations (peak and dip) in cross-sectional values for C, O and N can be

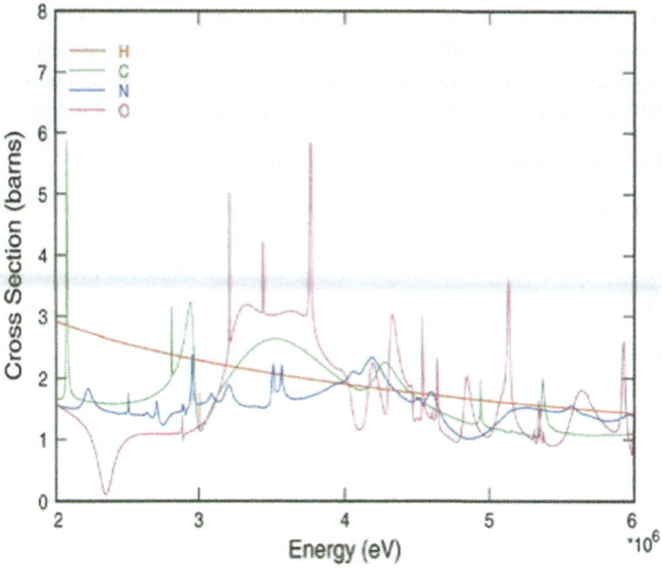

Fig. 12.10 Total neutron cross section for hydrogen, carbon, nitrogen and oxygen in the MeV energy region with the exception of hydrogen, clear and unique resonance structures exist for each element (Reproduced with permission from [65])

observed from Fig. 12.10 clearly. The element identification and density distribution/mapping of particular elements or say elemental sensitivity within an inspected object are possible if attenuation information is available at multiple neutron energies.

Two-dimensional projection image (as discussed earlier in this chapter) is the result of attenuations from all possible elemental contents present in the sample in the beam path under inspection. In FNRR, several such attenuation images are recorded at different neutron energies, to forms a set of linear equation which can be solved to map individual elemental contents [34].

Thus, compared to the conventional FNR, the advantages of using FNRR are the mapping of densities of multiple element (H, C, N, O and more), and based on the detected elemental density distribution, potential discriminations for various materials are possible. FNRR has main application for explosives detection, hidden within large volume of the transported items via truck, cargo etc.

Explosive detection in large volume of air/sea cargo employing FNRR method can be realized by measuring the elemental densities mainly of C, O, N and H elements and their relative ratio [13].

For example, one of the improvised explosives Tri-Acetone-Tri-Peroxide (TATP), as presented by Ilan Mor et al. [66], where FNRR is suitable and efficient for detection of small quantities of TATP. TATP is one of the high-grade potential explosives, such threat object can be simply made using commonly available household materials and chemicals like sulfuric acid, acetone, etc., which can be procured without any restriction. Detection of diamonds in kimberlite rock [67] is another attractive application of FNRR. The rock contains many elements such as O, Si, H, Fe andC; the presence of diamond is indicated via detecting carbon concentration/density in excess at a particular part of the rock. Different approaches have been adopted to realize FNRR technique for different applications, as categorized below:

FNRR using variable quasi-monoenergetic neutron beam

In this variant of FNRR, sequentially several transmission images are generated at various quasi-monoenergetic neutron energies [34, 35] using a continuous variable neutron energy, obtained by selecting/viewing different angles of the neutron source, for example, D-D neutron generator.

FNRR using switchable quasi-monoenergetic neutron beam

It utilizes switching between two discrete neutron energies suitable particularly for application of diamond detection in rocks.

FNRR using broad energy pulsed neutron beam

This variant of the FNRR selects different neutron energies from a broad energy (~1 to 10 MeV) pulsed neutron beam using time-of-flight technique. In this method, neutron transmissions through the object under inspection are measured at different preselected neutron energies. This method is also known as pulsed fast neutron transmission spectroscopy (PFNTS) [36]. Transmitted neutron energy spectra get modified depending upon the resonant features of neutron cross section exhibited by different elemental composition of the inspected sample. Thus, these transmitted neutron spectra are useful to reveal information about chemical composition of the inspected sample. A position-sensitive neutron detector can be used to measure the transmitted neutron spectra and time-of-flight techniques to determine the attenuation at preselected neutron energy. PFTNS has basic requirement of an imaging system with capability of time-of-flight measurements and precise information of the incident neutron energy which can be obtained via time-of-flight spectroscopy from a broad energy pulsed (ns) neutron beam.

Among above-mentioned three approaches of FNRR, the first two can be realized with the use of continuous or DC beam of neutron, while the last method requires an ns-pulsed neutron beam, which makes it more complicated compared to other two.

12.5.1.1 Key Challenges in FNRR

FNRR requires a neutron source of variable energy with minimal neutron energy spread at each selected energy, and for that accelerator is the only option. Secondly, FNRR imaging requires high neutron detection efficiency with high gamma rejection [68]. The presence of photons contaminates the cross-sectional measurements and probability of the detection of actual elements is reduced [34, 35]. Also, in order to perform FNRR measurements, one must measure the cross sections for all possible basis elements and based on that determine the number (> number basis elements) and neutron energies to be used. Despite its promising application and potential, practical demonstration of the FNRR has been difficult. Moreover, the reconstructed elemental density distribution depends on how precisely the least square fit has been performed to a measured neutron cross-sectional data set of the basic elements. The cross-sectional data measurement with detection/imaging system involves various steps and is not straightforward. However, few groups have demonstrated the perspective commercial applications of FNRR, for explosives and narcotics detection [69, 70], but FNRR-based systems are still out of reach at commercial scale for field applications because of its complex system design requirement demanding large size and high cost of accelerator-based high intensity neutron source.

12.5.1.2 Review of FNRR Techniques

FNRR using variable quasi-monoenergetic neutron beam

FNRR technique was first explored and studied by Lanza et al. [34, 65] demonstrating it as a viable technique providing elementally resolved images. RFQ-based ion accelerator producing high energy (~MeV) neutrons via D(d,n)He fusion reaction was used. The energy of the emitted neutron decided by its angle of emission, which is due to the kinematics of the D-D reaction. Viewing the neutron source from different angle, neutron beam of different energy can be obtained. This was realized by an object-detector assembly with the provision of rotating around the neutron source [35]. RFQ-based system with accelerated deuteron ion up to energy 2.3–2.5 MeV, striking on a 4-atm. gas target produces 5.6 and 2.0 MeV neutrons at emission angles (with respect to the deuteron ion beam direction) of 0° and 130°, respectively. Resonant radiography experiments were performed by placing sample and the detection system at various angles, which were selected corresponding to the energies of the resonances features in different C, O and N elements. Across the detector, there is an energy variation due to finite size of the detector; to reduce this, subtended angular aperture must be minimized and prefer to keep it less than 10 degree. This condition of angular aperture limits the maximum size of the object that can be inspected. Also, note that the neutron flux of $\sim 10^4$ n/s/cm^2 at 1 m distance from target decreases sharply as one move away from 0°. At 90°, it reduces almost 10% of neutron flux at zero degree. A thin slab of plastic scintillator in combination with low noise CCD camera is used for neutron detection. Plastics scintillators are also sensitive to gamma and cannot discriminate them from neutron as far as scintillations are concerned. As a result, the contrast of the experimentally obtained elemental images gets deteriorated, and the calculated attenuation coefficient differs largely from the expected or published data. Later on, a different detection approach was followed [66] using linear detector system, where detectors run in counting mode with gamma/neutron discrimination capability. Recently, Soubelet et al. [35] presented a feasibility study of FNRR using a compact deuterium-deuterium (D-D) neutron generator producing quasi-monoenergetic neutron of energy range from about 2.2 to 2.8 MeV. Perticone et al. [65] have also developed a prototype neutron resonance imaging system using a 3 MeV deuteron beam from a RFQ accelerator and provided the experimental evidence of the system's capability to identify the inspected object by reconstructing its chemical formula.

FNRR using switchable quasi-monoenergetic neutron beam

FNRR with use of switchable quasi-monoenergetic beam has emerged particularly useful for application of diamond detection in kimberlite rocks [68] and was first initiated by De-Beers Company, South Africa. In this application, rather than multiple energies, only two neutron energies at 6.8 and 7.8 MeV were considered for FNRR, which were selected as per the broad neutron cross-sectional structure of carbon element only. Neutron energy beams were produced by D-D fusion reaction, using 4 MeV RFQ accelerators to accelerate the deuterium ion beam with the provision of switching the deuteron energy either to 4 or 5 MeV. Neutron imaging detector consisted of thick scintillating fiber detector coupled to a CCD camera [68].

PFTNS: FNRR using broad energy pulsed neutron beam

This variant of FNRR was initiated long ago in 1985 by Oregon group [71, 72], which was improved to next level by Tensor Technology, Inc. [73]. As described earlier, PFNTS measures the transmitted neutron spectra by time-of-flight (TOF) technique. In this method, a short neutron burst (~1 to 2 ns) is produced by hitting pulsed accelerated ion-beam on the target. The individual neutron flight time is recorded at neutron detector using time-of-flight technique. The neutron detector is placed at a distance from the neutron source. Mathematically, it can be expressed as $T_{tof} = d\ (m/2E_n)^{1/2}$, where T_{tof} is the neutron flight, d = source-detector distance, m = neutron mass and En = neutron energy. The relation is valid for the nonrelativistic energy range. Convolution of detection instruments' time response and beam burst duration decides the overall time resolution of the system. Based on the FNRR instrument developed, one can choose either single event time-of-flight mode or time-of-flight method in integrative mode. First mode of TOF, that is the single event time, is the most commonly used conventional method of time recording, where individual neutron's arrival time is recorded at detector from its origin time relative to neutron burst at the target. On the other hand, in the integrative technique rather than individual neutrons, it integrates the detector signal for a well-defined gate time at a preselected T_{tof} corresponding to a preselected neutron energy bin.

In this mode, by varying the delay of integrating gate relative to the time of the burst, TOF spectrum is obtained. Tensor Technology Inc. has studied single event TOF PFNTS method and continuously improved by performing several critical and trial experiments along with Argonne NL group [74]. Later on, a significant progress has been made in the data acquisition using TOF, detection reconstruction algorithms and reduction in the accelerator size. Further, it proposed an airport explosive detection system using a cyclotron as a pulsed neutron source. However, the system spatial resolution was effectively inadequate towards detecting thin sheet explosives. In order to resolve this challenge, Dangendorf [75] had proposed a novel detector concept utilizing scintillating fiber coupled to pulsed counting image intensifiers for TOF measurements. Soreq NRC, Israel [76], in collaboration with PTB, Germany, has developed a PFNTS system having resolution of ns time, tested with counting in integrating mode using a special type detector named as Time-Resolved Integrative Optical Neutron (TRION) detector and also using single event-counting mode with TRECOR-Time-Resolved Event-Counting Optical Radiation detector [76–78]. TRION has been developed for detection of fast neutron pulses in FNRR technique. Utilizing the time-of-flight (TOF) method, TRION detector can record simultaneously up to 8 or more images, each at different incident neutron energy.

12.5.2 Associated Particle Imaging: API

API is a sophisticated and advanced technique for detection and identification of low Z materials in bulk by determining their chemical/elemental composition. Using the time-of-flight technique, the detected elements can be localized in the inspected volume. Hence, this technique is a sort of hybrid imaging where detection plays an important role. This is quite different from the conventional projection imaging where the transmitted neutron intensity through the sample is recorded. It is based on the use of neutrons generated in D-T neutron generator by D-T fusion reaction.

$$^{2}\mathrm{D} + {}^{3}\mathrm{T} = {}^{4}\mathrm{He} + {}^{1}\mathrm{n},\ \left(\mathrm{E_n} = 14.1\,\mathrm{MeV},\ \mathrm{E}_{4_{\mathrm{He}}} = 3.5\,\mathrm{MeV}\right)$$

D-T neutron generator produces neutron and alpha particles of 14.1 MeV and 3.45 MeV energies, respectively. Emission of each neutron is associated with an emission of an alpha particle and in opposite direction to each other. Thus, these two particles are correlated in space and time with respect to their production site, and this correlation can be used to tag the neutrons. The direction of the emitted neutrons can be determined via detection of associated alpha article with the use a position-sensitive charge particle detector. This way, the neutrons are tagged by its emission time as well as its direction. Size of the alpha detector governs the tagged neutron cone size and is utilized for imaging. Thus, collimation of neutron beam is achieved electronically without use of shielding material. The neutrons within this "tagged beam" (named "tagged neutrons" determined by correlated alpha particles) are utilized for imaging. Tagged neutrons, on interaction with the nuclei (mostly via inelastic scattering) of the object under inspection, produce neutron-induced prompt gammas. Element-specific characteristic gamma signatures [79] provide way towards identification of the nuclear species that the neutron interacts with thus the type of material present within the object. Second important parameter measured is the time difference [or say time-of-flight (TOF)] between alpha detection and tagged neutron-induced gamma radiation detection. It provides the distance traveled by the tagged neutron before its interaction with an element nucleus in the inspected object. Thus, in comparison to conventional neutron imaging, API is a neutron-based self-collimating technique, where neutrons are collimated/tagged by determining neutron emission time and its emission direction with the use of correlated alpha particles [1]. Thus, in this technique, the background is reduced by accepting (alpha–gamma coincidence) the events that are produced only by interaction of the tagged neutrons

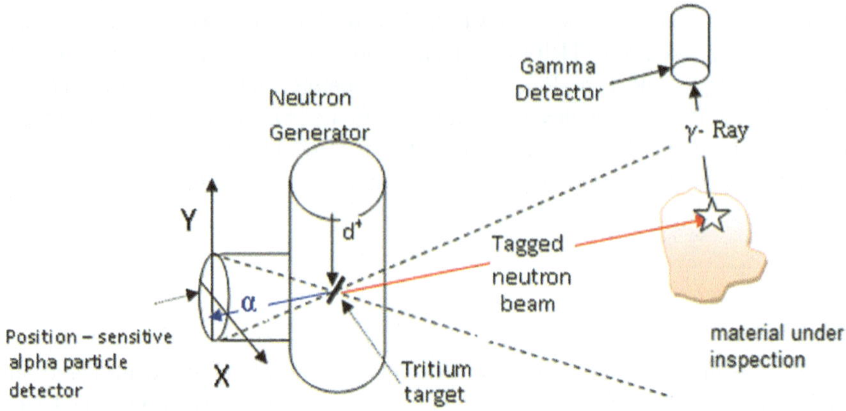

Fig. 12.11 Schematic view of tagged neutron-based associated particle imaging technique (Reproduced with permission from [80])

and rejecting all other events possibly caused by other neutrons as neutrons emitted isotropically from the source. Schematic view of the tagged neutron-based API is shown in Fig. 12.11.

12.5.2.1 Methodology of the API

Simulation is a primary first-hand tool in scientific research area. It can provide the first-hand information's on the required parameters simulating different experimental environment and configurations. Methodology of any technique before its implementation can be realized through simulation including estimation of necessary parameters and their interdependencies for a given configuration close to the possible planned experimental setup. In literature, various simulation studies performed by different groups related to the API-based system configuration have been reported. To broadly understand the methodology of API and concept of imaging, modeling of API by Saroj et al. [80] has been carried out using GEANT4 code [81]. Some other examples are modeling of a prototype tagged neutron system [82], LYSO detector response simulation study for tagged neutron system [83], generation of elemental spectra for EURITRAC system [79] and simulations studies on design of API method UNCOSS system, which is proposed for application of underwater object inspection [84] among many such studies. Briefly, we discuss here the major aspects of the API as demonstrated by Saroj et al. via simulation. In their simulation, a sample of benign (size 25 × 25 × 25 cm^3), or explosive (example, RDX –$C_3H_6N_6O_6$ sample of density 1.5 g cm^{-3}) is placed at a distance from neutron point source producing monoenergetic 14 MeV neutrons. The neutron-induced gamma energy spectra from sample and time of the gamma event (time spectrum) reaching at the detector were recorded. Next, by putting a suitable gate/window on time spectrum, respective correlated spectrum of gamma for a region of interest can be obtained. Now, using this approach, first of all, detector's response (experimentally measured) was incorporated in simulation model and spectral responses for range of pure elements such as carbon, oxygen, hydrogen and nitrogen, and many more were evaluated. These pure element spectra can be used as reference data set for extraction/determination of individual elemental's intensity/contribution from the gamma energy spectrum of any unknown sample which is composed of these elements. Further, a list of samples (benign as well as explosives) having composition of C, O, N and H was simulated and their C, O, N elemental fractions were extracted via unfolding sample's gamma spectra with use of reference data set of elemental spectra. To identify the material, these evaluated values were compared with theoretical values. These major steps of the API techniques are summarized in Fig. 12.12 taken from [85]. After these basic steps, identification of the material in presence of surrounding matrix was explored which is like closer to the realistic situation. In practice, a smaller volume containing threat objects is placed inside a larger volume of cargo, truck etc., that threat material gets covered/masked with the surrounding benign material and becomes difficult to detect. Thus, detection of such hidden threat material in presence of any surrounding material becomes more importance, because signals of the threat material have to be distinguished against the background signals generated from the surrounding environment material. The effect of matrix materials was studied on a RDX simulant hidden inside a large volume of a container filled with matrix material of either organic or metallic. It has been observed that RDX sample spectra got modified, due to overlapping of signals from surrounding matrices material. However, with the application of time window selection feature of API technique, it was still possible

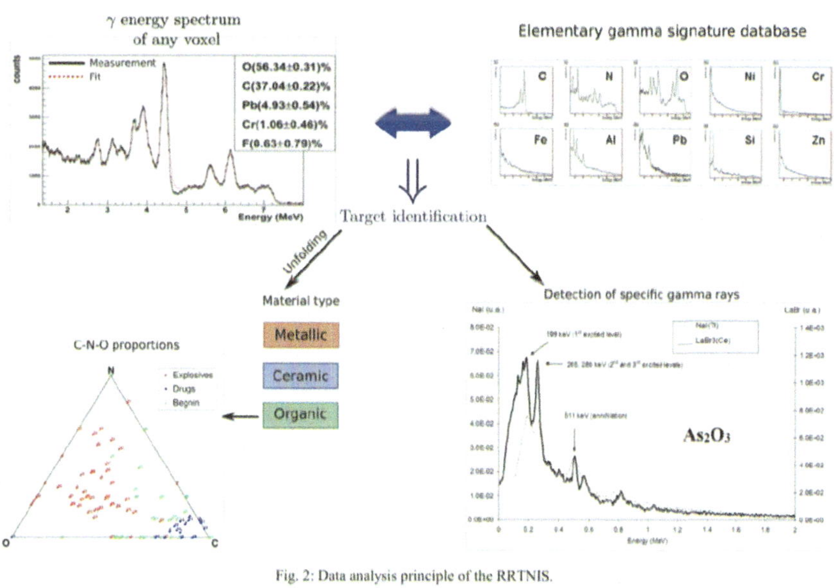

Fig. 12.12 Methodology of the API for material detection (Reproduced with permission from [85])

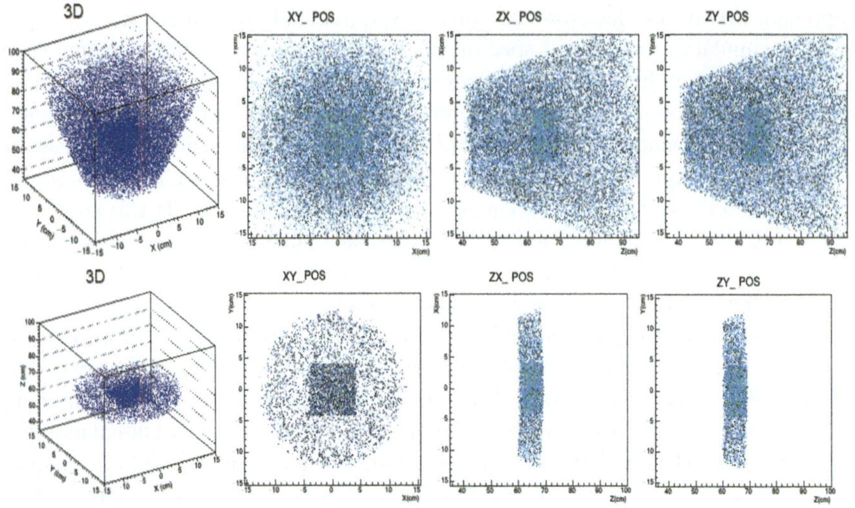

Fig. 12.13 With API approach, reconstructed images (3D, 2D) of neutron interactions for the case of RDX (1 kg) sample hidden inside container filled with iron matrix of density 0.1 gm/cc. (Top) when no time window selected (Bottom) with a condition of a 1.6 ns time window selected (Reproduced with permission from [80])

to identify/extract the sample signals from that of background clearly. Figure 12.13 shows the reconstructed 2D and 3D images of RDX surrounded with iron matrix material revealing the location of the interrogated object.

12.5.2.2 Key Components for API

The major components required for implementation of API technique and to produce the complete system are; a D-T neutron generator with a position-sensitive charge particle detector, set of gamma ray detectors and associated data acquisition hardware. The neutron generator irradiates the suspicious baggage, scintillation gamma detectors measure the neutron-induced γ-ray spectrum, and the hardware processes the measured data for baggage/large container material identification.

Towards improving signal to background ratio, API uses alpha–gamma coincidence and reduces the contribution of the random coincidences arising from surrounding/background. So, the primary aim is to track and record the true events of the alpha–gamma coincidence. If in a given coincidence window, more than one tagged neutron is produced from the NG, the system fails to recognize the neutron responsible towards producing the particular gamma ray. In this technique, the time from neutron production to gamma detection is around less than ~100 ns. The yield of neutrons which goes in the direction of the object under inspection must then be kept less than about 10^6 n s^{-1}. The reason can be accounted to the limitations in the alpha and gamma detector and their registration electronics. Therefore, the total D-T neutron yield (in 4 pi) must be kept in the 10^7–10^8 n s^{-1} range. Nowadays, many manufactures supply sealed tube neutron generator with built-in pixilated alpha particle tracker [86, 87], for tagged neutron-based applications. Tagged neutron-based EURopean Illicit Trafficking Countermeasures Kit (EURITRACK) project has used such sealed tube D-T neutron generator and reported various results [88] in which 64 tagged neutron beams are registered using an 8×8 multi pixel alpha detector.

Key characteristics required for alpha detectors are high count rate capability with good time resolution, mechanical robustness and resistance to high energy neutron-induced radiation damage. In particular, it must have minimum out gassing and it should survive prolonged radiation exposure. Semiconductor detectors are indeed primary choice as they are characterized by good energy resolution, good stability, excellent efficiency and timing characteristics. Some API system has been developed with built-in Si-based tagging detector [89] to D-T NG, but the long duration of exposure to neutrons and charged particles creates rapid performance deterioration due to radiation damage in them. Alternatively, scintillators of different types have been investigated for use in neutron generators for alpha charged particle detection. These were silver-activated zinc sulfide (ZnS:Ag), gallium-doped zinc oxide (ZnO:Ga), plastic scintillators and yttrium aluminum perovskite activated by cerium (YAP:Ce), etc. ZnS (Ag) has very high scintillation efficiency comparable to that of NaI (Tl), and it is polycrystalline nature. Its use is limited only to thin screens due to opacity of the multi-crystalline layer to its own luminescence. The coatings of the ultra-fast ZnO (Ga) scintillator are characterized by a decay time of about 1.5 ns, scintillation light peak at 390 nm [41] and its quite high light yield (~3300 photons/MeV). Thus, ZnO(Ga) has characteristics that make it suitable for charge particle (alpha) detection, and it has been explored as a phosphor thin coating for detection of alpha particle in API [86, 87]. Similarly, YAP:Ce (YAlO3:Ce), scintillator has various characteristics properties such as fast response (25 ns decay time constant), high light yield (40% relative to NaI (Tl)), non-hygroscopic, radiation hardness and low gamma radiation detection efficiency with its density of 5.37 g/cm^3 and atomic number of yttrium $Z = 39$, etc. In addition to these features, YAP: Ce crystal has emission spectrum with peak wavelength ~370 nm, which is in good match with typical characteristic of the photomultipliers. These characteristics of the YAP: Ce crystal makes it as a preferred choice for API-based system development. Various researches such as Euritrack project and Saroj et al. [88, 90] have successfully reported development of API-based system using YAP:Ce as tagging detector.

Next major component of the API is the selection of gamma ray detector, which is critical for gamma ray spectroscopy measurement and important component for efficient performance of API system. Efficient detection of characteristic high energy (up to ~10 MeV) gamma rays is the primary requirement and therefore the detector with material of high Z value preferred. Secondly, it must be able to resolve the energy peaks of interest and should be suitable for its use in mixed radiation field of neutron and gamma. These requirement narrows the search for this application to suitable crystals of large size and widely affordable such as NaI (Tl) (sodium iodide with thallium doped), or BGO ($Bi_4Ge_3O_{12}$—bismuth germanate). Other commercially available detectors like lanthanum bromide ($LaBr_3$ (Ce) or LYSO [83] even offer better energy resolution. For example, energy resolution of LaBr (Ce) at E = 662 keV is less than 3%. However, their development to larger size of 3 inch or more are still very expensive and this limits their wider usage. In EUITRACK system, large size NaI(Tl) gamma detectors in large numbers (20–22) were used for detection of neutron-induced gamma rays released from inspected container of cargo (Perot et al. 2008). TANGRA system [91] has used BGO detector for gamma and gallium arsenide as a pixilated alpha tracker coupled to a D-T neutron generator. Gierlik et al. [92] have performed a comparative study on properties of different sizes crystals of NaI (Tl) and BGO with high energy gammas for their suitability to API-based system towards large sea cargo inspections.

12.5.2.3 Review of the API Systems Developed for Different Applications

We briefly review the historical development of API and the detection systems that have developed or under field testing or have shown significant practical promise even at laboratory scale. We focus mainly on the API-based systems that have been developed for various applications including illicit material detection in cargo.

The concept of coincidence measurement of D-T neutron and its associated alpha charged particle was purposed long back 1969 by Valkovic et al. [93]. Using this concept, the API technique has been proposed quite long ago [94–96]. However, due to lack of advanced technology in the field of neutron generators, fast response detectors and fast data acquisition prevented its widespread use at that era. In the early 1990s, research work on API was limited to specific fields of nuclear and chemical treaty verification measurements. As discussed earlier, the D-T neutron generator with pixilated detector for alpha is a major component for API. It provides information precisely which voxel/part of the large volume is under inspection and enhances the signal to noise ratio via suppressing the background clutter. Initial reports on development and research progress of portable/sealed tube neutron generator with inbuilt associated particle detector for its use particularly for arms control, treaty verification and nonproliferation applications are available from Argonne National Laboratory (ANL) [97, 98]. These reports reveal that ANL has demonstrated the proof-of-principle experiments of API and analyses for identifying chemical and high-explosive munitions and related to the weapon dismantlement, detection of uranium and plutonium towards smuggling check, radioactive waste characterization, etc. These initial studies have proven the significant potential of API towards detection of illicit materials based on elemental compositions inside large volumes nondestructively. Further, research efforts were made on instrumentation development for data acquisition to improve the data quality. Chichester et al. [99] have developed a small STNG for field work with all digital control system and inboard diagnostic, platform independent control interface and safety interlocks. Nebbia et al. [86, 100] have developed neutron inspection system (TNIS) using STNG and performance tested successfully. It was focused mainly on investigating the neutron-induced prompt gamma using API methodology for locating and detection of threat material such as narcotics, explosives and special nuclear material for hold-up assay applications [101]. Though various experiments were carried out with fixed type D-T neutron generators particularly at laboratory environment, the advancement of technology in developing small portable/sealed associated particle detector embedded sealed tube neutron generator has shown potential towards transferring this technique out of the laboratory for field applications. This lately opened up various novel applications of API in different environments ranging from buried landmine, unexploded ordnance (UXO) [102], cargo scanning SENNA [103], EURITRACK [104] and under water threat detection UNCOSS [105]. A proof-of-concept study was performed for nondestructive identification of unexploded ordnance using API concept by Mitra et al. [102] at Brookhaven National Laboratory, USA. The system was designed with a portable neutron generator, inbuilt ZnO(Ga) alpha tagging detector and a NaI(Tl) detector of 12.7 cm × 12.7 cm diameter. A large API-based scanning system was developed and its capability for landmine detection has been demonstrated at Italy [106] by Lunardon and co-workers. The focus was on demining the post war activities in mine-affected countries. A concept of using an array of Parallel Plate Avalanche Counters for alpha detection and 10 BaF_2 scintillator detectors for coincident gamma rays has also been purposed. Applied Science and Technology Center, Russia [103] has developed a small portable device "SENNA" capable of finding small amount of explosive concealed or hidden, by scanning a large amount of surrounding material within a few minutes. In SENNA project, a nine-pixelated semiconductor-based alpha detector built in a portable neutron generator developed by VNIIA, was used and two BGO gamma detectors of size 6.3 cm × 6.3 cm. Thus, API method has been rapidly progressed and explored for various applications. In the last two decades, the focus on API has been towards national security applications in detection of illicit materials particularly explosives in sea/air cargo, small parcels and airline baggage etc. API-based EURITRACK inspection system is one of the most successful projects undertaken for sea cargo inspection. It has been developed for nondestructive inspection of suspect region inside large cargos, identified by existing photon (X-ray/gamma ray)-based scanner at a seaport. It is a fixed type inspection system implemented at Croatian sea port [104]. In EURITRACK system, an array of 8 × 8 pixels of YAP(Ce) scintillator coupled with multi-anode PMT of Hamamatsu designed for alpha particle detection and three separate arrays of gamma detector made of NaI(Tl) scintillator placed around the container (top, transmission and reflection position) for efficient collection of fast neutron-induced prompt gamma rays [107]. The read-out and data processing of all detectors in time and energy measurements are performed with front-end electronics using high-density VME modules and cards [108] interfaced to a PC. This makes data acquisition system compact with an option for ease of expansion to accommodate more numbers of gamma ray detectors if in case it is required in future. The application has been also extended for underwater threat detection. The aim was to develop systems towards detecting the significant munitions and unexploded ordinances lying at the base of coastal and inland waters, which might have been generated either during wartime or due to other activities like dumping, accidents, terrorist activities, training and munition testing. An underwater threat detection system "UNCOSS" (UNderwater COastal Sea Surveyor) was developed under research project of European Union aiming to improve maritime security for maritime/naval infrastructures of Europe and safety along important various sea routes which are vulnerable for terrorism attacks [109]. Scientists involved in UNCOSS project, have developed a compact neutron sensor, and its performance was evaluated with latest detector technology including high-speed data acquisition, processing and analysis system as well as advanced numerical analysis tools. UNCOSS system has a specially designed remote-controlled vehicle to carry or move the neutron generator with gamma detectors under the water to the location of unexploded ordnance or threat of interest lying at the base of sea. The capability of UNCOSS for field applications was demonstrated successfully at Punat Seaport, Croatia, via distinguishing explosive surrogates from sediments in metallic objects lying at the sea base in less than of 10 min measurements time. An API-based experimental system has also been developed at BARC (India), and its capability for the explosive type material detection was demonstrated with the use of a $3'' \times 3''$ size BGO gamma detector and VME-based data acquisition [90]. Neutrons were tagged using a multi pixel YAP:Ce detector incorporated with in-house developed D-T neutron generator. System performance and proof-of-concept experiments performed at laboratory environment with simulants of various materials including water, graphites, melamine, RDX,TNT etc. Experimental data bank of pure elemental spectra, analysis of the complex sample (RDX and melamine) and their comparison with the simulated results have proven the system's capability towards detection and localization of the samples under inspection.

Other than threat material detection, the tagged neutron-based API techniques has also been explored and been utilized for the search of diamonds from the kimberlite samples via detecting carbon signals present in excess at a particular location of the sample by Alexakhin and co-workers [110]. The high penetration capability of fast neutrons allows inspecting the appreciably large size of kimberlite samples. Thus, it is possible to identify and separate the diamond rich area before the crushing stage. For diamond detection from kimberlite ore, Rogov et al. [111] used 256 tagged neutron beams with 22 BGO gamma detector and gallium arsenide as position-sensitive alpha tagging detector built in a D-T neutron generator. API has also been configured for basic and applied research on fast neutron-induced nuclear reactions studies [91]. An initial study involving 14 MeV neutron-induced activation with API methodology performed by Garrett and Mitra [112] has shown its potential for in vivo body composition measurement of humans or farm animals. Mitra et al. have investigated C, N and O measurement of whole-body via irradiating a phantom of 41.4 kg meat. It was suggested that an API-based instrument having a small NG and four NaI(Tl) gamma ray detectors of size 15 cm $\times$ 15 cm $\times$ 45 cm can be designed for determination of in vivo body contents such as water, fat, protein etc.

The combinational use of transmission-based imaging methods such as X-/gamma ray (or FNR), FNRR and fast neutron-based API technique in providing first-hand and conclusive information regarding the suspected objects is no doubt the best that can be achieved for research, nondestructive testing, security and strategic applications.

12.6 Summary

Fast neutron radiography/imaging using neutrons of high energy (~1 MeV or higher) has been an excellent method towards inspecting high-density, thick or large objects. FNR has progressed over the years with the development of high intensity, high energy, neutron sources from accelerator and reactor. FNR has shown its capability in various fields such as NDT, security and strategic domain. Efficient detector development with very low sensitivity for gamma background radiation and high sensitivity for neutron is a challenging issue in FNR. It becomes more important in case of neutron sources of relatively low intensity such as D-T/D-D neutron generators. This has motivated initiation of new research frontier towards development of fast neutron detectors efficiently such as plastic scintillators, polypropylene embedded with ZnS (Ag). The major technical challenges associated with fast neutron fields are the presence of strong gamma background, high contribution of scattered neutrons and high energy neutron related damage to the detector matrix. These have been overcome to a certain extent using optimized collimator design and proper shielding around detectors. Collimator walls/shielding material must be chosen wisely or else it might contribute in enhancing the neutron scatter component. However, these issues are still unresolved and are under research to reach a matured stage. It is interesting to note that the scintillator type, its thickness and the L/D factors are more important in deciding the qualities of the FNR image. Various researchers have reported high-quality FNR images with mm order spatial resolution for thick samples. FNR has shown to be an effective inspection tool for various applications. FNR has achieved various milestones in field of scanning of the cargo for contraband such as narcotics, explosives and illicit drugs), gas/liquid flow and mixing, radiography and tomography of encapsulated heavy shielded low Z compound materials. Also, its extension to FNT has shown its ability in the reconstruction of sub mm size holes/cracks inside fuel bundle or holes drilled inside low Z which is masked or surrounded with high Z material,—as specific materials of interest.

FNRR and API are two modalities of the FNI for elemental imaging. These techniques are applicable particularly for detection of low Z materials such as drugs and explosives, hidden in large volume of car, truck, air cargo, etc. These fast neutron imaging techniques are currently being investigated worldwide to fulfill the demands for improved methods over the existing x-ray/γ-based inspection methods, for illicit and contraband substances. The preference is to directly image the contents of containers without unpacking and using neutron-specific properties identify the hidden illicit material. In this direction FNRR and API, two advanced modalities of the FNI, has been developed and explored as alternative interrogation techniques. However, their complex design and high cost of the system are not exactly in line with the stringent detection requirements of custom at port or check points like high through puts, small exclusion zone, low operational cost etc. As a result, these systems are less deployed for commercial applications. Alternatively, a multi-tier approach of scanning systems with X-ray/gamma system as primary scanning probe and fast neutron imaging techniques as secondary or confirmatory inspection tools can be developed in near future to meet adequate scanning requirements for effective inspection of illicit material/potential threat material and could provide increasing security in subsequent area.

References

1. Hassina Z, Bilheux, McGreevy R, Anderson IS, Neutron imaging and applications, ISBN: 978-0-387-78693-3; (a) IAEA-TECDOC-1604. Neutron imaging: a non-destructive tool for materials testing (2003–2006)
2. Berger H (1965) Neutron radiography. Elsevier Publishing Company, New York
3. Bai JH, Shin M, Whang JH, Monte Carlo A (2006) Calculation for neutron radiography facility using sealed-tube neutron generator, pp 5–10. 12th A-PCNDT 2006—Asia-Pacific conference on NDT, Auckland, New Zealand
4. Mikerov VI, Zhitnik IA, Barmakov JN, Zhitnik IA, Barmakov JN, Bogolubov EP, Ryzhkov VI, Koshelev AP, Soshin NP, Waschkowski W, Lanza RC, Hall JM (2004) Prospects for efficient detectors for fast neutron imaging. Appl Radiat Isot 61:529
5. Uncovering hidden defects with neutrons, Lawrence Livermore National Laboratory, S&TR 2001 May
6. Dietrich F, Hall J, Logan C, Conceptual design for a neutron imaging system for thick target analysis operating in the 10–15 MeV energy range
7. Bogolubov YP et al, Adelphi Technology Inc., http://www.adelphitech.com/
8. High yield neutron generator phoenix nuclear labs, http://phoenixnuclearlabs.com/product/high-yield-neutron-generator
9. Reijonen J (2005) Proceedings of particle accelerator conference, p 49
10. Bogoluov YP et al (2009) Nucl Inst Meth Phys Res A 605:62
11. Mor I et al (2015) Phys Procedia 69:304
12. Buffler A (2004) Contraband detection with fast neutrons. Radiat Phys Chem 71:853–861
13. Buffler A, Tickner J (2010) detecting contraband using neutrons: challenges and future directions. Radiat Meas 45:1186–1192
14. Seki Y et al (2017) Fast neutron transmission imaging of the interior of large -scale concrete structure using newly developed pixel type detector. Nucl Inst Meth Phys Res A 870:148
15. Otake Y et al (2017) Research and development of non-destructive inspection technique with a compact neutron source. J Disaster Res 13(3):585
16. Osterloh K, Bellon C et al (2015) Computed tomography with X-rays and fast neutrons for restoration of wooden artwork. Phys Procedia 69:472–477
17. Osterloh et al (2011) Proceedings of international symposium on digital industrial radiology and computed tomography, vol 69, p 472
18. Daniels GC et al (2015) Fast neutron radiography at an RFQ accelerator system. Phys Procedia 69:109–114
19. Bucherl T et al (2004) Radiography and tomography with fast neutrons at the FRM-II- a status report. Appl Radiat Isot 61:537
20. Ikeda Y et al (1989) Nucl Inst Meth Phys Res A 276:183
21. Zboray R et al (2017) Fast neutron radiography and tomography at a 10 MW research reactor Beamline. Appl Radiat Isot 119:43–50
22. Adams R et al (2016) Appl Radiat Isot 107:1
23. Takenaka N el al (1999) Application of fast neutron radiography to three-dimensional visualization of steady two-phase flow in a rod bundle. Nucl Instrum Methods Phys Res A 424:73
24. Zboray R et al (2014) Development of a fast neutron imaging system for investigating two-phase flows in nuclear thermal-hydraulic phenomena: a status report. Nucl Eng Des 273:10–23

25. Zboray R et al (2015) Rev Sci Instrum 86:075103
26. Rant J et al (1998) Proceedings of 7th European conference on NDT, p 2763
27. Mikerov VI et al (2004) J Appl Rad Isotopes 61:519
28. Matsubayashi M et al (2001) Nucl Instrum Meth A 463:324
29. Ikeda Y et al (2016) Nucl Inst Meth Phys Res A 833:61
30. Makowska M et al (2017) J Imaging 3:60
31. TRITEC. http://www.rctritec.com/en/scintillators.htm (2017)
32. Johnson MS et al (2017) Phys Procedia 90:47
33. Overley JC (1987) Element-sensitive computed tomography with fast neutrons. Nucl Instrum Meth B 24(25):1058–1062
34. Chen G (2001) Fast neutron resonance radiography for elemental imaging: theory and applications. Ph.D. Thesis, MIT, Apr 2001
35. Soubelet B et al (2019) Feasibility study of using a compact deuterium-deuterium (D-D) neutron generator for energy-selective transmission tomography. Radiat Phys Chem 156:292–299
36. Dangendorf V et al (2009) Multi-frame energy-selective imaging system for fast-neutron radiography. IEEE Trans Nuclear Sci 56(3):1135–1140
37. Evsenin AV, Kuznetsov AV, Osetrov OL, Vakhtin DN (2004) Detection of hidden explosives by nanosecond neutron analysis technique. In: Proceedings of the NATO ARW on "detection of bulk explosives advanced techniques against terrorism". Kluwer Academic Publishers, St.-Petersburg, pp 89–103
38. Mitra S (2011) Identification of UXO using the associated particle neutron time-of-flight technique, final report, BNL-96489-2011
39. Anderson BC, Holbert KE, Bowler H (2016) Design, construction, and modeling of a ^{252}Cf neutron irradiator. In: Science and technology of nuclear installations, vol 2016
40. Bishnoi S et al (2015) Simulation study of fast neutron radiography using GEANT4. JINST. https://doi.org/10.1088/1748-0221/10/02/P02002
41. Knoll GF, Radiation detection and measurement. Wiley
42. Oksuz I et al (2020) Characterization of a reactor based fast neutron beam facility for fast neutron imaging 2020, LLNL-PROC-814265
43. Fantidis JG et al (2010) Optimization study of a transportable neutron radiography unit based on a compact neutron generator. Nucl Instrum Methods Phy Res A 618:331–335
44. Bishnoi S et al (2019) Preliminary experimentation of fast neutron radiography with D-T neutron generator at BARC. J Nondestr Eval 38:13. https://doi.org/10.1007/s10921-018-0550-9
45. Sabo-Napadensky I et al (2012) Research and development of a dedicated collimator for 14.2 MeV fast neutrons for imaging using a D-T generator. JINST 7, C06005
46. Cremer JT (2012) Large area imaging of hydrogenous materials using fast neutrons from a DD fusion generator. Nucl Instrum Methods Phys Res A675:51–55
47. Adams R et al (2015) Development and characterization of a D-D fast neutron generator for imaging applications. Appl Radiat Isot 96:114–121
48. David LW (2020) A fast neutron radiography system using a high yield portable DT neutron source. J Imaging 6:128. https://doi.org/10.3390/jimaging6120128
49. Makowska MG, Walfort B, Zeller A, Grünzweig C, Bucherl T (2017) Performance of the commercial pp/zns:cu and pp/zns: ag scintillation screens for fast neutron imaging. J Imaging 3:60
50. Dietich F, Hall J (1996) Detector concept for neutron tomography in the 10–15 MeV Energy range, UCRL-ID-123490
51. Makowska MG et al (2017) Performance of the commercial PP/ZnS:Cu and PP/ZnS:Ag scintillation screens for fast neutron imaging. J Imaging 3:60
52. Chuirazzi WC et al (2018) Evaluation of polyvinyl toluene scintillators for fast neutron imaging, LLNL-JRNL-757419 J Radioanal Nuclear Chem
53. Ambrosi RM et al (1998) A Monte Carlo study of the effect of neutron scattering in a fast neutron radiography facility. Nucl Instrum Meth Phys Res B 139:286–292

54. Ambrosi RM, Watterson (2004) The effect of imaging geometry and the impact of neutron scatter on the detection of small features in accelerator-based fast neutron radiography. Nucl Instrum Meth A 524:340
55. Tang B et al (2013) The physics analysis and experiment study of zinc sulphide scintillator for fast neutron radiography. Nucl Instrum Methods Phys Res A729:327–333
56. Lehmann E et al (2005) Neutron radiography with 14 MeV neutrons from a neutron generator. IEEE Trans Nuclear Sci 52(1)
57. Babai RW et al (2015) Fast neutron tomography of low-Z object in high-Z material shielding. Phys Procedia 69:275–283
58. Brede HJ et al (1980) The Braunschweig accelerator facility for fast neutron research: 1: building design and accelerators. Nucl Instrum Meth 169(3)
59. Brede HJ (1988) Neutron yields from thick Be-targets bombarded with deuterons. Nucl Instrum Meth A274:332–344
60. Andersson P (2014) Fast -neutron tomography using a mobile neutron-generator for assessment of steam-water distributions in two-phase flows. Uppsala University, Digital comprehensive Summaries of Uppsala Dissertations from the faculty of science and Technology, p 1146
61. Oksuz I et al (2020) Characterization of a reactor based fast neutron beam facility for fast neutron imaging. LLNL-PROC-814265, 3 Sept 2020
62. Bucherl T et al (2011) NECTAR-A fission neutron radiography and tomography facility. Nucl Inst Methods Phys Res A 651:86–89
63. Bucherl T, Söllradl S (2015) NECTAR-Radiography and tomography station using fission neutrons. J Large-Scale Res Facil (JLSRF) 1–19
64. Fujine S et al (1999) Development of imaging techniques for fast neutron radiography. Jpn Nuclear Instrum Methods Phys Res A 424:190–199
65. Perticone D et al (2019) Fast neutron resonace radiography for elemental imaging. Nuclear Inst Methods Phys Res A 922:71–75; (a) Chen G, Lanza R (2002) Fast neutron resonance radiography for elemental mapping. IEEE Trans Nucl Sci 49(4):1919–1924
66. Mor I et al (2014) Reconstruction of material elemental composition using fast neutron resonance radiography. In: 10th world conference on neutron radiography, 5–10 Oct 2014
67. Watterson IJ, Ambrosi RM (2003) Some fundamental considerations in resonance imaging using fast neutrons. Nucl Instrum Meth A 513:367
68. Raas W et al (2005) Neutron resonance radiography for explosives detection: technical challenges. In: Proceedings 2005 IEEE nuclear science and medical imaging symposium, 2005, UCRL-CONF-217017
69. Blackburn BW, Chen G (2007) Fast neutron resonance radiography for homeland security. In: IEEE nuclear science symposium conference record (N19-2)2007
70. Overley JC et al (1995) Explosives detection through fast-neutron time-of-flight attenuation measurements. Nucl Instrum Methods Phys Res Sect B 99:728–732
71. Overley JC (1987) Element-sensitive computed tomography with fast neutrons. Nucl Instrum Meth B 24/25:1058
72. Overley JC (1995) Explosive detection through fast-neutron time-of-flight attenuation measurements. Nucl Instrum Meth B 99:728
73. National Materials Advisory-Board (NMAB) (1999) The practicality of pulsed fast neutron transmission spectroscopy for aviation security. National Academy Press, Washington, DC
74. Miller TG et al (1997) Contraband detection using neutron transmission. In: Proceedings of SPIE 2936, p 102
75. Dangendorf V et al (2004) Detectors for energy resolved fast neutron imaging. Nucl Instrum Meth A 535:93–97
76. Mor I et al (2015) Reconstruction of material elemental composition using fast neutron resonance radiography. Phys Procedia 69:304-313
77. Dangendorf V et al (2009) Multi-frame energy-selective imaging system for fast-neutron radiography. IEEE Trans Nucl Sci 56(3):1135–1140

78. Mor I et al (2015) High spatial resolution fast-neutron imaging detectors for pulsed fast-neutron transmission spectroscopy. Phys Procedia 69
79. El Kanawati W et al (2011) Acquisition of prompt gamma-ray spectra induced by 14 MeV neutrons and comparison with Monte Carlo simulations. Appl Radiat Isot 69:732–743
80. Bishnoi S, Thomas RG et al (2019) Modeling of tagged neutron method for explosive detection using GEANT4. Nuclear Inst Methods Phys Res A 923:26–33
81. Agostinelli S et al (2003) Geant4-a simulation toolkit. Nucl Instrum Methods A 506:250–303
82. Donzella A et al (2006) Simulation of a tagged neutron inspection system prototype. J Phys Conf Ser 41:233–240. https://doi.org/10.1088/1742-6596/41/1/024
83. Batyaev VF, Belichenko SG, Bestaev RR (2015) Simulation of non-organic scintillation detector response for the problems of active interrogation by tagged neutron technology. WIT Trans Model Simul 59. WIT Press, www.witpress.com. ISSN 1743-355
84. Eleon C et al (2009) Preliminary Monte Carlo calculations for the UNCOSS neutron based explosive detector. Nucl Instrum Methods A. https://doi.org/10.1016/j.nima.2009.10.128
85. Sardet A, Pérot B, Carasco C et al (2018) Gamma signatures of the C-BORD tagged neutron inspection system. EPJ Web Conf 170:07011. *ANIMMA 2017*. https://doi.org/10.1051/epjconf/201817007011
86. Wu Y et al (2009) Development of a RF-driven neutron generator for associated particle imaging. IEEE Trans Nuclear Sci 56(3)
87. Nebbia G, Pesente S et al (2004) Performance of a tagged neutron inspection system (TNIS) based on portable sealed generators. Nucl Instrum Meth A 533:475–480
88. Carasco C et al (2008) In-field tests of the EURITRACK tagged neutron inspection system. Nucl Instrum Meth A 588:397–405
89. Bogolyubov EP et al (2017) Neutron generators and DAQ systems for tagged neutron technology. In: Proceedings of XXVI international symposium on nuclear electronics & computing (NEC-2017)
90. Bishnoi S et al (2020) Study of tagged neutron method with laboratory D-T neutron generator for explosive detection. Eur Phys J Plus 135:428
91. Ruskov I et al (2017) TANGRA-an experimental setup for basic and applied nuclear research by means of 14.1 MeV neutrons. EPJ Web Conf 146
92. Gierlik et al (2006) Comparative study of large NaI(Tl) and BGO scintillators for the EURopean Illicit TRAfficking countermeasures kit project. IEEE Trans Nuclear Sci 53(3)
93. Valkovic V, Miljanic D, Tomas P, Antolkovic B (1969) Neutron-charged particle coincidence measurements from 14.4 MeV neutron induced reactions. Nucl Instrum Methods 76:29–34
94. Ussery LE, Hollas CL (1994) Design and development of the associated-particle three-dimensional imaging technique. Report LA-12847-MS (UC-706 and UC-700)
95. Beyerle A, Hurley JP, Tunnel L (1990) Design of an associated particle imaging system. Nucl Instrum Meth A 299(1–3):458
96. Rhodes E et al (1992) IEEE Trans Nucl Sci 39:1041
97. Rhodes E, Dickerman CE, Brunner T et al (1995) Associated particle sealed tube neutron generator studies for arms control. Report ANL/ACTV-95/1
98. Hollas et al (1990) Report LA-11876-MS
99. Chichester DL, Lemchak M, Simpson JD et al (2005) The API-120: a portable neutron generator for the associated particle technique. Nucl Instrum Meth Phys Res B 241:753–758
100. Pesente S, Nebbia G et al (2005) Tagged Neutron Inspection system (TNIS) based portable sealed generators. Nucl Instrum Meth Phys Res B 241(1–4):743–747
101. Nevada B (1998) Associated particle imaging. DOE/NV11718 -223, Special Technologies Laboratory, CA
102. Mitra S et al (2011) IEEE Nucl Sci Symp Conf Record, pp 285–289
103. Vakhtini DN, Gorshkov YU et al (2006) SENNA portable sensor for explosive detection based on nanosecond neutron analysis. In: Detection and disposal of improvised explosives. Springer, Dordrech
104. Perret G, Perot B et al (2006) EURITRACK tagged neutron inspection system. J Phys: Conf Ser 4(375):1742–6596

105. Eleon C, Perot B et al (2011) Nucl Instrum Meth Phys Res A 629:220
106. Lunardon M et al (2004) Appl Radiat Isot 61:43–49
107. Pesente S, Nebia G, Viesti G et al (2007) Nucl Instrum Meth B 261:268
108. Lunardon M et al (2007) Nucl Instrum Meth Phys Res B 261:391
109. Underwater Coastal Sea Surveyor 'UNCOSS', Project Final Report-218148, http://www.uncoss-project.org
110. Alexakhin VY et al (2015) Detection of Diamonds in kimberlite by the tagged neutron method. Nucl Instrum Meth Phys Res A 785:9–13
111. Rogov Y et al (2020) Application of tagged neutron method for detecting diamonds in Kimberlite. Instruments 4:33. https://doi.org/10.3390/instrumnets4040033
112. Mitra S, Wolf J et al (1995) Application of the associated particle technique for the whole-body measurement of protein, fat and water by 14 MeV neutron activation analysis-a feasibility study. Phys Med Biol 40:1045–1055